Goulden
Heritage Resources Centre
U. of Waterloo
Waterloo Ontario Canada
N2L 3G1
Phone (519) 885-1211 (2072 or 4555)
Fax (519) 746-2031.

NATO ASI Series

Advanced Science Institutes Series

A series presenting the results of activities sponsored by the NATO Science Committee, which aims at the dissemination of advanced scientific and technological knowledge, with a view to strengthening links between scientific communities.

The Series is published by an international board of publishers in conjunction with the NATO Scientific Affairs Division

A	Life Sciences	Plenum Publishing Corporation
B	Physics	London and New York
C	Mathematical and Physical Sciences	Kluwer Academic Publishers
D	Behavioural and Social Sciences	Dordrecht, Boston and London
E	Applied Sciences	
F	Computer and Systems Sciences	Springer-Verlag
G	Ecological Sciences	Berlin Heidelberg New York
H	Cell Biology	London Paris Tokyo Hong Kong
I	Global Environmental Change	Barcelona Budapest

PARTNERSHIP SUB-SERIES

1.	Disarmament Technologies	Kluwer Academic Publishers
2.	Environment	Springer-Verlag/Kluwer Acad. Publishers
3.	High Technology	Kluwer Academic Publishers
4.	Science and Technology Policy	Kluwer Academic Publishers
5.	Computer Networking	Kluwer Academic Publishers

The Partnership Sub-Series incorporates activities undertaken in collaboration with NATO's Cooperation Partners, the countries of the CIS and Central and Eastern Europe, in Priority Areas of concern to those countries.

NATO-PCO DATABASE

The electronic index to the NATO ASI Series provides full bibliographical references (with keywords and/or abstracts) to about 50000 contributions from international scientists published in all sections of the NATO ASI Series. Access to the NATO-PCO DATABASE compiled by the NATO Publication Coordination Office is possible in two ways:

- via online FILE 128 (NATO-PCO DATABASE) hosted by ESRIN, Via Galileo Galilei, I-00044 Frascati, Italy.

- via CD-ROM "NATO Science & Technology Disk" with user-friendly retrieval software in English, French and German (© WTV GmbH and DATAWARE Technologies Inc. 1992).

The CD-ROM can be ordered through any member of the Board of Publishers or through NATO-PCO, Overijse, Belgium.

Series G: Ecological Sciences, Vol. 40

Springer
*Berlin
Heidelberg
New York
Barcelona
Budapest
Hong Kong
London
Milan
Paris
Santa Clara
Singapore
Tokyo*

National Parks and Protected Areas

Keystones to Conservation and Sustainable Development

Edited by

James Gordon Nelson

University of Waterloo
Environmental Studies Building 1
Waterloo, Ontario, Canada N2L 3G1

Rafal Serafin

Polish Environmental Partnership Foundation
6/6 Bracka St.
31-005 Krakow, Poland

With 35 Figures

Proceedings of the NATO Advanced Research Workshop on Contributions of National Parks and Protected Areas to Heritage, Conservation, Tourism and Sustainable Development, held in Krakow, Poland, August 26–30, 1996

Cataloging-in-Publication Data applied for

Die Deutsche Bibliothek - CIP-Einheitsaufnahme

National parks and protected areas : keystones to conservation and sustainable development ; [proceedings of the NATO Advanced Research Workshop on Contributions of National Parks and Protected Areas to Heritage, Conservation, Tourism and Sustainable Development, held in Krakow, Poland, August 26 - 30, 1996] / ed. by James Gordon Nelson ; Rafal Serafin. - Berlin ; Heidelberg ; New York ; Barcelona ; Budapest ; Hong Kong ; London ; Milan ; Paris ; Santa Clara ; Singapore ; Tokyo : Springer, 1997
  (NATO ASI series : Ser. G, Ecological sciences ; Vol. 40)
  ISBN 3-540-63527-0

ISSN 0258-1256
ISBN 3-540-63527-0 Springer-Verlag Berlin Heidelberg New York

This work is subject to copyright. All rights are reserved, whether the whole or part of the material is concerned, specifically the rights of translation, reprinting, reuse of illustrations, recitation, broadcasting, reproduction on microfilm or in any other way, and storage in data banks. Duplication of this publication or parts thereof is permitted only under the provisions of the German Copyright Law of September 9, 1965, in its current version, and permission for use must always be obtained from Springer-Verlag. Violations are liable for prosecution under the German Copyright Law.

© Springer-Verlag Berlin Heidelberg 1997
Printed in Germany

Typesetting: Camera ready by authors/editors
Printed on acid-free paper
SPIN 10532415   31/3137 - 5 4 3 2 1 0

Acknowledgements

We would like to thank the NATO International Scientific Exchange Program for financial support for the Advanced Research Workshop of which this book is a result. We are also grateful to NATO for a grant in aid of publication of this book. We also offer thanks to the Social Sciences and Humanities Research Council of Canada (SSHRC) for grants for related research. The Environmental Partnership for Central Europe (EPCE) also provided funding for the Workshop. We are of course very grateful to all the participants in the Workshop, most of whom are noted in this volume. Many people assisted with organization and other arrangements in Kraków, including Wies³aw Krzemiñski and his staff at the Polish Foundation for Nature Conservation and the Natural History Museum of the Institute of Evolutionary and Animal Systematics at the Polish Academy of Sciences which hosted the Workshop in Kraków, Andrzej Biderman and Wojciech Bosak of Ojców National Park, Ró¿a Biderman of the Kraków Development Forum and Dorota Mech of Wye College, University of London. Stephen Woodley of Parks Canada and Andrzej Biderman of Ojców National Park served on the Workshop Advisory Committee. Many participants also received some travel and other support from their organizations. Lisa Weber and Patrick Lawrence of the Heritage Resources Centre at the University of Waterloo assisted with the preparation of the book manuscript. We offer our sincere thanks to all those not mentioned here who helped make the Workshop a success.

J. Gordon Nelson
University of Waterloo
Canada

Rafal Serafin
Polish Environmental Partnership Foundation
Kraków, Poland

Contents

Keys to Life 1
Gordon Nelson and Rafal Serafin

Part I Broadening the Scope

Science and Protected Area Management: An Ecosystem-Based Perspective 11
Stephen Woodley

Landscape Ecology as a Basis for Protecting Threatened Landscapes 23
Almo Farina

Landscape Approaches to National Parks and Protected Areas 31
Adrian Phillips

Land Use and Decision-Making for National Parks and Protected Areas 43
Gordon Nelson, Rafal Serafin, Andrew Skibicki and Patrick Lawrence

A Clash of Values: Planning to Protect the Niagara Escarpment
in Ontario, Canada 65
Anne Varangu

The Economic Pitfalls and Barriers of the Sustainable Tourism Concept
in the Case of National Parks 81
Jan van der Straaten

Environmental Education in Protected Areas as a Contribution to
Heritage Conservation, Tourism and Sustainable Development 93
Andrzej Biderman and Wojciech Bosak

Extending the Reach of National Parks and Protected Areas:
Local Stewardship Initiatives 103
Jessica Brown and Brent Mitchell

The Role of Banff National Park as a Protected Area in the Yellowstone to
Yukon Mountain Corridor of Western North America 117
Harvey Locke

The Danube Challenge: Protecting and Restoring a Living River 125
Philip Weller

Exploratory Planning for a Proposed National Marine Conservation Area
in Northeast Newfoundland 133
Paul McNab

Experience in Cross-Border Cooperation for National Park and
Protected Areas in Central Europe 145
Zygmunt Denisiuk, Stepan Stoyko, and Jan Terray

Parks for Life: An Action Plan for the Protected Areas of Europe 151
Marija Zupancic-Vicar

Part 2: Information and Communication

Usable Knowledge for National Park and Protected Area Management:
A Social Science Perspective 161
Gary Machlis and Michael Soukup

Integration of Information on the Kraków-Czêstochowa Jura for
Conservation Purposes: Application of CORINE Methodology 175
*Anna Dyduch-Falniowska, Malgorzata Makomaska-Juchiewicz,
Róza Kazmierczakowa, Joanna Perzanowska, and Katarzyna Zajac*

The Potential Role Of Biosphere Reserves In Piloting Effective Co-operative
Management Systems For Heritage, Landscape and Nature Conservation 187
Henry Baumgartl

The Berezinski Biosphere Reserve in Belarus: Is Ecotourism a Tool
to Support Conservation in the Reserve?.. 191
Sylvie Blangy

Canada-Hungary National Parks Project: 1992-1995 195
Tom Kovacs

A Sustainability Appraisal of the Work of the Yorkshire Dales
National Park Authority.. 199
David Haffey

Forest Management and Planning and Local Populations:
Assessing the Case of Niepolomice Forest, nr. Kraków, Poland 207
*Jerzy Sawicki, Rafal Serafin, Bogumila Kuklik,
Katarzyna Terlecka, and Tomasz Terlecki*

Part 3: Learning from the Experience of Others

Conflict Between Skiers and Conservationists and an Example of its Solution:
The Pilsko Mountain Case Study (Polish Carpathians) 227
Adam Lajczak, Stefan Michalik, and Zbigniew Witkowski

Research, Conservation, Restoration and Eco-tourism in National Parks:
Experience from Hungary 235
Mihaly Vegh and Szilvia Gori

Questions Related to the Rehabilitation of the Egyek-Pusztakocs Marshes 243
Szilvia Gori and Aradi Csaba

Bulgarian Experience in Nature Protection: Contributing to
Sustainable Development 247
Elizaveta Matveeva

The Emerging System of Protected Areas in Lithuania: A Review 253
Tadas Leoncikas

Landscape Protection in Estonia: The Case of Otepää 257
Monika Prede

Protected Areas and Plant Cover Diversity in the Ukrainian Carpathians:
An Assessment of Representativeness 265
Lydia Tasenkevich

Sultan Marshes, Turkey: A New Approach to Sustainable
Wetland Management 269
Nilgül Karadeniz

Conservation of The Mediterranean Environment: The Vital Need to Protect
and Restore the Quality of National and Cultural Resources in Libya and the
Role of Parks and Protected Areas. 277
Amer Rghei and Gordon Nelson

Workshop Participants: contact addresses and affiliations.................................. 285

Keys to Life
Contributions of National Parks and Protected Areas to Heritage Conservation, Tourism and Sustainable Development.

J.G. Nelson[1] and Rafal Serafin[2]

[1] Heritage Resources Centre, University of Waterloo
Waterloo, Ontario Canada

[2] Polish Environmental Partnership Foundation
Bracka 6/6
31-005 Kraków, Poland

1 Setting the Agenda

Heritage conservation, tourism, and sustainable development are the ideas or concepts which we initially chose to use in thinking about the wide-ranging and fundamental role of national parks and protected areas. However, as the title, and the rest of this introduction and book show, these concepts are not far-reaching enough to capture the full significance of protected areas. These areas are actually best thought of as keys to the well-being of human and other life on earth. In saying this we are thinking not only of national parks, wildlife refuges, biosphere reserves and other protected areas which are largely of western origin and usually set up by governments. We are also thinking of protected areas set up by local people over the centuries for conservation of forests, pastures, hunting grounds or other purposes. The stress in the book is on the more formal governmental protected areas. However we recognize that these need to be linked with private stewardship efforts more fully than in the past in order to achieve greater overall effectiveness.

The papers and reports which make-up this book are the basis of the foregoing fundamental conclusions - which may be seen as extreme or radical in some scientific, scholarly, professional and civic quarters. The book consists of many of the papers presented at a NATO Advanced Research Workshop held in Krakow, Poland, August 26-30, 1996. In seeking financial and general support from NATO we put forward a set of ideas to justify the holding of the workshop. These ideas are presented below more or less as they were to NATO in spring, 1996. Some minor changes have been made in the text but these are predominantly for clarification. The following ideas and justification will therefore provide a picture of our thinking before the workshop and show how our thoughts evolved or elaborated as a result of the workshop.

2 Justification for the NATO Advanced Research Workshop: April, 1996.

2.1 Need

New scientific, educational and planning approaches are needed to make it possible for national parks and protected areas to fulfill their vital role in maintaining and enhancing significant environments, resources and sustainable development options in the NATO countries, their neighbours and other parts of the world, especially in the light of growing tourism pressures.

2.2. Importance

Many people view national parks and protected areas as having mainly been established to protect highly significant environments or ecosystems, landscapes, plants, animals, water, archeological, cultural and other resources for conservation, scientific, educational, aesthetic and recreational purposes. Since their inception in North America in the nineteenth century, national parks have been strongly associated with the development of tourism. Visits by rising numbers of local, national and international tourists have generated revenue for many interested parties, as well as having recreational, educational and other benefits to society. Visits by recreationists and tourists have made people more aware and supportive of the values, benefits and roles of national parks and other protected areas in economy and society.

Since the first World Conservation Strategy in 1980, the Brundtland Report in 1987, the IVth World Parks Congress in Caracas and the World Environment Summit in Rio in 1992, national parks and protected areas have also been seen as playing a fundamental role in sustainable development. In this regard, national parks and protected areas have been recognized as providing a range of ecological services which are necessary for sustainable development. Among these services are those which: preserve essential ecological processes, for example nesting of staging areas for migratory birds; protect biodiversity, for example habitat for large predators; and maintain productivity of resources, for example water quality and quantity in parts of otherwise settled watersheds. Parks and protected areas are also increasingly seen as offering important opportunities for science-based educational and interpretation programs for teachers as well as students and for enhancing environmental understanding on an ongoing basis among citizens. In the broad sense, parks and protected areas are also seen as essential elements in sustainable regional and national conservation strategies.

At the same time that their fundamental importance has received greater recognition, increasing land use and resource development pressures have been placed upon national parks and protected areas by accelerating tourism, agricultural, forestry, fishing, industrial, urban and other development ensuing from economic and population growth. As a result the basic wildlife, soil, water and other ecosystem services offered by national parks and protected areas are threatened in many parts of the world, as are their related values for sustainable development.

New approaches are needed in many countries to deal with these challenges. Some promising approaches have been developed which could help in this regard in fields such as ecology, social science, education and planning. An example of an ecological advance is landscape ecology. An example of a social science advance is economic incentives or other aspects of ecological economics. An example of an educational approach is citizen monitoring and participatory and social learning. An example of a planning advance is transactive or interactive and adaptive planning. Also important is the increasing role of citizens in private stewardship and co-management arrangements. Such advances or approaches are better known in some areas than others and it is a major purpose of this workshop to make them and other new approaches more widely understood as a basis for improving both theory and practice in the future.

More specifically, a number of relatively new ideas and methods are available for improving understanding and practice in ways that can help deal with the stresses. Examples include: greater park ecosystems; conservation biology; bioregional theory; comprehensive resource assessment systems such as the Abiotic-Biotic-Cultural (ABC) resource inventory method; environmental, social and technology assessment; economic methods such as contingency analysis; and mediation, conflict resolution, and visioning in participatory and adaptive planning. The study, discussion, communication and application of such ideas and approaches in a workshop providing for exchange of international learning and experience, would assist considerably in dealing with pressures on national parks and protected areas not only in Poland and Canada, but also in other parts of the world.

2.3 Timeliness

The workshop is seen as very timely for a number of reasons. At the international level new conventions such as those on biodiversity and climatic change are very relevant and should be built into national park and protected area research and education, as well as overall planning and management. At the national and local level, many land use and socio-economic pressures are threatening the ecological character and the quantity of the services provided by national parks and protected areas. It is very important to deal with these pressures as soon as possible before many unwanted changes proceed too far. Among the most relevant of the pressures are those arising from tourism, a development option which is being stressed in many countries as a means of compensating for economic decline and providing for economic renewal. Ecotourism, green tourism and other forms of tourism fall in this category and deserve careful assessment in terms of their possible large scale implementation and effects in or near national parks and protected areas.

2.4 Follow-up

This workshop is strategic in the sense that it is intended to assess current knowledge and its possible application in decision-making as well as educational and research needs. The workshop is also strategic in promoting development of a network of researchers and professionals interested in enhancing the contribution of national parks and protected areas to heritage conservation, tourism and sustainable development. This network could communicate by electronic mail and other means and could convene other meetings in the future to follow up on key issues and topics. Indeed, following on from the conceptual focus of this first workshop, future meetings would be better positioned to address issues relating to the development and use of new tools in protected area, heritage conservation, tourism and sustainable development planning, such as geographic information systems (GIS) and remote sensing. In this way, the workshop would feed into and complement efforts of other organizations such as the World Conservation Union and its Parks for Life action plan for the protected areas in Europe. The workshop would also support the work of relevant government and private organizations, such as park and protected area agencies, regional and municipal planning authorities, tourism and business interests, university departments and research groups.

2.5 Organization of the Workshop

A basic idea in the motivation and organization of the workshop was the potential gains to be derived from new and emerging theory and methods as well as research and protected area experiences in different countries, notably in North America, Western Europe and the Central and Eastern European NATO Co-operating Partner (CP) countries. In this respect the workshop arose from several years of co-operation among scientists, scholars, professionals and knowledgeable citizens in Canada and Poland, notably the Heritage Resources Centre at the University of Waterloo, Canada and the Polish Foundation for Nature Conservation in Kraków, Poland. In organizing the workshop speakers and participants were invited who were scientists and scholars, planners and managers and also active concerned citizens.

With the foregoing background and motivation in mind, the workshop program was designed to improve understanding and make research recommendations by : (1) critically assessing existing knowledge of the contributions of national parks and protected areas to heritage conservation, tourism and sustainable development; (2) identifying research needs and directions; and (3) identifying ways of promoting closer relations among scientists, scholars and professionals, especially from NATO and Cooperating Partner (CP) countries.

Another underlying objective was to provide information which would assist the Central and Eastern European Countries with work on national parks and protected areas during the transition from a centrally-planned communist to a free market system. As noted earlier however, the expectation was that all involved would learn from the experience of the others.

To meet these objectives, the organizers used the following means and procedures: (1) invited a mix of participants involved in research and practice related to protected areas as scientists, managers, planners, members of non-governmental organizations and other groups to prepare keynote theme and case study papers, posters, and other materials as a basis for discussions intended to help meet the objectives; (2) developed and organized two illustrative case study field tours and guides; and (3) invited and arranged exhibits, for example from Georgian Bay Islands National Park, Canada, and from Ojców National Park and the Directorate of the Jura Landscape Parks System near Kraków, Poland.

The organizers solicited applications for workshop participation through several relevant networks, including the participant list of the University of Waterloo Heritage Resources Centre, Federation for Nature and National Parks of Europe (FNNPE), the Commission for National Parks and Protected Areas (CNPPA) of the International Union for the Conservation of Nature (IUCN or the World Conservation Union), the International Society for Ecological Economics (ISEE), World Wide Fund for Nature (WWF) and the Environmental Partnership for Central Europe (EPCE). The organizers interacted with workshop authors in order to ensure that a workable program could be developed that addressed both theory and practice from the perspectives of North America, Western and Central Europe. Forty-nine speakers and participants from 21 countries attended the workshop. Evaluations from them indicated that the approach proved very successful in generating communication and information exchange opportunities - and indeed proved itself as a research tool for clarifying the complex nature of interaction between theory and practice and between researchers, professionals or practitioners, and concerned citizens.

2.6 Summary of Workshop Results

The workshop report submitted to NATO in October 1996 was organized under three major headings: Current Knowledge; Research; and Linkages and Relationships.

2.6.1 Current Knowledge

The workshop reviewed scientific, educational and planning approaches used by national parks and protected areas to fulfill their vital role in maintaining and enhancing significant environments, resources and sustainable development options in the NATO and Cooperating Partner countries. The workshop presentations and discussions showed that current knowledge of and perspectives on the roles and contributions of national parks and protected areas in various countries is quite uneven. Differences include the U.S. and Canadian tendencies to place greater emphasis on: ecosystem approaches; comprehensive nature conservation; wildlife; social science; and public and private stewardship programs. The European countries tend to place greater emphasis on: landscape approaches; the natural sciences; and environmental education. Environmental education programs seem to be especially well developed in the U.K. and Poland. Landscape parks and protected areas are well developed or emerging at the local government level in many European countries such as the United Kingdom and Germany, as well as Eastern European countries such as Poland and the Czech Republic. Central and Eastern European countries also are leading in the establishment and development of cross-border parks and protected areas and related environmental programs in boundary areas, including those where military and security considerations formerly dominated. These include the Carpathian Mountain Region (Ukraine, Slovakia, Czech Republic, Romania, Poland) and the Danube basin (Austria, Czech Republic, Hungary, Romania, Slovakia and Ukraine). These cross-border approaches to parks and protected areas can be linked to North American experience with bioregional approaches, for example, in the Rocky Mountains.

Similarities among countries include growing interest in: patterns and rates of land use change and their effects in and around national parks and protected areas; improved monitoring and assessment of natural and human changes; the nature and effects of tourism, including eco and heritage tourism; a stronger role for parks and protected areas in economic planning and environmental and international security; better development of ecologically-based approaches to planning, management, and decision-making (biosphere reserves, greater park ecosystems, bioregions); and improved scientific and civic methods of communicating and working with people in planning, management and decision-making at the community, state, regional, and international levels. The differences and similarities among countries and regions in North America and Europe present excellent opportunities to learn from one another and to identify important research needs.

2.6.2 Research Needs

The theme and case study papers analyzed at the workshop show that many internal and external land use and socio-economic pressures are threatening the ecological character and the level of the services provided by national parks and protected areas. Conventional management and planning procedures that focus on action within protected area boundaries are insufficient to contend with these development pressures and so the degradation of resources. Successes have been achieved however, by emphasizing linkages among parks and protected areas and the surrounding lands and waters of greater park ecosystems or planning regions.

A fundamental conclusion is that a very useful overall research approach arises from the broad field of human ecology. This is especially so where research involves groups with mixed backgrounds, interests, and expertise - such mixed pluralistic groups being characteristic of national park and protected area situations.

In this context, scientific and scholarly studies are needed, especially with regard to six broad areas:
1) ecological approaches to parks and protected areas and their role in the larger land use matrix (Greater Park Ecosystems, Bioregions, Watersheds);
2) land use and land cover changes in and around parks and protected areas (remote sensing, field mapping, GIS, ecological mapping systems, and monitoring);
3) theory and methods in regard to the role of parks and protected areas in environmental education;
4) assessment and study of existing and potential institutional arrangements and their effectiveness in national parks, protected areas, and surrounding regions (zoning, use of environmental, social and economic assessment, buffers, nodes and corridors, trusts, easements and related tenure arrangements, co-management, conflict resolution);
5) improved understanding of the role of parks and protected areas in river, lake, marine and coastal situations; and
6) the interrelationships among the foregoing broad areas of research, especially as they relate to environmental conservation, sustainable development, and public and private stewardship.

2.6.3 Linkages and Relationships

Greater development of electronic and other interactive communication systems received much support at the workshop and a lot of activity is already underway in this regard. Enhancement of linkages with and among IUCN (World Conservation Union), the Federation for Nature and National Parks, the World Bank and other international agencies was strongly supported. A proposal for a series of linked National or Regional Park and Protected Area Institutes - along the lines of the U.S. National Park/University Centres - was also of interest to the workshop participants. Considerable stress was placed on the need to develop civic linkages involving scientists, government personnel, and people of many backgrounds who benefit from and can contribute to research and decision-making for parks and protected areas and their role in nature conservation and sustainable development. These civic linkages can make a strong contribution to the greater development of public and private stewardship. As an initial attempt to do this, the Kraków Workshop was seen as quite useful. Plans were made for follow-up, first with the European Union Parks for Life Program in regard to cross-border parks and former military land, and second, in establishing an ongoing International Park and Protected Area Civic Forum. A subsequent meeting is tentatively planned for Canada, with a focus on Greater Park Ecosystems.

The Greater Park Ecosystem theme provides an excellent context in which to bring together research, mutual learning, and co-operation along the lines noted earlier in this report.

2.7 Overview

Following the foregoing report to NATO, the papers from the workshop were edited in detail and 29 of them are included in this book. During the detailed editing much more thought occurred about the papers and their significance as well as about how they should be organized in this book. In this respect, we believe that three fundamental themes run through most of the papers and the workshop discussions. These themes are related to but go beyond the ideas contained in the October 1996 report to NATO. Thinking about these three general themes led us beyond the initial concepts of heritage conservation, tourism, and sustainable development as a framework for evaluating the contributions of national parks and protected areas to the much bigger idea that protected areas were actually key or essential to the well-being of human and other life on earth. We owe the development of this larger idea to all of the papers and participants although the concept of Parks for Life as summarized by Maria Zupancic-Vicar in this volume, brought the idea home to us in a forceful way.

The three general themes around which the papers are organized are:
(1) Broadening the Scope: Toward a wider and more fundamental understanding of the contributions and significance of national parks and protected areas.
(2) Information and Communication: The need to expand and improve upon information and communication in understanding, planning, managing, and deciding upon protected areas.
(3) Learning from the Experience of Others: Case studies which show what can be learned by regular monitoring and assessment of the work and experiences of people in other places.

The papers are organized individually under these three headings in accordance with our judgment about where they seem to fit most appropriately. However, most of them include information which is relevant to all three themes.

Most of the papers in this book contain suggestions or recommendations for research - in the spirit of the NATO Advanced Research Workshop Program. We have chosen not to summarize these in detail here for practical reasons. Our final editing of the papers has led us to make only a few amendments to the research findings sent to NATO in October; these findings stand essentially as they were submitted at that time.

In our view if one fundamental finding arises from this book and the NATO workshop it is that parks and protected areas must be understood, planned, managed and decided upon as an integral part of the local as well as the larger regional, national or international context in which they find themselves. The national parks and protected areas are part of their locality or region in ecological, land use, economic, social, educational, stewardship and other senses. They are also part of the larger national and international scene for similar reasons. Just as local and regional economies are linked with global ones so too with protected areas.

This book shows that national parks and protected areas play a necessary role in economic, social and environmental well-being at the local, national, regional and international levels. Without well-designed protected areas the prospects for heritage conservation, tourism, sustainable development and local, regional and global well-being are dim indeed. Much more recognition of the fundamental importance of protected areas is therefore necessary not only among scientists, scholars, planners, managers and politicians but also among business people and citizens generally. Scientists, scholars, and protected area professionals in particular have to make more intense, wide-ranging and collaborative efforts to promote better understanding, planning, management and decision-making about protected areas - in the broad context of land use, socio-economic, environmental and regional planning - if the conditions essential to human well-being and diversity of life on earth are to be maintained for this and future generations.

Further to this point when taken as whole the papers may make a case for another broad perspective on the contribution or role of parks and protected areas in society and earth generally. This perspective seems to lie between nature conservation and sustainable development and can be termed environmental security. From this perspective, parks and protected areas not only serve to conserve nature for biodiversity and other more strictly ecological purposes but also for protection of plants, animals, watersheds, water quality and other processes, elements or resources useful for human welfare and sustainable development. This security role includes protection or buffering against flooding and other hazards. If this thinking is useful then conservation, environmental security and sustainable development can be thought of as interelated and vital concepts in understanding, planning and implementing policies and programs for a more desirable earthly and human future -- and parks and protected areas play a key role in all of them.

The papers in this book show that understanding, planning, managing and making decisions about national parks and protected areas is a highly complex matter not only for citizens, but also for researchers, professionals, administrators and politicians. The understanding of parks and protected areas involves drawing upon many ways of knowing - local or indigenous, scientific or scholarly. This understanding also involves many specialties or disciplines, such as geology, hydrology, biology, soils, archaeology, economics, sociology, anthropology, geography, law and planning. This understanding also involves many different perspectives or approaches, such as ecology, landscape, history, land use, governance and human ecology. Understanding, planning, managing and deciding upon parks and protected areas in the broad context of nature and culture requires the work of many different kinds of knowledge, interests and values.

This complexity makes national parks and protected areas an unusually demanding and yet revealing way of learning about ourselves and the world around us. This complexity is also a great challenge to people who wish to promote better understanding of the role and values of protected areas and provide the means of making better decisions about them at a time when they are under increasing threat of often unwanted change.

Do the papers in this book provide useful information and insights in this regard? Each paper makes at least some contribution to better understanding of parks and protected areas - as well as larger concerns about conservation and sustainable development. However, it is the papers taken as a whole that are most instructive about what we need to think about in understanding and making decisions on protected areas and their role in the world. Here we are concerned about the various kinds or types of information that seem to be needed as a broad framework or context for addressing specific protected area issues.

Major elements of a broad framework for understanding and deciding upon parks and protected areas are presented throughout the book, but especially in Part 1: Broadening the Scope. These elements include - understanding the concepts and methods of the earth and life sciences - of ecology and related theory - of human and natural history and landscape and land use - of values, ethics and culture - of economics, technology and development - of social systems and social guides - of regulations, governance, organizations and institutions.

We do not wish to press this matter any further than we have. There are many ways of organizing the concepts and frameworks to varying levels of detail and scale. We do wish to stress, however, the need to take a broad view, even in understanding, planning and dealing with seemingly focussed or narrow concerns, such as fire in the ecosystem or the cultivation of rice and the formation of landscapes. With these thoughts we urge you to explore the many thought-provoking papers contained in this book.

Part I
Broadening the Scope

The fundamental message here is that parks and protected areas should not be thought about separately, but rather as an integral part of the socio-economic and environmental context in which they find themselves.

- National parks and protected areas and surrounding lands and waters are closely linked in environmental, economic, land use, educational and social terms and so must be understood, planned, managed and decided upon in relation to one another.
- Science - especially ecosystem science - can provide valuable information for understanding, planning, managing and deciding upon protected areas and surrounding lands and waters by: 1) explaining cause and effect relationships; 2) predicting future events; and 3) providing models of how systems operate, for example ecosystems.
- The field of landscape ecology can provide tools such as nodes and corridors approaches to help deal with fragmentation of forests, wetlands and other environments by agriculture, transport and other land uses.
- Clash of values is an inevitable and fundamental part of understanding, planning, managing and making decisions on protected areas and related concerns; values have not been addressed explicitly and systematically in the past and do require direct and immediate attention in future work.
- The idea of landscape - of the interaction of nature, human activity and culture over the centuries - can play a much stronger role in conservation and sustainable development along with the wilderness or wildlands ideas stressed internationally in recent decades.
- An understanding of land use history and landscape change - of land use and landscape dynamics - is essential to understanding, planning, managing and deciding up parks, protected areas and conservation and development; yet these dynamics are still frequently neglected by scientists, professionals, managers and citizens.
- National parks and protected areas are part of the economy of the region in which they are located and this needs to be taken into account in planning, management and civic decision-making generally.
- Although economic theory has limitations in dealing with protected areas and environmental affairs generally, contingency analysis and other methods can be used to show that people value protected area tourism experiences highly and are prepared to pay fees and other charges to the benefit of protected areas and local people.
- National parks and protected areas offer opportunities for education programs that will increase awareness and understanding of the protected areas and surrounding regions in environmental, economic, social, land use and public policy terms; such education programs can contribute to the building of civic capacity in the region as well as among visitors from farther away.
- Residents and visitors from farther away bring a range of values to bear on land use, wildlife, plants and other aspects of protected areas; the different values lead to ongoing controversies and conflicts which need to be addressed by more interactive and adaptive planning, negotiation and civic approaches.
- The concept of stewardship - of individual and group responsibility and care - offers much potential to bring government, business and citizens together in planning, managing and deciding upon land use in protected areas as well as in the lands and waters around them. Considerable success has been achieved in public and private stewardship in recent years in North America. The stewardship approach holds promise in balancing the economic aspirations of new private land owners with the need for conservation in formerly communist countries in Europe.

- More stress is needed on effective planning, management and decision-making for marine parks and conservation areas. Stewardship approaches seem especially appropriate here, for example in association with fishing and related activities.
- The beginnings of large scale bioregional and watershed approaches to protected areas and sustainable development are emerging in North America (Rockies) / and Central Europe (the Danube); these approaches deserve careful study and support because of their comprehensive nature and potential for wide-ranging effectiveness
- Cross-border parks and protected areas are growing in number in Central and Eastern Europe in particular, because they can build ecological, economic, social and political bridges among places and people and also promote international peace and security; the progress in Europe deserves careful study for application in other parts of the world.
- Planning for and deciding upon national parks and protected areas on a continental scale has been introduced in Europe in the form of the Parks for Life program; this approach holds promise for other areas such as North America. The approach parallels free trade and other economic planning now underway, continentally and globally.
- The scope of the national park and protected area domain is constantly expanding as the services that these areas provide and the needs that they address are more widely recognized. Interest in national parks, state or provincial parks, wildlife refuges, and historic sites has expanded to include marine parks, local conservation areas, cross-border parks and a range of cooperative government and private stewardship programs at the local, regional, national and international levels; these include biosphere reserves, World Heritage Sites and Parks for Life which addresses parks and protected areas on a continental or European scale.

Science and Protected Area Management: An Ecosystem-based Perspective

Stephen Woodley[1]

[1] Forest Ecologist, Natural Resources Branch,
Canadian Parks Service, Hull, Quebec, Canada K1A OM5

Abstract. Recently the value of protected areas as conservation tools has been questioned. Protected areas are often too small to protect viable populations of many organisms, especially large vertebrates and carnivores. Most protected areas are also too small to allow large-scale ecological processes, such as wildfire or insect epidemics, to occur. If protected areas are to act as one of humankind's main tools to conserve biodiversity, they must be managed from new perspectives and with revised management goals. In this context an ecosystem based perspective is presented for protected area management. Such a perspective or approach requires newly defined goals based on ecological integrity. While this approach is normative, it must be based on a comprehensive understanding of the state of a greater regional ecosystem, which includes the protected area and its surrounding landscape. A science program to achieve ecological integrity is discussed and examples given. The science program must provide an ongoing assessment of the state of the greater park regional ecosystem as well as a set of clear directions for management intervention.

Keywords. Protected areas, national parks, ecosystem management, ecological integrity, biodiversity, ecosystem stress, greater park ecosystem

1 Introduction

Citizens, political leaders, public interest groups and even scientists place an amazing amount of faith in protected areas. The establishment of protected areas is generally seen as the end point of a conservation struggle, a victory for protecting nature from the ravages of human exploitation. Indeed protected areas can be seen as the most common human response to human induced ecosystem degradation, as evidenced by the fact they are found in virtually all countries worldwide. The dramatic worldwide increase in both the total number of parks and the total park area has paralleled an increase in the human understanding of the global environmental crisis.

However, recent evidence from conservation biology has questioned the value of protected areas as conservation tools. Protected areas are often too small to protect viable populations of many organisms, especially large vertebrates and carnivores. Most protected areas are also too small to allow large-scale ecological processes, such as wildfire or insect epidemics, to occur. For example, fire in the boreal forests of Canada is highly variable, with an average fire size of about 500 ha. Over 30 % of fires cover more than 1000 ha and large fires regularly burn over 100,000 ha (CIFFC 1996). With fire return intervals of 100-150 years in the boreal forest (Dansereau and Bergeron 1993), it is difficult for most protected areas to sustain patterns of natural burning.

Even after establishment, protected areas are still subject to a host of other human-caused stresses, including: habitat fragmentation and edge effects; disease transmission from domestic animals; introduced species; long-range transport of pollutants; the impacts of tourism; poaching; and, global climate change. As a result, the ecological integrity of many protected areas is already severely degraded or is at risk.

If parks and protected areas are to play a key role in addressing global environmental degradation, they require new approaches to management. In this paper, the argument is for an approach that is based on ecological integrity as a management endpoint, rather that the more traditional endpoints of wilderness, or natural. Moreover, ecological integrity needs to be based on a strong program of ecosystem science. Without a science-based program to inform management action, protected areas are unlikely to meet their goals of conserving nature.

2 The Evolution of the Rationale for Protected Areas

Three philosophical streams can be envisioned in thinking about humanity's relationship with nature. Until recently, the dominant paradigm in natural resource management has been utilitarian, notably in terms of commercial harvest of natural resources. Management concerns were aimed at maximizing the harvest of individual desired species, be they fish, trees or ecosystem products like fresh water. There has been, however, an alternative philosophy that was inherent in at least part of the park and protected areas movement. This philosophical stream can be termed romantic. Traditionally, and still commonly today, parks and protected areas were and are managed for management goals like wilderness, natural or pristine nature. These terms have been used worldwide to describe human desires for protected areas, where humans have had minimal impact and nature reigns supreme. Given our current ecological understanding of protected areas, it is highly problematic to apply these concepts. There are virtually no protected areas in the world that are unaffected by long range pollutants, exotic species, species loss or an alteration in historical disturbance regimes (Machlis and Tichnell 1985).

In recent years the utilitarian and romantic themes have converged. The terms ecological integrity, ecosystem health, and biodiversity have all been suggested as new end-points for managers, whether they are managing commercial forests or parks. While there is some disagreement about what constitutes ecological integrity or ecosystem health, there is at least an increased reliance on ecosystem science in the development of this new class of goals. More importantly, this new approach to protected areas is a major departure from the traditional reliance on very general goals such as wilderness. The new paradigm is embodied by the term, ecosystem management, or, ecosystem-based management (see Halvorson and Davis 1996; Agree and Johnson 1988). Ecological integrity, or similar wording, is always stated as the endpoint for management based on the ecosystem approach.

With the adoption of an ecosystem management approach, definitive ecosystem goals become essential to the management of protected areas. However, it is often extremely difficult to set definitive ecosystem goals. The very act of setting such definitive goals takes the manager away from the traditional safe reliance on wilderness, natural or pristine nature, where god or some other forces of natural regulation determine the goals. Despite the difficulty of setting goals, there is a surprising degree of convergence in the literature on a set of goals. Grumbine (1994) describes 5 common ecosystem management goals used under the banner of ecological integrity:

1. Maintain viable populations of all native species in situ.
2. Represent, within protected areas, all native ecosystem types across their natural range of variation.
3. Maintain evolutionary and ecological processes (i.e., disturbance regimes, hydrologic processes, nutrient cycles etc.)
4. Manage over periods of time long enough to maintain the evolutionary potential of species and ecosystems.
5. Accommodate human use and occupancy within these constraints.

The foregoing goals, or variants of them, have been used by a variety of agencies attempting to employ ecosystem management in protected areas. They are, in fact, value statements that are at least rooted in our understanding of ecosystem science. These new goals imply a departure from the idea that protected areas contain self-regulating systems, where the best management is to always leave them alone. An ecological integrity approach to management implies the possibility that management actions will be taken when the system moves away from a desired state.

The natural regulation versus interventionist approach to park management has been the subject of intense debate, especially in the United States National Parks (Kay 1990). It should be noted that the interventionist approach does not imply that ecosystems are not self-regulating. It simply means that the area contained within the boundary of a protected area will likely not maintain native biodiversity and ecological processes. There is another side to this debate that argues that many of the desired characteristics of protected areas were human-modified or human-created. Natural regulation ignores humans as keystone species that controlled ecosystems, primarily through hunting and the use of fire. An ecological integrity approach forces managers to define precise goals for the system and manage for those goals, which often means intervention.

The term, biodiversity, has been also invoked in recent years as a new goal for protected areas. The Convention on Biological Diversity was signed by over 90 percent of United Nations member countries at the 1992 Earth Summit in Rio de Janeiro, Brazil. According to the declaration of the international agreement for conserving biodiversity, in-situ conservation of biodiversity, through protected areas, is one of the key strategies of biodiversity conservation. Biodiversity is most commonly defined as the diversity of life. The term includes both ecosystem structures such as species richness or landscape-level community diversity, and ecosystem functions such as productivity or nutrient cycling. Thus the goal of conserving biodiversity, is really identical to the goals advanced by Grumbine, and listed previously. In other words, biodiversity, as broadly defined by the Biodiversity Convention, is similar in intent to ecological integrity. In this respect, the biodiversity convention has subsections that require a park to:
(1) maintain viable populations of species in natural surroundings;
(2) ensure environmentally sound and sustainable development in areas adjacent to protected areas with a view to furthering protection of these areas;
(3) prevent the introduction, control or eradication of alien species which threaten ecosystems, habitats or species;
(4) Endeavour to provide the conditions needed for compatibility between present uses and the conservation of biological diversity and the sustainable use of the system at hand.

3 The Greater Ecosystem Approach to Protected Area Management

Ecosystem management, or ecosystem-based management, is not a new idea. It has origins in a call to integrate biological, physical and sociological information. Ecosystem management in protected areas was discussed as early as 1932, by the Committee for the Study of Plant and Animal Communities of the Ecological Society of America (Shellford 1932). Committee members recognized that a comprehensive system of sanctuaries in the United States must: protect ecosystems as well as particular species; represent a wide range of ecosystem types; manage for ecological fluctuations (i.e. natural disturbances); and employ a core reserve and buffer approach. The committee also discussed the need for inter-agency cooperation and public education to make the approach successful. These components remain as foundations for more recent approaches toward ecosystem management. More recently, Agee and Johnson (1988) published an edited volume on ecosystem management in protected areas. The modern application of ecosystem management was pioneered in Yellowstone National Park and the Greater Yellowstone Ecosystem has been the subject of much writing and debate (see Keiter and Boyce 1991). In Canadian National Parks, the ecosystem approach developed from the extensive use of biophysical inventories in the 1970s. These biophysical inventories involved an integrated examination of the natural world, including topography, soils, and vegetation.

Ecosystem-based management is a term applied to the activities of many different agencies, and has been interpreted in a variety of ways. In current practice in protected areas, ecosystem management is recognized as a way to (i) integrate parks and protected areas into their surrounding landscapes so that parks do not function as isolated islands; (ii) account for the range of interactions that occur at spatial and temporal scales beyond the traditional scales used in park management; and (iii) incorporate a range of human values into the protection and use of the landscape (Woodley and Forbes 1995). Ecological integrity is an endpoint for ecosystem-based management.

4 Issues of Scientific Accuracy and Value

It is a difficult exercise for scientists to translate the results of quantitative science into detailed sets of management guidelines for park managers. There are several reasons for this difficulty. First and foremost, ecosystems are far more complex than any other system that humans have tried to understand or manage. Ecosystem science has many informing concepts that are useful in a general sense, but fail to qualify as analytical concepts. Ecosystem science is especially limited by the simple fact that studies have traditionally taken place in short temporal and small spatial scales. Brown and Roughgarden (1990) noted that 60% of all ecological studies had been conducted on a spatial scale less than one square meter and 70% on a time scale less than one year. Thus, it is not surprising that ecosystem scientists understand a lot about individuals, less about populations and little about communities and ecosystems. The problem is that there are few long-term, large-scale studies that are directly relevant to management.

Since ecosystems are so complex, scientists and resource managers are forced to measure only parts of the system. A common method is to use the notion of indicator species in making generalizations about the larger systems. This approach has limitations but is necessary because not all species can be monitored and studied. The science/manager linkage is critical in choosing these indicator species.

Ecosystems are far more complex than financial systems, yet society spends billions monitoring, assessing and tracking financial systems, often with unsuccessful results in terms of predicting future changes. It is not surprising then, that there is extreme difficulty in trying to predict responses to forest management activities within highly variable and complex ecosystems. Scientists are trained always to be aware of levels of accuracy and precision. Thus scientists are often reluctant to specify exact prescriptions when uncertainty exists. Management guidelines are therefore generally based on the best available science, and represent the professional judgment of the scientists and resource managers.

For reasons discussed above, predicting the behaviour of an ecosystem almost always involves some level of uncertainty. Conversely, traditional resource management, such as a forest harvesting system, operates to minimize uncertainly and maximize predictability of the resource. There is almost always a gap in precision between the two approaches. A forest manager can easily predict the impacts of a 75 metre versus 100 metre stream-side buffer on the allowable cut. However, researchers cannot easily predict the varying effects on biodiversity, wildlife movement, or water quality between the same two buffers. Research can indicate with some certainty that buffers are important, but it is more difficult to specify the influence of 50 or 75 m widths. This precision gap is often a source of misunderstanding between researchers, and managers of resources. The only solution to the precision gap, short of more research, is to rely on best professional judgment and the precautionary principle. The precautionary principle simply means that in an absence of sound information, the best policy is to err on the side of conservation.

Any consideration of an ecosystem ultimately depends upon both spatial and temporal scale. Ecosystems are composed of hierarchies of organization which exhibit both similar and unique properties at different levels. Any scientific management of protected area ecosystems must therefore consider scale as a central principle. When time horizons for ecosystem response stretch into decades or longer, the ability to solve a given environmental problem is severely limited. Both ecological theory and environmental management have tended to deal with issues that are shorter term and thus related to individuals and populations rather than longer term and related to community and ecosystem structure and function. For the management of spruce budworm, Holling (1986) has calculated the time horizon for analysis of management prescriptions must be in the order of 200 years. This estimate is based on the notion that the time span must be a small multiple of the slow variable, that is the 70 year rotation age of the spruce. In another example, Peterson (1984) has calculated that populations of wolves and moose will exhibit regular cyclic fluctuations of 30-40 years. This calculation is based on the body size and the notion that body size regulates the period of time required to complete a population cycle.

5 Providing Answers to Public Questions

Science is perhaps most valuable as a support tool to communicate to the public the value of protected areas. As implied earlier, three general arguments can be advanced for protecting biodiversity: utilitarian; ecological; and romantic. The utilitarian argument lists the potential uses for humanity of biodiversity and generally cites insignificant species that have conferred some human benefit such as a medicinal plant. The ecological argument sees humans as part of the interconnected web of life, with our own survival depending on maintaining the web. The romantic argument ascribes existence values to other species that are independent of human values. Different individuals and publics will focus on one or a combination of these reasons in their support for protected areas.

Science cannot provide information or support for all of the foregoing reasons, but it can provide answers to many relevant questions such as: If this area is set aside, what biodiversity is it likely to protect in the long term? This is a complex question and science can offer much insight with understandings of species-area relationships or the persistence of small populations. Consider the issue for example, of whether small, remnant wolf populations are likely to persist in Polish National Parks. Conservation science can make projections on the probability of survival, based on the present population size, habitat availability and life table data such as age-specific mortality, natility, and dispersal.

Another commonly asked question is: Why is this particular species important to conserve?. This question is often asked in connection with protected areas, where large sums of money are spent to protect or restore populations of a given species. Again science can provide part of the answer to questions of this type. Science can provide information on the specific role of species in an ecosystem or on the more general question of the relationship between species richness and ecosystem function. Although the relationship between ecosystem structure and function is unresolved in ecology, there are now some specific insights that support protected areas. Based on recent evidence, the capacity of ecosystems to resist changing environmental conditions, and to rebound from unusual climatic or biotic events is positively related to species numbers. Thus keeping all native species in protected areas is supported by science.

6 A Science Program for Protected Areas

If protected areas are to play the roles discussed above, namely to maintain ecological integrity by conserving native biodiversity and ecological processes, the role of science becomes clearer. There are several critical areas where ecosystem science can inform protected areas managers and citizens. These include protected area design as well as management of existing protected areas, including the restoration of existing systems. For an existing protected area to maintain ecological integrity, it must have an ecosystem science program that provides for the following: (1) characterizing the ecosystem, including its past and dynamics and its ecological condition; (2) predicting the future of the system; and (3) understanding the nature of ecosystem stresses affecting the protected area. All of these elements are required to contribute to a larger ecosystem-based management approach that will maintain ecological integrity.

6.1 Design of Protected Area Networks

Increasingly, research is leading to the conclusion that protected areas will only function to conserve native biodiversity if they are managed as integrated networks rather than individual, relatively small reserves. As an example, the IUCN category II protected areas in Poland (N=13) average only 103.6

km² in area (range 17.3 to 354.9). This is not a large enough area for the long term persistence of most area-requiring species, such as mammalian carnivores, or populations restricted to specialized habitats. It is essential that individual protected areas be integrated into a larger scale bioregional approach to conservation. This can take two forms. One way is for protected area staff to work with adjacent land owners and managers to ensure land management is compatible with conserving all native biodiversity. The other, and not necessarily exclusive, approach is to integrate protected areas into functionally connected networks as illustrated in Figure 1 (see Noss, 1995). Theoretically, these networks connect a range of large and small protected areas, which may include all IUCN classes. The concept is illustrated in Figure 1. A program of ecosystem science can calculate the probability of success of an integrated network. It can compare the utility of an integrated network approach against traditional management scenarios.

Figure 1. (source Bill Stephenson, Parks Canada). A theoretical framework for an integrated network of various classes of protected areas. The network must be functionally connected allowing movement of individuals between sub-populations The network consists large core reserve areas (large circles), smaller classes of protected areas (squares and small circles), functional corridors and buffer zones of compatible landuse.

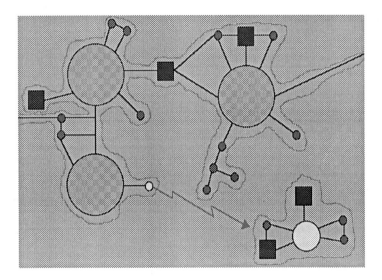

6.2 Characterizing the Ecosystem

Consideration of any ecosystem must begin with a basic description and understanding of the components of the system. In all cases, data characterizing the ecosystem should to be kept in a common up-datable data base. The main needs for ecosystem management are a characterization of past and present landscape levels. This should include landscape scale patterns of biological community types, including size, and distribution. Such information should include human land use information such as roads, buildings, visitation patterns and densities

At a community level, chronosequences of both natural and anthropogenic origin should be determined. Communities should be described for species present, structure of the vegetation, soils, topography and drainage.

At a species level detailed population dynamics information is required on indicator species. An indicator species is an organism whose characteristics are used as an index of ecological attributes that are too difficult, inconvenient or expensive to measure otherwise (Landres et al. 1988). Ideally, selected indicators should be hypersensitive to stress, have ubiquitous natural distribution, be easy to collect and assess, and be a population that is not harmed by sampling for assay purposes. Examples of useful indicator species are large carnivores, such as wolves, or k-selected species, such as eagles.

6.3 Understanding the dynamics of the ecosystem

Management can be defined as a choice between different predicted futures. If science is to inform management, it must be able to provide predictions of the future of protected areas. This is most relevant in regard to patterns of vegetation that are driven by dynamic elements such as fire, wind, floods, and extreme weather events. Most protected area managers tend to ignore or downplay the fact that ecosystems are dynamic. For example, in most world protected areas, fire is suppressed, rather that allowed to burn unimpeded. Biodiversity is adapted to a range of seral stages and individual species often make use of different seral stages. An example of changes in fire regimes is shown in Table 1. This table illustrates how the suppression of fire in the last 50 years in five Rocky Mountain National Parks in Canada has dramatically altered the fire regime. Area burned for the five parks has decreased from an historical average of 1083 to 29.5 km^2 per decade. As a result the average age of the forest has increased and early successional classes are virtually eliminated.

Understanding the dynamics of an ecosystem also means understanding the changes in community structure and species richness. An example of a dramatic change (decline) in species richness is presented in Table 2 from a small national park (Point Pelee) in southern Canada. Clearly, this table shows an impairment of ecological integrity. In addition to assessing the state of the protected area, such data can be used to model which species are likely to be at risk in the future and thus be applied to conservation efforts.

The ability to predict species extirpation and the probability of species survival is currently under development in ecosystem science. A collection of techniques is termed 'population viability analysis' and it can be of great value to protected area managers. The probability of persistence of a given population of a species is a function of age specific natality and mortality, immigration and emigration. These are in turn, controlled by habitat availability and suitability and other factors such as pollutants, hunting, and interaction with other species. Science can predict with certainly that small, isolated populations have a higher probability of extinction than large, interconnected populations. This has profound implications for protected area design and management.

Table 1. An example of alteration of the fire regime in 4 national parks in Canada
(source Pengelly 1995)

National Park (historic period)	Historic burned area km^2 per decade	1940-1995 burned area km^2 per decade
Jasper (1510-1930)	590	5.8
Banff (1488-1928)	267	8.6
Kootenay (1491-1931)	91	4.5
Yoho (1700-1980)	43	10.6
Waterton (estimate)	92	nil
TOTAL	1083	29.5

6.4 Research on Mitigation and Avoidance of Known Stressors

In some cases, research needs to be conducted on the specific mitigation and avoidance techniques that are required to ensure that external land use, tourism and other activities are compatible with the maintenance of ecological integrity. Often, the major stressors impacting on protected areas arise outside the national park. For example, in Northern American national parks, intensive logging is often carried out around the park boundaries. Some parks are working with timber managers and harvesters to determine what patterns of forest harvest on the landscape are best suited to the maintenance of native biodiversity. Native biodiversity cannot be maintained if intensive forestry dramatically alters species composition, stand age-class structure, or the patterns of stands on the landscape. Forestry, as a stressor, must be studied at both the community and landscape levels in order to plan for its compatability with protected areas. In some cases stressors originate within the park, from such things as inadequate sewage treatment, tourism developments or simply sheer numbers of visitors disturbing wildlife. A 1992 survey of stressors in 34 Canadian national parks listed 29 different stresses reported to be causing significant ecological impact. (Parks Canada, 1994) The most commonly reported impact was visitor/tourism facilities, with exotic vegetation being second. Surprisingly only four of 34 parks reported significant impacts from sewage. A scientific assessment of these stressors can provide managers with part of the understanding required to mitigate the problem.

Table 2. Changes in species richness (presence) of amphibians in Pt. Pelee National Park, Canada from historical (c1900) to 1994.

Species	Historical	1972 survey	1994 survey
Mudpuppy	x	extirpated	extirpated
Tiger salamander	x	extirpated	extirpated
American toad	x	x	x
Fowlers toad	x	extirpated	extirpated
Blanchard's frog	x	x	extirpated
Spring peeper	x	x	x
Western chorus frog	x	x	x
Gray treefrog	x	x	extirpated
Bullfrog	x	x	extirpated
Green frog	x	x	x
Leopard frog	x	x	x
Total Species Present	**11**	**8**	**5**

6.5 Assessing the State of Ecological Integrity

One critical role for a science program in protected areas is to assess regularly the state of ecological integrity of the area. Monitoring is an integral part of any field of endeavor. Monitoring the state of the protected area is an as essential management step, rather that an optional element. Much has been written in recent years on monitoring the state of ecosystems in general and protected areas specifically through the use of ecological indicators (see Noss 1990; McKenzie at. al. 1990; Woodley et al. 1993; Noss 1995). Sometimes indicators follow medical jargon (i.e. diagnosis, prognosis) and are dealt with according to integrity, threat and recovery. In my view, these are similar models. It is generally accepted that a suite of carefully chosen indicators can be used to assess the state of an ecosystem, providing indicators to measure both structure and function at a range of ecological scales.

There is also general agreement on the need to monitor measures of the ecological stressors acting on the system.

Table 3 shows one approach to developing such a of indicators. This list requires considerable investment in data collection, analysis and management to be successful. Other lists can be assembled that are less costly financially, but are not as rigorous scientifically. For example road densities, traffic volumes, levels of deer browsing and presence of large carnivores have all been suggested as simple indicators of ecological integrity (Noss 1995).

Table 3. An example of a suite of ecological indicators chosen to assess ecological integrity. This suite was chosen by Parks Canada for their State of Parks Report

Ecosystem Structure (Biodiversity)	Ecosystem Functions	Stressors
Species richness changes in species diversity number & extent of exotic species	Succession retrogression disturbance, frequencies and size (fire insect, flooding ..) vegetation age class distribution	Land-use patterns (human) Land-use maps, roads, building, development areas...
Population Dynamics reproduction, mortality, emigration and immigration rates of indicators species population viability of indicator species	Productivity Remote sensing - large scale Site specific monitoring program	Habitat fragmentation patch size, inter-patch distance, forest interior remaining
Trophic structure size and distribution of all taxa predation levels	Decomposition Site specific monitoring program	Pollutants sewage, petrochemicals long range transport of toxics
	Nutrient retention Site specific monitoring program for Ca, N,...	Other park specific issues

6.6 Ecosystem Restoration

Another role for ecosystem science is in the restoration of ecosystems, both inside and outside protected areas. Restoration may range from small scale species reintroductions to massive landscape-level projects, such as planned restoration of water flow patterns in the Florida Everglades (Boucher 1996). The present day Everglades is only half its original area and the colonies of wading birds have shrunk by 95 percent since the 1930s. This has made the Everglades a candidate for the biggest ecosystem recovery project in history, a project aimed at reinstating the hydrological complexity of the original natural system. It is estimated the project will require billions of dollars of mostly public funds. Its success depends on a high degree of ecological understanding of the ecosystem as well as the development of new technologies, and the forging of new political alliances.

Ecosystem restoration also means the restoration of ecological processes, of which wildfire is perhaps the most difficult. The vast majority of protected areas in the world protect fire-adapted ecosystems. Indeed, because of concerns for park visitors, built facilities and neighboring lands, fire exclusion is now practiced in the majority of protected areas. Yet biodiversity in these areas has evolved with fire, and fire is required for the persistence of many organisms. Developments in fire suppression, fire behaviour prediction and fire ecology now give managers the tools to restore fire to protected area ecosystems by using prescribed fire. However, despite scientific advances, there is still a reluctance to use fire in most protected areas.

As a result of the changes that have already taken place in many protected areas, there is an enormous need for ecosystem restoration. Based on the Florida Everglades example, the scope of restoration projects is mostly limited by our ability to set meaningful goals and build political alliances. However, restoration cannot be done without a sound program of ecosystem science.

7 Conclusion: Science and Decision Making

Science can realistically do only three things; (1) explain cause and effect relationships, (2) predict future events, and (3) provide a model for how systems operate. Thus, from a management perspective, the most that can be expected is that the science program will inform and support management, inform public opinion and create support, critique or assess policy, and guide actions. However, these are not trivial outcomes. They are essential parts of managing a protected area to conserve biodiversity.

A program of science is essential for parks which have discarded the old management paradigm of unspoiled nature, where ecosystems evolve independent of humans. As a result of the changes discussed in this paper, the natural regulation - or naturally evolving - model of protected areas is not workable, at least in the majority of world national parks. If protected areas are to fulfill a key role of conserving in-situ biodiversity, then the goals for protection must be explicitly stated, based on ecological integrity or a similar notion. In many cases, conservation will require careful human intervention.

Human intervention in the functioning of protected areas is a brave new world. In thinking this way, we are guilty of thinking that we understand more about the system that we really do. Our interventions must be cautious and we must be willing to admit mistakes and adapt to changing situations. Management actions must be considered experimental and they will often be risky. The use of prescribed fire in protected areas is a case in point. However, based on what we know of the functioning of ecosystems, to never intervene will result in protected areas that do not conserve native biodiversity.

A program of ecosystem science has an essential role to play in guiding protected area management. However, management must be willing to make decisions in the absence of a full understanding. Thus, management must always be precautionary, erring on the side of conservation. In a world of growing human pressures, including massive visitation to protected areas, this is increasingly difficult. However, if countries are to live up to the letter and spirit of the international Biodiversity Convention, then conservation must dominate management, in actions and in rhetoric.

8 References

Boucher, N. 1996. *Back to the Everglades*. Internet Article. URL http://web.mit.edu /afs/athena/org/t/techreview /www/articles/aug95/Boucher.html.

Canadian Interagency Forest Fire Centre (CIFFC). 1996. *Canadian Fire Statistics*. Internet URL. http://www.nofc.forestry.ca/fire/cwfis/stats/index.html

Dansereau, P.-R. and Y. Bergeron. 1993. Fire history in the southern boreal forest of northwestern Quebec. *Can. J. For. Res*. Vol. 23:25-32.

Grumbine, R. E. 1994. What is ecosystem management? *Conservation Biology*. 8(1):27-38.

Halvorson, W. L. and G. E. Davis (eds). 1996. *Science and Ecosystem Management in the National Parks*. The University of Arizona Press, Tucson, Arizona.

Holling, C.S. (1986). The resilience of terrestrial ecosystems: local surprise and global change. In. W.C. Clark and R.E. Munn (eds.) *Sustainable Development of the Biosphere*. Cambridge University Press, Cambridge: 292-317.

Kay, C. E. 1990. *Yellowstone's northern elk herd: A critical evaluation of the natural regulation paradigm.* Ph.D. Dissertation, Utah State University, Logan, Utah.

Keiter, R.B. and M.S. Joyce (eds.). 1991. *The Greater Yellowstone Ecosystem*. Yale University Press, New Haven.

Machlis, G. E. and D.L Tichnell. 1985. *The State of the World's Parks*. Westview Press, Boulder, Co.

McKenzie, D. H., D. E. Hyatt, and J. McDonald (eds.) 1990. *Proceedings of the International Symposium on Ecological Indicators.* Ft. Lauderdale, Fl. October 16-19, 1990. Elsevier, Amsterdam.

Newmark, W.D. 1985. Legal and biotic boundaries of western North American national parks: A problem of congruence. *Biological Conservation,* 33:197-208.

Noss, R. 1990. Indicators for monitoring biodiversity: A hierarchical approach. *Conservation Biology* 1: 159-164.

Noss, R. 1995. *Maintaining Ecological Integrity in Representative Reserve Networks.* A World Wildlife Fund Canada/ World Wildlife Fund United States Discussion Paper. World Wildlife Fund Canada, Toronto.

Peterson, R.O., Page, R.E. and Dodge, K.M. (1984). Wolves, moose and the allometry of population cycles. *Science* 224:1350-1352.

Rapport, D.J., Regier, H.A. and Thorpe, C. (1981). Diagnosis, prognosis, and treatment of ecosystems under stress. In. G. Barrett and R. Rosenberg (eds.) *Stress Effects on Natural Ecosystems.* John Wiley Ltd. , Chichester: 269-80.

Shellford, V.E. 1932. Ecological Society of America: A nature sanctuary plan unanimously adopted by the Society, December 28, 1932. *Ecology* 14(2):240-245.

Woodley, S. 1992. *A survey of ecosystem stresses in Canadian National Parks.* Internal research report. Parks Canada, Ottawa.

Woodley, S.J. 1993. *Assessing and monitoring ecological integrity in parks and protected areas.* Ph.D. Thesis. Faculty of Environmental Studies, University of Waterloo, Waterloo, Ontario.

Woodley, S. 1993. Tourism and sustainable development in parks and protected areas. In: Nelson, J.G., R. Butler and G. Wall (eds.). *Tourism and Sustainable Development: Monitoring, Planning, Managing.* University of Waterloo, Dept. of Geography Publication Series Number 37, Waterloo: 83-96.

Woodley, S. and Forbes, G. 1995. Ecosystem Management and Protected Areas. In: Herman, T.B., S. Bondrup-Neilsen, J.H.M. Willison and N.W.P. Munro (eds.) *Ecosystem Monitoring and Protected Areas.* 1994 Proceedings of the Second International Conference on Science and the Management of Protected Areas, Dalhousie University, Halifax, Canada. pp. 50-58.

Landscape Ecology as a Basis for Protecting Threatened Landscapes

Almo Farina[1]

[1] Laboratory of Landscape Ecology, Lunigiana Museum of Natural History
Aulla, Italy

Abstract. In human dominated environments it seems more and more clear that the mitigation of the human impact and the conservation of the ecodiversity of the systems are priority goals to assure sustainable development. Most of the principles and paradigms used in landscape ecology may be applied to preserve threatened landscapes inside and outside parks and protected areas. The hierarchical and scaled approach of this discipline, combined with the study of the land mosaic, are essential to track processes that occur at different spatial and temporal scales in a complex environment in which different disturbance regimes occur. The capacity to take into account natural as well as human processes appears to be the distinctive attribute of landscape ecology.

Keywords. Threatened landscapes, landscape ecology, protected areas, scale

1 Introduction

We can consider a landscape to be any piece of earth, water and air nested in a hierarchy of spatial scales. Landscapes are generally large portions of a region and may be coincident with the watershed or they may be larger and include watersheds of different order. Scaling properties characterize each level and the spatial arrangement of the components of the land mosaic is important to predict the functioning level of each component. The surrounding system controls and is conditioned by the target system. A disturbance regime maintains dynamism in the system and in its sub-components (Figure 1).

Figure 1. General model illustrating the relationship between biodiversity, ecosystem processes and landscape dynamics. This new paradigm considers two-way interactions between biodiversity, landscape and ecosystem when we move to a broad scale. The dashed arrows indicate: influence, solid lines; direct interactions and hourglasses; controls (from Turner et al. 1994).

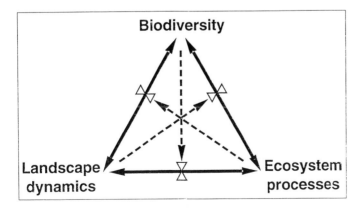

A landscape is a very complex system in which abiotic and biotic components are strictly connected in space. The study of this complexity is the preferred theme of landscape ecology (Naveh and Lieberman 1984; Risser et al. 1984; Forman and Godron 1986). The conceptual framework of landscape ecology is based on the paradigm that a system is composed of a hierarchy of sub-systems and is part of a superior system. Many features of the landscape mosaic such as: size distribution; boundary form; perimeter to area ratio; patch orientation; context; contrast;connectivity; richness; evenness; dispersion and predictability; can be measured (Wiens et al. 1993). The strength of landscape ecology mainly depends on the capacity to link natural processes with the human sphere (Naveh and Lieberman 1984) despite the common trend to ignore the human ecological needs (Grizzle 1994) and seems a formidable tool to approach complexity in parks and protected areas.

The cultural landscape that is the result of cultural evolution in the land, is a very complex system. Cultural evolution produces a high cultural diversity largely connected with natural diversity (Haber 1995; Farina 1995). During recent decades accelerating biological and cultural landscape degradation has forced ecologists to investigate more fully the interactions between landscape and cultural forces to assure sustainable management (Naveh 1995; Plachter and Rossler 1995). We can protect a landscape more efficiently when we understand how this system functions across a range of spatio-temporal and ecological scales. Landscape ecology appears to be a very efficient approach to the study, management and protection of valuable threatened landscapes because this approach can scale and manage information for the selected system and the hierarchy of other systems of which it is part (Mladenoff et al. 1994).

2 Scaling Approach and Management

The comprehension of processes that produce patterns is the essence of the science. Many of the changes in pollutant and green house effects have origins at a fine scale although they are perceived at broad scale (Levin 1992). Ecological processes are controlled by factors that may be different at a fine as opposed to a broad scale (Turner et al. 1994). Tracking patterns and processes is important for any management strategy but often they change across a scale. Consequently scaling is one of the first actions to take into consideration in studying a landscape. This means deciding at which spatial and temporal scale the patterns or processes show or give the maximum information. Working at various scales with the shaping factors allows us to understand the more comprehensive spatial and temporal realm in which the factors are moving and how the information is transferred from one level of the system to another. The management consequences appear extremely important; for example in moving from fine to broad scale.

Managing a landscape means applying ecological models to human activity. These models should take in account the maintenance of the ecological fluxes and avoid being dazzled by ephemeral species oriented patterns. The methods to achieve this goal are mainly shared among geostatistics, GIS procedures and fractal geometry (Isaaks and Srivastava 1989; Burrough 1986; Milne 1991). Choice of sample size and the resolution of quantitative analysis are two faces of the same coin. For instance variograms and neighboring analysis of the matrix are two of the main tools to separate processes across scales. Fractal geometry is widely applied to investigate broad scale phenomena in terms of patch mosaic complexity (Krummel et al. 1987) and also to study the movement and diffusion of organisms in the landscapes (Johnson and Milne 1992; With 1994; Wiens et al. 1995). GIS procedure is used to implement and to test models, and uses a data base to indicate or map the geographical character of landscapes (Burrough 1986) (Figure 2).

3 Ecotone Perspectives

An important issue in landscape ecology and mainly in nature conservation, is the study of fluxes of materials and energy across ecosystems. If isolated, a system rapidly declines and sensitive species become extinguished. The maintenance of fluxes is essential to assure healthy conditions in the system. Often landscapes are severely threatened by the isolation produced by human infrastructure and land use (examples, concrete barriers of highways, large stands of intensive croplands separating forests, urban development separating cultural landscapes). Focusing studies on ecotones allows us to investigate mosaic complexity and dynamics (Risser 1995).

Figure 2. The human and natural shaping effects on the landscape have been emphasized by the fractal analysis by Krummel et al. (1987). D= fractal dimension (calculated regressing log of perimeter on log of area of forested patches of the Mississippi flood plain). The change in D value moving from small forested patches to larger patches represents the threshold between human influence (cultivation, hedgerows) that simplifies the land mosaic and natural processes that occur at broader scale, producing a more complicated mosaic.

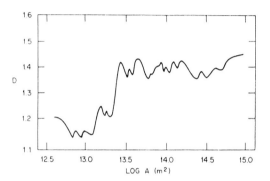

In a heterogeneous landscape the patches composing the mosaic are in contact with each other. The surface of contact is considered to be an ecotone. The ecotone concept is extremely important to understanding the dynamic of the systems (Wiens et al. 1985). Each component of the mosaic is not an island that receives energy from the atmosphere and that preserves or degrades energy and biomass in the interior of the patch. Rather the system communicates with the neighboring patches across a semipermeable membrane, the ecotone. Ecotone is located where the rate and the dimension of the ecological transfers - solar energy, nutrient exchange - undergo abrupt change. The extent of ecotones, and their complexity, are important factors conditioning the presence and the persistence of species. They are therefore, relevant structures to support high biodiversity. Ecotones exist across a broad range of spatial and temporal scales and are produced by a hierarchy of tension factors such as air mass dynamics, megatopography, local geomorphology, disturbance, competition, plant growth and development (Hansen and di Castri 1992).

Size, shape, biological structure, structural constraint, internal heterogeneity, density, fractal dimension and porosity are some of the variables that can be used to study these structures. The extension of ecotones is an efficient index of landscape fragmentation; when the landscape appears very fragmented the ecotone length is very high.

4 Applying the Metapopulation Concept

A landscape that by definition is heterogeneous, may be seen as a chessboard in which suitable cells are intercalated to hostile ones. Plants and animals living in such a system tend to be present in a cell or group of cells creating a clumped distribution. Each group maintains contacts with the other groups of sub-populations; this system has been called by Levins (1970) metapopulation. Extinction and recolonization are two essential components of the metapopulation model. The distance of a sub-population to another is the key factor determining survival or the extinction (Figure 3).

Figure 3. Examples of metapopulation models. Habitat patches are indicated by closed circles; filled=occupied, unfilled=vacant. Dashed lines represent the boundaries of population. Migratory movements are indicated by arrows. A, Levins metapopulation. B, Core-satellite metapopulation. C, Patchy population. D, Non-equilibrium metapopulation. E, Intermediate case of B and C (from Harrison 1991)

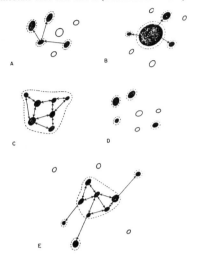

This model may be applied to many real situations and explains genetic variability at broad scale and contemporarily the habitat adaptation of a sub-population at fine scale. Dispersion, connectivity and fragmentation are usual processes for the organisms living in a land mosaic. The dispersion rate of a species depends on the energetic convenience to move, the intraspecific competition and the reduction of inbreeding risk. In this respect the metapopulation paradigm has been usefully employed in planning bird conservation in fragmented landscapes (Opdam et al. 1994).

5 Source-Sink Model to Assess Population Dynamics

In heterogeneous systems the quality of the patches depends on many factors and is species-specific. Some patches play a fundamental role in conserving a population or a species. Pulliam (1988) calls a source: a patch in which the rate of offspring exceeds the number of local extinctions. From these source patches individuals disperse into the surroundings.

On the contrary a sink is a patch in which the death rate is higher than the offspring and a species will be extinguished without a continuous immigration of individuals from other patches. Location of source patches is essential to maintain a population in an area. The assessment of the suitability of a habitat or patch may lead to disastrous results if we cannot distinguish source or sink characteristics of a patch, because sometimes sink patches can support a very high population density.

6 The Application of Percolation Theory

Percolation theory - which was created to model fluid diffusion (Gefen et al. 1993; Stauffer 1985) - is extensively used in landscape ecology to create what are called neutral models (Gardner et al. 1987; Gardner et al 1992). The behavior of a mosaic composed of a matrix in which each cell represents the spatial distribution of a variable can be efficiently studied by applying this theory. A variable - which can be a land cover type, a vegetation type, a structuring factor, and so forth - is percolating when it occupies at least 59% of the matrix as a web of contiguous cells from one side to another of the matrix. This percolation threshold or critical probability, has been calculated using a high number of simulations on a computerized matrix. Number, size, and shape of patches changes as a function of the fraction of landscape occupied by a specific land cover or habitat. Interesting relationships exist between percolation threshold and movements of animals and the structure of ecotones. Neutral models are useful to simulate the spreading behavior in the mosaic landscape of influential factors such as fire, diseases, fragmentation rate, and animal movements (Gardner et al. 1991).

7 Disturbance and Fragility in Biological and Cultural Landscape

Disturbance is a common process in nature but has not received much attention until recently (Pickett and White 1985). Many valuable threatened landscapes are at risk through modification of this process. Disturbance is an essential factor and a system may be more or less exposed to it. To protect landscapes it is often essential to understand the value and the importance of the disturbance. An exemplar case is represented by Yellowstone National Park. The fire repression policy of the past decades modified this system and created a high inflammable system. Ecodiversity was depressed by this management policy (Romme and Despain 1989). After the big fire of 1988 the dynamics of the system were rejuvenated and spatial heterogeneity has been increased, with benefits for plants and animals .

Recently the fragility concept has been included in protected area perspectives. Fragility refers to sensitivity to change and is an important concept in landscape management. Many protected landscapes, for example riparian ones, are fragile systems. For these systems the size of the area assumes a special importance. In fact the more fragile the system, the more undisturbed surroundings are needed to guarantee its healthy status (Nilsson and Grelsson 1995). From this perspective cultural landscapes have to be considered fragile systems. In fact as human stewardship is reduced by land abandonment and other processes, species diversity and ecodiversity in general tend to decline (Farina 1991).

8 Spatial Arrangement of the Land Mosaic

According to island biogeography theory (MacArthur and Wilson 1967) the size and isolation of an area are fundamental characteristics influencing species diversity. The landscape ecology approach involves this theory and takes in account size, shape, spatial position and similar attributes of the patches in the mosaic. The patches may be considered as islands although they are not completely isolated but more or less surrounded by a hostile environment that may represent a barrier for some species or habitat gradient for others. The shape of patches has great implications for species diversity and distribution, and more generally for the dynamics of the system (water, nutrient fluxes, resource allocation) and for the movement of animals (Gutzwiller and Anderson 1992). Patches elongated as peninsulas have more edges than semicircular patches (Forman 1995). Spreading or clumping are important attributes of patches and can be monitored by using contagion indexes (O'Neill et al 1988).

9 Fragmentation and Corridors

Fragmentation is one of more relevant patterns arising from disturbance processes. Fragmentation means the reduction of size and increase in isolation of patches from a main core or cores. Largely produced by human disturbance regime, this pattern severely affects many species of plants and animals (Saunders et al. 1991) in tropical forests (Lovejoy et al. 1987)as well as in boreal forests (Harris 1984) (Figure 4). Intensive agriculture increases fragmentation of forests, menacing habitat exigent species, and inner species, and favoring edge and opportunistic species.

Figure 4. Bird species richness plotted against the area of grassland fragments. The number of bird species is positively correlated with the fragment size (from Herkert 1994).

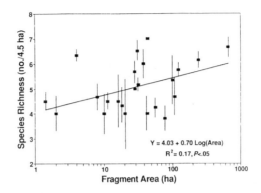

To calculate the level of fragmentation - expressed in terms of patch connectivity - many landscape indexes are available. Some are weakly correlated with dispersal processes (Schumaker 1995). In many cases the possibility of reducing forest fragmentation and mitigating the consequent effects is not yet workable and the maintenance of the connectivity among the fragments is the only possible strategy (Shafer 1995). In this respect landscape ecology again assumes a primary role.

Corridors are narrow strips of a cover type surrounded by different habitats. Corridors are important structures that can mitigate habitat loss. Plants and animals can preferentially use these corridors. Corridors play a fundamental role in dispersion and migratory movements of plants and animals and in maintaining regional biodiversity, especially along rivers (Knopf et al. 1988; Naiman et al. 1993).

10 Conclusions

Landscape ecology is a promising discipline for studying the ecological processes occurring in landscapes across a hierarchy of scales. It can lead to workable management tools. The transdisciplinary and anticipatory character of landscape ecology is essential to face the challenges involved in studying the complexity of systems. Taking in account that the influence of humanity on the landscape processes will increase in the future, new tools will be necessary to mitigate the impact, protect ecological diversity and provide for sustainable development in human societies. In this respect the Red Book Program first launched by Naveh (Naveh and Lieberman 1984) has elicited interest and support among landscape ecologists (Grove and Rackham 1993). The main goal of the Red Book is to describe in formal and detailed ways the ecological conditions of threatened landscapes. They can be used to create awareness and interest among decision-makers, politicians and the media. They can be used to build scenarios for multi-beneficial and sustainable land use.

11 References

Burrough, P.A. 1986. Principles of geographical information systems for land resource assessment. Clarendon Press, Oxford.

Farina, A. 1991. Recent changes of the mosaic pattern in a montane landscape (North Italy) and consequences on vertebrate fauna. Option Méditerranéennes: Serie seminaires, n. 15, Zaragoza, Spain, pp. 121-134.

Farina, A. 1995. Uplands farming systems of the Northern Apennines. The conservation of biological diversity. In: Halladay, P. and Gilmour, D.A. (eds.), Conserving biodiversity outside protected areas. The role of traditional agro-ecosystems. IUCN , Gland, Switzerland, and Cambridge, UK, pp. 133-135.

Forman, R.T.T. 1995. Land Mosaics. The ecology of landscapes and regions. Cambridge University Press, Cambridge.

Forman, R.T.T. and Godron, M. 1986. Landscape ecology. Wiley and Sons, New York.

Gardner, R.H., Milne, B.T., Turner, M.G., O'Neill, R.V. 1987. Neutral models for analysis of broad scale landscape patterns. Landscape Ecology 1: 19-28.

Gardner, R.H., Turner, M.G., O'Neill, R.V., Lavorel, S. 1991. Simulation of the scale-dependent effects of landscape boundaries on species persistence and dispersal. In: Holland, M.M., Risser, P.G., Naiman, R.J. (eds.), Ecotones. The role of landscape boundaries in the management and restoration of changing environments.Chapman and Hall, New York, pp.76-89.

Gardner, R.H., Turner, M.G., Dale, V.H., O'Neill, R.V. 1992. A percolation model of ecological flows. In: A.J. Hansen and F. di Castri (eds.) , Landscape boundaries, consequences for biotic diversity and ecological flows. Springer-Verlag, New York, pp 259-269.

Gefen, Y., Aharony, A., Alexander, S. 1983. Anomalous diffusion on percolating clusters. Physical Rev. Lett. 50: 77-80.

Grizzle, R.E. 1994 -Environmentalism should include human ecological needs. BioScience 44: 263-268.

Grove, A.T. and Rackham, O. 1993. Threatened landscapes in the Mediterranean: examples from Crete. Landscape and Urban Planning 24: 279-292.

Gutzwiller, K.J. and Anderson, S.H. 1992. Interception of moving organisms: influences of patch shape, size, and orientation on community structure. Landscape Ecology 6: 293-303.

Haber, W. 1995. Concept, origin and meaning of Landscape. In: Droste, B.V., Plachter, H., Rossler, M. (eds.), Cultural landscapes of universal value. Gustav Fisher, Jena.

Hansen, A.J. and di Castri, F. (eds.) 1992. Landscape boundaries. Consequences for biotic diversity and ecological flows. Springer-Verlag, New York.

Harris, L.D. 1984. The fragmented forest. University of Chicago Press, Chicago.

Harrison, S. 1991. Local extinction in a metapopulation dynamics. Biological Journal of the Linnean Society 42: 73-88.

Herkert, J.R. 1994. The effect of habitat fragmentation on midwestern grassland bird communities. Ecological Applications 4: 461-471.

Isaaks, E.H. and Srivastava, R.M. 1989. An introduction to applied geostatistics. Oxford University Press, New York.

Johnson, A.R. and Milne, B.T. 1992. Diffusion in fractal landscapes: simulations and experimental studies of tenebrionid beetle movements. Ecology 73: 1968-1983.

Knopf, F.L., Johnson, R.R., Rich, T, Samon, F.B., Szaro, R.C. 1988. Conservation of riparian ecosystems in the United States. Wilson Bull. 100: 272-284.

Krummel, J.R., Gardner, R.H., Sugihara, G., O'Neill, R.V., Coleman, P.R. 1987. Landscape patterns in a disturbed environment. Oikos 48: 321-324.

Levin, S.A. 1992. The problem of pattern and scale in ecology. Ecology 73: 1943-1967.

Levins, R. 1970. Extinction. In: Gerstenhaubert, M. (ed.) Some mathematical questions in biology. Lectures in mathematics in the life science. American Mathematical Society, Providence, Rhode Island, pp. 77-107.

Lovejoy, T.E. Bierregaard, R.O., Rynalds, A.B., Malvom, J.R., Quintela, C.E., Harper, L.H., Brown, K.S., Powell, A.H., Powell, G.V.N., Shubart, H.O.R., Hays, M.B. 1987- Edge and other effects of isolation on Amazon forest fragments. In: Soulé, M.E. (ed.), Conservation Biology. The science of scarcity and diversity. Sinauer Associates, Sunderland, Massachusetts, pp. 257-285.

MacArthur, R.H. and Wilson, E.O. 1967. The theory of island biogeography. Princeton University Press, Princeton.

Milne, B.T. 1991. Lessons from applying fractal models to landscape patterns. In: M.G. Turner and R.H. Gardner (eds.), Quantitative methods in landscape ecology. Springer-Verlag, New York, pp. 199-235.

Mladenoff, D.J., Crow, T.R., Pastor, J. 1994. Applying principles of landscape design and management to integrate old-growth forest enhancement and commodity use. Conservation Biology 8: 752-762.

Naiman, R.J., Decamps, H., Pollock, M. 1993. The role of riparian corridors in maintaining regional biodiversity. Ecological Applications 3: 209-212.

Naveh, Z. 1995. Interaction of landscapes and cultures. Landscape and Urban Planning 32: 43-54.

Naveh, Z. and Lieberman,A.S. 1984. Landscape ecology: theory and application. Springer-Verlag, New York.

Nilsson, C. and Grelsson, G. 1995. The fragility of ecosystems: a review. Journal of Applied Ecology 32: 677-692.

Opdam, P., Foppen, R., Reijnen, R., Schotman, A. 1994. The landscape ecological approach in bird conservation: integrating the metapopulation concept into spatial planning. Ibis 137: S139-S146.

O'Neill, R.V., Krummel, J.R., Gardner, R.H., Sugihara, G., Jackson, B., Deangelis, .L., Milne, B.T., Turner, M.G., Zygmunt, B., Christensen, S.W., Dale, V.H., Graham, R.L. 1988. Indices of landscape patterns. Landscape Ecology 1: 153-162.

Pickett, S.T.A. and White, P.S. 1985. The ecology of natural disturbance and patch dynamics. Academic Press, New York.

Plachter, H. and Rossler, M. 1995. Cultural landscapes: reconnecting culture and nature. In: von Droste, B., Plachter, H., Rossler, M. (eds.), Cultural landscapes of universal value. Gustav Fisher, Jena. pp.15-18.

Pulliam, R. 1988. Sources, sinks, and population regulation. American Naturalist 132: 652-661.

Risser, P.G. 1995. The status of the science examining ecotones. BioScience 45: 318-325.

Risser, P.G., Karr, J.R., Forman, R.T.T 1984. Landscape ecology. Directions and approaches. Illinois Natural History Survey, Special Publication, number 2, Champaign, Illinois.

Romme, W.H. and Despain, G. 1989. Historical perspective on the Yellowstone fires in 1988. BioScience 39: 695-699.

Saunders, D., Hobbs, R.J., Margules, C.R. 1991. Biological consequences of ecosystem fragmentation: a review. Conservation Biology, 5: 18-32.

Schumaker, N.H. 1995. Using landscape indices to predict habitat connectivity. Ecology 77: 1210-1225.

Shafer, C.L. 1995. Values and shortcomings of small reserves. BioScience 45: 80-88.

Stauffer, D. 1985. Introduction of percolation theory. Taylor and Francis, London.

Turner, M.G., Gardner, R.H., O'Neill, R.V. 1994. Ecological dynamics at broad scales. Ecosystems and landscapes. Supplement to BioScience, pp 29-35.

Wiens, J.A., Crawford, C.S., Gosz, R. 1985. Boundary dynamics: a conceptual framework for studying landscape ecosystems. Oikos 45: 421-427.

Wiens, J.A., Stenseth, N.C., Van Horne, B., Ims, R.A. 1993. Ecological mechanisms and landscape ecology. Oikos 66: 369-380.

Wiens, J.A., Crist, T.O., With, K.A., Milne, B.T. 1995. Fractal patterns of insect movement in microlandscape mosaic. Ecology 76: 663-666.

With, K.A. 1994. Using fractal analysis to assess how species perceive landscape structure. Landscape Ecology 9: 25-36.

Landscape Approaches to National Parks and Protected Areas

Adrian Phillips[1]

[1] Department of City and Regional Planning, University of Wales, Cardiff, UK and Chairman of IUCN's Commission on National Parks and Protected Areas

Abstract. Wilderness protection was for long the driving force behind protected areas. However, the concept of wilderness can be misleading, and many humanized areas (i.e. landscapes) are also rich in natural and cultural values. Though a concern with landscapes has tended to be associated with Europe, it is of broader relevance. This is apparent in: the IUCN Protected Area Management Category V - Protected Landscape/Seascape, (there are >2,200 such sites around the world); Cultural Landscapes under the World Heritage Convention; the European Strategy for Landscape and Biological Diversity; and the proposed European landscapes convention. There is a need for research in landscape typology, landscape evaluation and landscape management; the last has to address the key question of how to manage landscape when it is the product of human use.

Keywords. Landscape, parks and protected areas, wilderness, International Union for Conservation of Nature (IUCN)

1 Introduction

My thesis is that conservation philosophy and practice was for many years driven primarily by concern for the protection of wilderness and pristine nature: but that gradually a complementary concept has taken root which recognizes the value of more humanized environments - or landscapes. As evidence of the growing importance given to landscape issues, four important recent policy developments are discussed here. The paper concludes by highlighting some of the research questions which face the academic and policy communities in applying the concept of landscape to protected areas policy and practice.

2 The Roots of Wilderness Concern

Much has been written about the origins of human obsession with the wilderness (see for example Thomas 1983; and Schama 1996): of the close affinity between a primeval fear, and near religious awe of wild places. It was however, in the United States, during the last century, that both the philosophy of wilderness worship and the practice of wilderness protection became a real political force.

The reasons for this were several. There was a reaction against the plunder of wild nature by many of the early North American settlers; there was too a search for the soul of a new nation which had no Gothic cathedrals, great castles nor fine palaces to elevate into national monuments. There was an element of the deliberately democratic in American practice, aiming to protect nature so as to make it available to all. Henry Thoreau was among a number of American writers who saw in this opening up of nature for human pleasure and enjoyment a contrast with the European practice of closing off areas of land for the exclusive enjoyment of the wealthy and powerful. And of course the wonders of the west of the United States were truly splendid: the greatest trees ever seen, the greatest canyon in the world, the mightiest geysers.

The Yosemite Valley in California, and the Big Trees (sequoias) of the Mariposa Grove were the first focus of attention. President Lincoln and the US Congress were not too preoccupied with the

American Civil War to take action in 1864 to protect the area, and to grant it to the State of California for the benefit of the people, for their resort and recreation, to hold them inalienable for all time. The Big Trees were regarded as some sort of botanical pantheon, a living American monument (Schama 1996). The idea of there being a pantheon was reinforced when the mightiest sequoias were named after great American heroes: General Sherman lives on to this day:

> 'The sequoias seem to vindicate the American intuition that colossal grandeur spoke to the soul. It was precisely because the red columns of this sublimely American temple had not been constructed by the hand of man that they seemed providentially sited, growing inexorably more awesome until God's new Chosen People could discover them in the heart of the Promised West'. (Schama 1996).

Following the designation of the Yellowstone National Park in the State of Wyoming as the world's first true national park in 1872, a number of other new nations followed suit: Australia (its first park was the Royal National Park, designated in 1879), Canada (Banff National Park 1885) and New Zealand (Tongariro National Park 1894). In 1916, the United States National Parks Service was established, and in more recent years its strong international division has carried the North American vision of national parks to the furthest corners of the world.

Americans are rightly proud of their world-wide influence. Roderick Nash (1967) wrote 'since Yellowstone, we have exported the National Park idea round the world. We are known and admired for it, fittingly, because the concept of the national park reflects some of the central values and experiences in American culture'. Everhardt (1972) talks about all the other parks and reserves created since, in the United States and elsewhere, as 'Yellowstone's Children'. However, he is sensitive enough to acknowledge that it would be 'a disservice to the world parks' movement and to the contributions that many other nations have made to the development of the national park idea, to suggest that other countries simply copied the American experiment'.

So although the American model of a national park provided the stimulus, the concept took root in rather different ways. In Africa and to a lesser extent in the Indian sub-continent, for example, the colonially-led steps towards conservation were focused upon protecting large mammals, often those which had previously been ruthlessly hunted. And in some of the more long settled parts of Europe the emphasis was more upon protecting humanized landscapes - and as this happened the words, national park, were gradually replaced by the broader term: protected area.

By the time that IUCN, the International Union for the Conservation of Nature, now known widely as the World Conservation Union, held its 10th General Assembly in New Delhi in 1969 there were some 3,000 or so protected areas around the world, reflecting a great diversity of approaches to protected area concepts. Indeed the variety of types of protected areas had become a source of concern to the traditionalists, who wished to hang on firmly to the North American ideal of a national park as essentially a wilderness area available for recreation use by the public. Accordingly the 1969 General Assembly adopted a Resolution which defined national parks as places:

'1. Where ... eco-systems are not materially altered by human exploitation and occupation,
2. the highest competent authority of the country has taken steps to prevent or eliminate as soon as possible exploitation or occupation in the whole area and enforce .... respect for ecological, geomorphological and aesthetic features,
3. and where visitors are allowed to enter under special conditions ....'

The Resolution went on to request governments not to designate as national parks scientific reserves from which all but scientists are excluded, nor parks and reserves managed by private institutions or by lower levels of government, nor wildlife reserves like those of Africa. Most of all, they asked that this term should not apply to inhabited and exploited areas where landscape planning is in place and measures are taken for the development of tourism.

The Resolution is instructive to read in association with the 1969 report of what was then known as IUCN's Commission on National Parks. This makes it clear that the Commission members wished to maintain the integrity, as they saw it, of the last scenic areas and natural habitats of an increasing number of threatened animal and plant species. The Commission's report also made clear that the protection of wilderness values was to be its central responsibility in future.

3 Wilderness versus Landscape

Nearly 30 years later we can see that this approach to the protection of nature was seriously deficient in two important respects:

1. The concept of wilderness areas of pristine nature is largely discredited, and
2. The importance of protecting other areas in which nature lives alongside humanity is increasingly recognized.

3.1 There is No Pristine Nature Left

The mythical pristine environment exists only - or very largely - in human imagination. It is now known, for example, that Australia's plants and animals, and indeed its whole landscape, have been dominated by human-induced land use practices (notably fire) going back perhaps 50,000 or even 60,000 years (Hill and Press, 1994). Whereas Europeans arriving in the island continent assumed that they saw a pristine environment, in fact humankind had shaped it for far longer than many parts of the Asian continent from which the new Australians came. Or again, 95% of the great forests of Amazonia were influenced by a human population of at least 8 million at the time of Christopher Columbus's first voyage. By 1492, these forests were already subject to centuries of management, with tree species deliberately moved around to create forest fields where plants, used by indigenous people, were collected together as food depots (Denevan 1992). There are many other such examples from around the world of where apparently natural environment turns out to have been subject to human influence over a long period of time.

3.2 The Importance to Conservation of Landscapes in Which People Live and Work

Human manipulation can enhance biodiversity. For example, in Britain, a traditional woodland management practice known as coppicing provides ideal conditions for woodland flora. Coppicing involves the periodic cutting back of trees to their stumps, from which they sprout again, thereby permitting the sunlight to reach the woodland floor for a few years at a time. A study of native agriculture in North America has shown that it can enhance biodiversity and that species richness is greater than in adjacent or analogous habitats that are not cultivated (Reichardt et al. 1994). Various studies in Asia show that shifting cultivation, long considered damaging to conservation interests, can help enhance or protect nature, providing it takes place at levels of low intensity. Traditional shifting agriculture can favour elephants, wild cattle, deer, wild pigs and so on; and the herbivores attract tigers, leopards and other predators (Pretty and Pimbert 1995).

Also, farmers can be agents of biodiversity. Traditional farming methods make a great contribution to biodiversity within species of domesticated crops or livestock. Thus there are thousands of land races, or crop populations which have, through centuries of cultivation, adapted to particular regions, even individual small valleys or hillsides. These are now being replaced by modern highly-bred crop cultivators, which may lead to improved yields but result in more dependence on chemicals and all the other problems of monoculture cultivation. Similarly, many of the numerous animal breeds which have developed through traditional farming are being lost. Over 220 breeds of cattle worldwide have become extinct, and 126 breeds of pig (WCMC 1992).

Of particular relevance in these post-Rio days, as we search for models of sustainable resource use, many human-altered landscapes contain important lessons for resource management. Such areas reflect ways in which people's use of the land has evolved along sustainable lines, conserving water, soils and vegetative cover. This is not true in every case, of course, but it is now widely recognized that there is much to be learnt from the study of traditional land use systems. Agenda 21 (chapter 10), for example, refers to the need to strengthen land management systems for land and natural resources by including appropriate traditional and indigenous methods: examples of these practices include pastoralism, Hema reserves (traditional Islamic land reserves) and terraced agriculture.

Many such humanized landscapes have other values too: they are important for their scenic qualities and can support a thriving tourist industry in areas which are often better suited to accommodate large numbers of people than more natural sites. They help to protect watersheds and perform other environmental services. Also landscapes are often important from a cultural perspective, since many are not only a physical expression of a community's long interaction with the land but are often seen as the embodiment of that community's culture, celebrated in song and writing, and dreamt about in exile.

3.3 The Growing Acceptance of Landscapes in Protected Areas Policy

The idea of important landscape areas - used here to mean places which have been shaped by human use over the centuries and which contain valuable natural and cultural resources - has become increasingly accepted in recent years. For example, particularly in Europe many such areas have been designated under national or provincial law as protected landscapes, nature parks, regional parks and even national parks. Indeed, not only in Europe, but more generally, our view of what we are trying to protect, and why, has broadened out. From a position thirty years ago when there was a simple-minded equation between the degree of naturalness and the importance of conservation, there has developed a far more sophisticated strategy to use a variety of approaches in conservation as well as a range of protected area tools. In this new vision, landscape is recognized as an important resource in its own right, as worthy of protection as wilderness, great natural features and grand wildlife spectacles.

4 Four Examples of the Growing Place of Landscape in Protected Areas

4.1 IUCN Protected Area Management Categories

IUCN, and more particularly its Commission on National Parks and Protected Areas (CNPPA), has undergone a complete revolution of thinking since 1969 in New Delhi. In 1978, in order to reflect the existence of many ideas wrapped up in the term protected area, and not only the original one of a national park, CNPPA produced for IUCN an initial attempt at a protected area categories system for international use. After a major review, culminating in debate at the Fourth World Congress on National Parks and Protected Areas in Caracas, Venezuela, 1992, CNPPA recommended the adoption of a revised categories system, based on management objectives, in 1994 (IUCN 1994) (Table 1). Category VI is a recent addition, and statistics are not yet available on the number and extent of such areas. However, they do exist for the other categories, as follows (Table 2).

Although Category V, the Protected Landscape or Seascape, first appeared in the 1978 IUCN system of protected area management categories, it was regarded as very much a European concept, and a poor relation beside the more highly protected categories, such as category II national parks. Gradually, however, IUCN has come to take a greater interest in protected landscapes. This stems from the convergence of two lines of thought: that conservation of species and habitats cannot be achieved in strict nature reserves and national parks alone; and that conservation depends upon the involvement of people, and therefore places where people co-exist with nature are worthy of special attention. In 1987, this interest was brought to a head when a UK government agency, the Countryside Commission, with the Council of Europe and IUCN, convened an international symposium on protected landscapes. The main output was The Lake District Declaration, which sought to reflect the importance of protected landscapes to IUCN (Countryside Commission 1987; 1988).

Table 1. Categories of Protected Areas

I	Strict Nature Reserve/Wilderness Area: protected area managed mainly for science or wilderness protection
II	National Park: protected area managed mainly for ecosystem protection and recreation
III	Natural Monument: protected area managed mainly for conservation of specific features
IV	Habitat/Species Management Area: protected area managed mainly for conservation through management intervention
V	Protected Landscape/seascape: protected area managed mainly for landscape/seascape conservation and recreation
VI	Managed Resource Protected Area: protected area managed mainly for the sustainable use of natural resources.

source: IUCN, 1994

Table 2. World Data on Protected Areas: 1993

Category	Number	Area (ha)
I	1460	86,473,325
II	2041	376,784,187
III	250	13,686,191
IV	3808	308,314,011
V	2273	141,091,932
Total	9832	926,349,646

source: WCMC, 1994

Another result was a decision of the IUCN General Assembly held in 1988 in Costa Rica, to adopt a resolution on the topic. This stressed the importance of the protected landscape approach, recognized that such places were important from an economic, environmental, social and cultural point of view, and called on IUCN to advise on the protected landscapes approach. The next step was the publication of Protected Landscapes: a Guide to Policy Makers and Planners (Lucas 1992) and the adoption of the revised management categories in 1994, which included this definition of a protected landscape/seascape:

'an area of land, with coast and sea as appropriate, where the interaction of people and nature over time has produced an area of distinct character with significant aesthetic, ecological and/or cultural value, and often with high biological diversity. Safeguarding the integrity of this traditional interaction is vital to the protection, maintenance and evolution of such an area' (IUCN 1994)

IUCN identified the following values within protected landscapes/seascapes:
- conserving nature and biodiversity
- buffering more strictly controlled areas
- conserving human history in structures and land use patterns
- maintaining traditional ways of life
- offering recreation and inspiration

- providing education and understanding
- demonstrating durable systems of use in harmony with nature (Lucas 1992).

The objectives of management, in terms of guidance for selection and notes on the organizational responsibility for their administration, are given in Table 3. The inclusion of phrases such as: 'harmonious interaction of nature and culture', supporting 'lifestyles ... in harmony with ... the preservation of the social and cultural fabric of the communities concerned', and 'manifestations of unique or traditional land-use patterns and social organizations as evidenced in human settlements and local customs, livelihoods and beliefs' all recognize the value of a positive interaction between humankind and nature (see also Countryside Commission 1988). Some 2,273 Protected Landscapes/Seascapes are recognized by IUCN in the 1994 UN List of National Parks and Protected Areas. Their distribution between the regions of the world is shown in Table 4. While these figures confirm that there are many such areas in Europe, it also shows that there are protected landscapes in many other countries around the world. There is a particular concentration in Australia; most of this is accounted for by the Great Barrier Reef, which is clearly a special case and one which may soon be re-assigned to Category VI.

4.2 Cultural Landscapes Under the World Heritage Convention

The World Heritage Convention was adopted in 1972. Operated under the auspices of UNESCO, it provides for the designation of areas of outstanding universal value as World Heritage sites, with the aim of fostering international co-operation for their protection. Sites may be natural or cultural, or mixed i.e. they are considered worthy of inclusion on the list on both natural and cultural grounds. However, until recently, the Convention was not well designed to recognize landscapes of outstanding merit, because these were usually too modified to be acceptable as natural sites, and too natural to be accepted as cultural sites. So a convention which uniquely seemed to bring together cultural and natural perspectives singularly failed to make the all-important connection.

Following on from a resolution at the Caracas World Parks Congress, a review took place of the operational guidelines for the Convention, which led - some two years later - to the adoption of a new type of cultural site, cultural landscapes. The guidelines distinguish between three categories of cultural landscape:
- landscapes designed and created intentionally by man, such as gardens and parks,
- organically evolved landscapes, i.e. those resulting from human responses to the natural environment, which may be either relict (i.e. fossil) landscapes, or continuing and evolving landscapes where the evolutionary process is still in progress,
- associative landscapes, whose significance lies in the powerful religious, artistic or cultural associations of the natural elements rather than in any material cultural evidence.

Although cultural landscapes are treated as a type of cultural site, most will clearly have important natural qualities too. Therefore their inclusion in the Convention requires collaboration between the two advisory bodies: IUCN for natural values and ICOMOS for cultural ones. These are as yet early days. Three cultural landscapes have been designated. Two were already recognized as World Heritage natural sites i.e. Tongariro in New Zealand; and Uluru Kata Tjuta National Park - in the Ayers Rock area of central Australia.

These are now also inscribed for their associative values to the Maori and Aboriginal peoples. The third, the Philippines rice terraces of Luzon, has been recognized as an organically evolving landscape, with very important cultural and natural qualities (see also Droste et al. 1995). The importance of this development is not confined to the relatively few sites which will be recognized under the Convention. Just as important in the long run is the encouragement that this will give to the conservation of landscapes generally and to collaborative work among experts in cultural conservation and the conservation of natural values.

Table 3. Guidance on Protected Landscapes/Seascapes

Objectives of Management

- to maintain the harmonious interaction of nature and culture through the protection of landscape and/or seascape and the continuation of traditional land use, building practices and social and cultural manifestations
- to support lifestyles and economic activities which are in harmony with nature and the preservation of the social and cultural fabric of the communities concerned
- to maintain the diversity of landscape and habitat, and of associated species and ecosystems
- to eliminate where necessary, and thereafter prevent, land uses and activities which are inappropriate in scale and/or character
- to provide opportunities for public enjoyment through recreation and tourism appropriate in type and scale to the essential qualities of the area
- to encourage scientific and educational activities which will contribute to the long term well-being of resident populations and to the development of public support for the environmental protection of such areas, and
- to bring benefits to, and to contribute to the welfare of, the local community through the provision of natural products (such as forest and fisheries products) and services (such as clean water or income derived from sustainable forms of tourism)

Guidance for Selection

- the area should possess landscape and /or coastal and island seascape of high scenic quality, with diverse associated habitats, flora and fauna along with manifestations of unique or traditional land-use patterns, and social organizations as evidenced in human settlements and local customs, livelihoods and beliefs
- the area should provide opportunities for public enjoyment through recreation and tourism within its normal lifestyle and economic benefits

Organizational Responsibility

The area may be owned by a public authority, but is more likely to comprise a mosaic of private and public ownerships operating a variety of management regimes. These regimes should be subject to a degree of planning or other control, and supported, where appropriate, by public funding and other incentives, to ensure that the quality of the landscape/seascape and the relevant local customs and beliefs are maintained in the long term.

Source: IUCN (1994)

Table 4. Distribution of Protected Landscapes/Seascapes

CNPPA region Category V sites:			
	Number	Area (ha)	% of land surface
N. America	507	25,793,725	1.10%
Europe	1307	33,748,761	6.61%
N. Africa and Middle East	54	5,111,744	0.39%
East Asia	72	4,297,156	0.36%
N. Eurasia	10	151,217	0.01%
Sub-Saharan Africa	20	2,299,947	0.10%
South Asia	56	799,052	0.09%
Pacific	7	14,571	0.03%
Australia	32	48,273,354	6.28%
Antarctic/New Zealand	1	1,000	<0.01%
C. America	4	6,671	0.01%
Caribbean	28	708,257	2.97%
S. America	175	19,886,467	1.10%
Total	2273	141,091,932	0.95%

From data compiled by the WCMC (1994)

4.3 European Strategy for Biological and Landscape Diversity

The third example of the growing acceptance of landscape protection as a suitable subject for international co-operation relates to the European Strategy for Biological and Landscape Diversity (UN/ECOSOC 1995) adopted by European Environment Ministers at their meeting in Sofia, Bulgaria in October. This strategy is an important means by which European countries, including those in the European Union, can collaborate to implement the Convention on Biological Diversity.

The inclusion of landscape in the title and the measures proposed, suggest an appreciation that it is through landscape management and protection that much European biodiversity can best be safeguarded. Based on a thorough assessment of the state of the European environment (Stanners and Bourdeau 1995), the strategy contains proposals to: compile a comprehensive guide to Europe's landscape resources; develop guidelines for landscape management; undertake special measures for landscape parks and landscapes of geological importance; promote awareness of landscape values among landowners and other concerned persons; and work on the positive relationship between the protection of traditional landscapes and the strength of the rural economy.

4.4 Proposed European Convention on Landscape

The last initiative is as yet at a formative stage. Following a proposal in the IUCN action plan for protected areas in Europe, Parks for Life (IUCN, 1994a), a group of local and regional authorities in Europe, working under the auspices of the Council of Europe, are currently developing a draft Convention for the Protection of European Landscapes. At present, a number of questions remain to be resolved and governments are non-committal. Nonetheless, the idea has gathered support from a broad spectrum of organizations. If it comes into being, the convention seems likely to: encourage governments to record their landscape resources; understand better the changes taking place in the landscape; and put in place measures designed to protect the landscape while at the same time supporting traditional land uses which sustain landscape character.

The draft convention contains proposals to recognize certain landscapes as of European importance. Should this be acted upon, there would then be a three-tier hierarchy of landscape protection in Europe:

- sites of global importance (i.e. World Heritage sites),
- sites of European importance, and designated under the European landscapes convention, and
- sites of national importance, designated nationally and recognized by IUCN as protected landscapes/seascapes.

This framework could be an important one in relation to the research requirements set forth in the last part of this paper.

5 Some Research Priorities [1]

By comparison with biological resources and to some extent cultural ones, landscape - the meeting ground between the natural and the cultural world - is poorly understood. If, as is argued in this paper, landscape issues are increasingly playing a role in shaping protected areas policy, then there are a number of gaps in knowledge which will need to be addressed. These are broadly of three kinds:

- the need to develop a typology of landscapes, so that there is some frame against which to organize analysis,

---

[1] The rest of this paper draws on an unpublished article by the author, prepared for a UNESCO Regional Thematic Study of the Asian Rice Culture and its Terraced Landscapes: Manila, Philippines, 28 March - 4 April 1995, entitled 'The Nature of Cultural Landscapes - an IUCN Perspective'.

- the need for methods of evaluating landscapes in order to compare one landscape with another, and
- most important of all
- the need to find ways to manage landscapes in protected areas so that the qualities which merit designation can survive.

Some preliminary ideas are offered in respect of each of these.

5.1 A Landscape Typology

The development of a typology requires the identification and classification of landscapes by their types, taking account of such factors as those listed in Table 5. Such an exercise, leading to the development of a landscape typology, is desirable as the first stage in identifying individual landscapes for their suitability for designation. The resulting categorization of different landscape types can be used to identify and compare individual landscapes. At its simplest, this exercise is analogous to the classification of ecosystems or biotopes (e.g. wetlands and mountains), or species (on a taxonomic basis).

Table 5. Factors Contributing to the Identification of Landscape Types

---

Physical factors
eg   geology
       landform
       drainage
       soils

Natural Factors
eg   ecosystems
       species (fauna and flora)

Human use of land
eg   farming systems
       forestry and other land uses
       settlements
       transport systems

Cultural factors
eg   aesthetic (visual etc.)
       associations (historic, artistic)

The question of typology is complicated by that of scale. Landscapes form a hierarchy, occurring at every scale from the local to the global. Thus the landscape of a small valley of a couple of kilometres width may be a component of a local landscape unit of undulating topography and mixed farming covering 200 square kilometres. This in turn, may be a part of a regional or national landscape of lowland agriculture, covering many thousands of square kilometres; and this too, may be part of world landscape unit.

This is a very brief introduction to a complex topic, but it is nonetheless one which policy makers will need guidance upon from the research and academic community as they seek to apply the growing interest in landscape issues to the development of policy and practice.

5.2 Evaluating landscapes

The development of a typology is descriptive and analytical: the evaluation of landscapes involves the exercise of value judgment. While it is in the nature of evaluations of landscape that there will always be a large element of subjective judgment, the extent to which that is an informed judgment should increase over time. This requires, however, that experience is consolidated in the form of guidance and assessment criteria.

A possible starting point is to be found in some work done in the UK (Cobham Resources 1993) on the factors to be taken account of in evaluating landscapes (Table 6); this check list is not exhaustive.

5.3 The Management of Landscapes

The survival of natural values in landscape and indeed the continued presence of healthy life-support systems (clean water, clean air, productive soils etc.) in the rural environment, often depends upon maintaining a relatively low-intensity form of land use. However once inputs begin to be applied in large amounts to increase production, the natural qualities of the area will almost always decline. Even if the appearance of the landscape maintains - for a while at least - a superficial similarity to that which was produced through less intensive farming methods, the value of the landscape from the perspective of biodiversity conservation will be reduced. A similar process will occur where rural populations may persist in traditional land use practices but are forced by increased human numbers to work the land more intensively, removing remaining pockets of natural vegetation or adopting shorter cycles of fallow or forest regrowth. Paradoxically, natural values can also be put at risk when land is abandoned; for example, if upland herb rich grasslands are encroached upon by secondary forest growth when grazing ceases, or if traditional systems for periodic flooding of farmland are neglected, the biodiversity value of landscapes can be reduced.

Table 6. Check List of Items for Evaluating Landscapes

Landscape as a Resource: the importance of the resource in terms of rarity and representativeness
Scenic Quality: the scenic quality of the landscape (the presence of pleasing or dramatic patterns and combinations of landscape features, and aesthetic or intangible qualities)
Unspoilt character: the degree to which the landscape within the area is unspoilt by large scale, visually intrusive or polluting industrial or urban development, or infrastructure
Sense of place: how far the landscape has a distinctive, including topographic and visual unity. It is helpful if local people have a perception of the value of the area
Harmony with Nature: how far the landscape demonstrates a harmonious interaction between people and nature, based upon sustainable land-use practices, thereby maintaining a diversity of species and ecosystems
Cultural Resources: whether the landscape contains buildings and other structures of historical and architectural interest; the integrity of these features should be apparent
Consensus: the presence of a consensus among professional and public opinion as to the importance of the area; reflected, for example, through associations with writings and paintings about the landscape
Management: not part of the analysis as such, but there should be a legal/administrative framework for the managment of the area, which ensures that its qualities are protected.

Herein lies the central dilemma in landscape protection: since landscapes are the product of a particular human society living in a particular way at a particular population density, changes in that society - and especially in the land use practices which it follows - will bring about changes in the landscape itself, and thus often affect its value for biodiversity. It is not enough therefore to attempt to protect the landscape as such: attention must be given as well to the ways of life of those who are the architects of the landscape, and upon whom the survival of the biodiversity within it depends.

This is not an argument for seeking to fossilize the way in which communities use the land, although in some instances it may be appropriate to support certain traditional land use practices because these sustain certain key aspects of the landscape. Rather the aim should be to encourage communities to adopt more sustainable patterns of living, so that they can both improve their prospects of economic and social progress, and continue to maintain the landscape that they have created. There can be little doubt that this will be the management challenge facing many of the landscapes recognized as of global, regional and national importance.

The issue can be put in these terms: it is often assumed that there are only two choices open to traditional land users, the architects of many valued landscapes. These are to stick to the traditional way of life; or to adopt those of the dominant society and abandon all tradition. In fact there is a third option: to modify subsistence systems, combining the old and new in ways that maintain and enhance distinctive cultural identity, while allowing the society and economy to evolve. This calls for an approach which recognizes the rights of people to have control over their own resources - especially land - in the belief that this will ensure that the timing, pace and manner of development minimizes harmful social, cultural and environmental impacts. The approach also requires policy makers to work with local people as partners who have much to offer by way of experience and knowledge in regard to sustainable rural development. Some of the issues which have to be addressed are set out in Tables 7 and 8.

Table 7. Requirements for Management of Protected Landscapes

- a sound legal basis for the management of the area, based in national law, but reflected also in site specific regulations;
- a national authority with expertise and resources to oversee policy and implementation for the protection of landscapes;
- a managing body at the local level, able to call on a range of professional expertise;
- a good understanding of the social, economic and environmental conditions prevailing in the area, and how they inter-relate;
- ways of providing two-way communication between the people living in, and/or working within the protected landscapes; other interests such as visitors and commercial concerns; and the managing body;
- a continuing monitoring and feed-back process which ensures that policies are kept under review at the national land local levels - and revised should this be required.

Table 8. Principles for Management of Protected Landscapes

- Landscape protection requires the presence of a vital and sound local economy. It is also true that landscape resources are needed to ensure that development can be sustainable. Thus the management of a protected landscape is, in fact, the management of the local economy and of change
- Landscape protection requires the support and involvement of the local people. Thus protection must be seen to be in their interests, using educational and financial incentives, and local powers of decision
- The basic resources of the area (natural and cultural) should be recorded, examined and protected
- Planning and management in the area should involve the public discussion of options
- Regulatory measures are necessary, but they should be flexible and respect the rights, interests and needs of local people
- The traditional knowledge of local people in sustainable land use should be respected and supported
- No protected landscape can survive in isolation from the areas around it.

Against this background, the research challenge is to help develop from empirical evidence the principles and guidelines for the management of valued landscapes, and to ensure that these are available to the managers of protected areas. A powerful tool in this regard will be case studies of what has worked well - and what has not - and why.

6 Conclusion

Historically, protected areas policy has been driven for many years by a simplified model in which wilderness stood at the top of the scale of values and interest declined as human presence and intervention increased. This view is now being widely challenged. As a result, IUCN, especially through its Commission on National Parks and Protected Areas, is promoting a new perspective in which there is a growing interest in the ways in which rural and indigenous communities around the world interact with nature. The physical expression of the relationship between people and nature is the landscape; and the policy expression of the desire to safeguard that relationship is the protected landscapes approach to protected areas.

7 References

Cobham Resources. 1993. *Landscape Assessment Guidance.* Countryside Commission, Cheltenham, UK.

Countryside Commission. 1987. *The Lake District Declaration.* Countryside Commission, Cheltenham, UK.

Countryside Commission. 1988. *Protected Landscapes: Symposium Proceedings* Countryside Commission, Cheltenham, UK.

Denevan W. 1992. The Pristine Myth: the landscape of the Americas in 1492 *Annals of the Association of America Geographers.* 82: 369-385.

Droste B., Rossler M. and Plachter H., 1995 *Cultural Landscapes of Universal Value* Gustav Fischer, Jena.

Everhardt W. 1972. *The National Parks Service.*Praeger, New York.

Hill M.A. and Press A.J. 1994. Kakadu National Park - a History. In *Biodiversity Broadening the Debate,* ed. Longmore R., Australian Nature Conservation Agency, Canberra.

IUCN - the World Conservation Union. 1994. *Guidelines for Protected Area Management Categories* IUCN, Gland.

IUCN - the World Conservation Union. 1994a. *Parks for Life: Action for Protected Areas in Europe.* Gland.

Kemf E., 1993. *Law of the Mother.* Sierra Club Books, San Francisco.

Lucas P.H.C. 1992. *Protected Landscapes.* Chapman and Hall, London and New York.

Nash, R. 1967. *Wilderness and the American Mind.* Yale University Press, New Haven.

Pretty J. and Pimbert M. 1995. *Parks, People and Professionals: Putting Participation into Protected Areas Management.* Discussion Paper 57, United Nations Research Institute for Social Development, Paris.

Reichardt K., Mellink E., Nabham G., and Rea A. 1994. *Biodiversity in a Sea of Pasturelands: Indigenous Resource Management in the Humid Tropics of Mexico.* Etnoecologica, vol. II, no.3, 1994, 37-52.

Schama, S. 1996. *Landscape and Memory.* Vintage Books, Random House of Canada, Toronto.

Stanners D. and Bourdeau P. (eds.). 1995. *Europe's Environment: the Dobris Assessment.* European Environment Agency, Copenhagen.

Thomas K. 1983. *Man and the Natural World.* Penguin Books, London and New York.

United Nations Economic and Social Council. 1995. *Pan-European Biological and Diversity Strategy.* Economic Commission for Europe, Paris.

World Conservation Monitoring Centre (WCMC). 1992. *Global Biodiversity.* Chapman and Hall, London and New York.

Land Use and Decision-Making for National Parks and Protected Areas

Gordon Nelson[1], Rafal Serafin[2], Andrew Skibicki[1], and Patrick Lawrence[1]

[1] Heritage Resources Centre, University of Waterloo
Waterloo, Ontario, Canada N2L 3G1
[2] Polish Environmental Partnership Foundation
Bracka 6/6,31-005 Kraków, Poland

Abstract. Sustainable development, ecotourism, landscape ecology and other new concepts and approaches, have been advanced to help deal with the growing land use pressures threatening the ecological and social services national parks and protected areas offer to society. Yet, relatively little research is being done on the nature and character of the land use changes. Land use changes are the most important challenge to planning, managing, and deciding upon parks and protected areas on an ecosystems basis. Land use changes pose the greatest threats to the sustainability of water quality, air quality, wildlife, research, education, ecotourism, and other services that parks and protected areas offer to communities and societies in all parts of the world.

Serious historical and evaluative research is also lacking on the nature and effects of land use planning, management, and decision-making. Without such information we are in a poor position to make informed choices about how land use changes could be planned, managed and decided upon more effectively in future. We recommend that park and protected area personnel, university and other researchers, and concerned government officials and citizens, strongly support more research of this kind. Remote sensing, Geographical Information Systems (GIS), and other methods are available to carry out land use studies more effectively and efficiently than in the past. Comprehensive ecologically based land use mapping and analysis systems such as the ABC Resource Survey Method, are also being developed. New concepts, and methods for analyzing and assessing experience with planning, management, and decision-making systems also have been developed in recent years. These ideas have mainly been generated for other than national parks and protected areas but seem applicable to them. They deal with such relevant topics as environmental and civic science.

Keywords. Parks and protected areas, land use, sustainable development, ecotourism, landscape conservation, ecosystems, planning, management, decision-making, civics

1 Contributions, Stresses, New Concepts and Approaches

The many important contributions of national parks and protected areas to local, national, and global societies are being increasingly recognized by more and more people (Nelson 1991). These services include: conservation of water, forest, and other resources, wildlife habitats, ecosystems, and biodiversity; protection of archaeological, cultural, and historic resources; education; research; tourism and other economic activities. National parks and protected areas offer support not only to human kind, but to all life on earth. Yet at the same time that awareness of their contributions is rising, national parks and protected areas are under growing and often damaging pressure from accelerating tourism, recreation, forestry, mining, and other land use activities (Ross and Saunders 1992).

2 Sustainable Development

The idea of sustainable development is seen by many people as one important response to the land use pressures upon parks and protected areas (IUCN 1980; World Commission on Environment and Development 1987). Sustainable development is planned in such a way as to: 1) preserve essential ecological processes such as animal migration routes, or staging areas for birds; 2) protect the range or diversity of plants, animals, and natural communities; and, 3) maintain productivity of soil, wetland, and other resources. According to the World Commission on Environment and Development, or Brundtland Commission (1987), sustainable development also provides for equity, or for comparable access to environmental opportunities for present and future generations. The application of the concept of sustainable development to national parks and protected areas seems highly promising. On the other hand some observers believe that the concept will open the way for more environmentally damaging growth in the longer run. A basic challenge with the concept of sustainable development is to determine what it actually means in terms of land use changes and their effects 'on the ground'. Assessments of the land use, economic, social and environmental effects of proposed changes are therefore needed to help decide whether they are compatable with ideas on sustainable development.

3 Ecotourism

The concept of ecotourism is seen as another basic way of responding to growing recreation and tourism pressures upon national parks and protected areas (Nelson et al. 1993). Various definitions of ecotourism have been put forward. These are generally similar in stressing that ecotourism builds on the ecological qualities of an area and is conducted in such a way as not to damage or destroy these valued qualities. Ecotourism also conveys the idea that tourism activities are, to a considerable degree, under local control and provide a high level of local benefits. Some observers have noted that the idea of ecotourism is still largely perceptional. Much lies in the eye of the beholder. The meaning of the term can vary from person to person, place to place, and group to group. An understanding of the kinds of land uses involved in ecotourism and an assessment of their ecological, social and economic effects is therefore, necessary for sound decision-making.

4 Theory and Method

Advances in theory and method are also seen as a way of reducing economic and land use pressures on national parks and protected areas. These advances in theory seem to be particularly important in North America where for many decades notions of pristine environments and wilderness have been fundamental guiding concepts in planning and managing national parks and protected areas (Nash 196; McKibben 1990).

Among the more important new fields of thought are landscape ecology and conservation biology (Foreman and Godron 1986; Woodley 1996; Nelson and Serafin 1992). In general, these fields build on concepts such as maintaining viable wildlife populations, biodiversity, ecological integrity and corridors and connectivity among parks and protected areas and surrounding greater park ecosystems (see both Woodley and Farina, this volume).

It is from the lands and waters around national parks and protected areas that many, if not most, of the stresses on them generally arise. Interest is growing, therefore, in identifying and delimiting the greater park ecosystems that are tied by river flow, watershed, or other processes to the national parks and protected areas, and visa versa. The ecological integrity or health of the national parks and protected areas is seen as mainly being a function of the ways in which the lands, waters, and atmosphere in the greater park ecosystem are planned, managed, and decided upon. The ecological health and sustainable development of the greater park ecosystem is, in turn, seen as dependent upon the ecological and social services offered by the national parks and protected areas located within the system's boundaries.

The basic problem is that, while new concepts such as sustainable development, ecotourism, and landscape ecology can be helpful, they are not in themselves sufficiently precise to provide specific guidance for land use planning, management and decision making. Any proposed land uses and their social and environmental effects need to be described, analyzed, and assessed before we can decide whether they represent ecotourism or sustainable development.

5 Land Use

In spite of the obvious importance of land use processes, patterns, and changes to planning, management, and decision-making for national parks and protected areas, the topic has not received attention commensurate with its ecological and social significance (Turner et al. 1990). This statement is especially applicable to North America, although it also is relevant in Europe and other parts of the world. This is not to say that illustrative examples of the value of land use studies and information are not available to us. We can think of Colin Tubbs' (1968) work on the natural history of the New Forest in southern England, with its detailed account of the more than thousand year role of domestic stock grazing in maintaining the character and distribution of the ancient forests, the heath, and other communities.

Oliver Rackham (1986) has used historic land use studies to explain the character and distribution of trees and woodland in British landscapes. In *A Natural History of the Hebrides*, J. Morton and Ian L. Boyd (1990) have interwoven historic land and resource use with natural changes to explain the evolving character of the remote western islands of Scotland. In the United States and Canada, a smaller amount of work has been done on land use changes in national parks and protected areas, for example, in the Everglades (Tebeau 1986).

Various approaches to land use studies on protected areas can be made, depending on the situation and the available resources. Simple historical mapping of changes in recreational land use and technology can be quite instructive for planning, managing, and decision-making, and for education and interpretation. Figures 1, 2, 3, and 4 are simple maps estimating increases in the kind and extent of recreational facilities and technology in Banff National Park from 1930 to 1992 (Nelson 1994).

The Banff maps show that recreational land use has intensified and spread cumulatively over the years and suggest that many associated changes have taken place in the park environment. From a planning standpoint, it is important to prepare simple historical land use and recreational change maps and analyses for all major national parks in order to approximate the extent of land use changes changes and estimate their environmental and socio-economic effects. This basic mapping can be used strategically to decide whether more detailed research is needed on any ongoing or proposed changes and their effects.

A fundamental reason for preparing such maps and studies is that people are generally unaware of the growth in intensity and extent of recreation, tourism and related land uses over the years. They are therefore, not as concerned as they might be about further recreational or other land use changes and their effects on forests, animal habitat, and other aspects of park and protected area ecosystems. This lack of awareness extends not just to visitors but also to planners, managers, and decision-makers, including politicians.

Information on changes in land use in and around national parks and protected areas is something that people of varied backgrounds, experience, and interests can understand. Appreciation of these changes can promote interest in and concern about the effects of land use changes on economy, society and environment. For this fundamental reason land use changes should be mapped, monitored and reported upon regularly in national parks and protected areas.

6 Land Use and the ABC Resource Survey Method

A comprehensive, dynamic, and interactive approach to land use mapping and research is ultimately desirable and useful for learning and general understanding, as well as for planning, management, and decision-making. The land use, resource and environmental survey and assessment system known as the ABC method has been developed over the last fifteen years in Canada with these desired

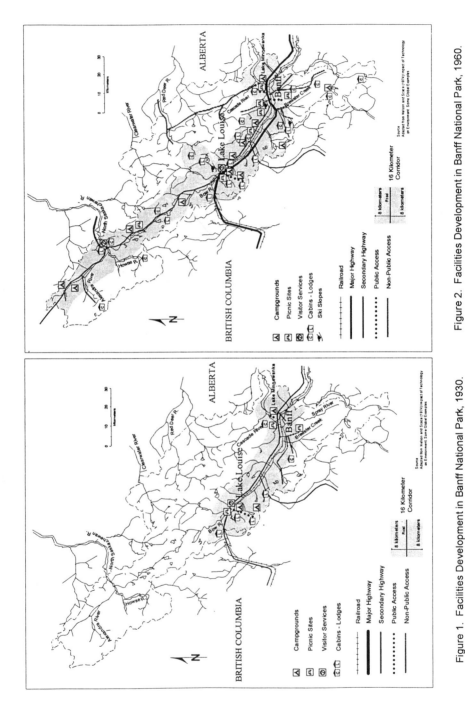

Figure 1. Facilities Development in Banff National Park, 1930.

Figure 2. Facilities Development in Banff National Park, 1960.

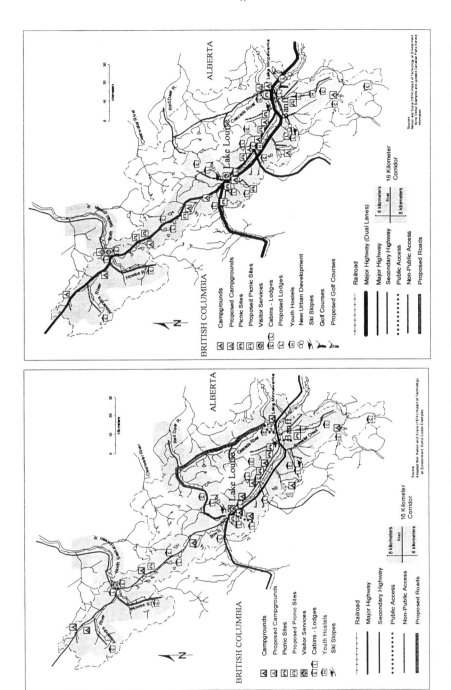

Figure 3. Facilities Development in Banff National Park, 1971.

Figure 4. Facilities Development in Banff National Park, 1992.

characteristics of comprehensiveness, dynamism, and interaction in mind (Bastedo et al. 1984; Nelson et al. 1988). The method is basically one of applied human ecology. The ultimate goal is to map, analyze, and assess human relations with the environment. Working toward this goal requires studies not only of geologic and biological aspects of the environment, but also the values, ideas, technology, policies, and land use activities which humans use to adapt to and change the world around them.

The ABC resource survey method (Figure 5) is comprehensive in that it covers: Abiotic, including geologic, landform, and hydrologic information; Biotic, including plant, animal, and soils information; and Cultural, including land use, economic, human heritage, land management, institutional, and other human information. The method is dynamic in that it can be used to prepare historic maps and analyses and to link abiotic, biotic, and cultural (or human) patterns, processes, and changes. It is also dynamic in that it provides for both historic and current mapping and analyses.

The method is interactive in that it can be used in a cross-disciplinary way to link scientists, scholars, planners, and managers of different training, backgrounds, and interests. It can also be used to involve businessmen and concerned citizens and gain information from them for mapping, research and planning. In particular, people in a study area can be consulted about what is important to them and the results can be used in selecting topics for theme, significance and constraint mapping.

For the purposes of this paper, the important thing is that the method can be used to undertake land use mapping as part of the C component of the ABC system. This land use mapping can then be linked to land tenure and management, to commercial and economic activity, to bird watching and other recreational activities, to forest and wildlife patterns, and to other social and environmental elements, processes and effects of interest in planning, management, and decision-making. The ABC method is flexible in terms of the topics and issues or themes to be mapped as well as in terms of scale and research detail. Decisions on selection of information and mapping ultimately depend on the purposes of the study. The criteria for deciding on significance and constraints mapping (level 2) can be selected in terms of the goals and objectives of the study, as specified at the fourth level in the method (Figure 5).

In this paper we are using land use in the general sense that is often understood for this term, i.e. human activities that are revealed on the surface of the earth as residential areas, croplands, forests, and other patterns. However, these patterns have been referred to by some professionals as land cover or landscape patterns with the term, land use, then referring to the human activities that occur within and are reflected to various degrees, in such land cover types (Meyer and Turner 1994). Mapping cultivated areas or forest cover types does not, for example, normally reveal what logging, recreational, or other uses are actually occurring within the land cover types. In this context, what we have mapped in our use of the method has often been land cover changes.

7 Results

At the first stage of mapping (level 1) information can be mapped in two basic ways. The first way is by preparing a set of theme maps on topics or issues of concern to planners, managers and citizens. The use of theme maps is especially appropriate where a considerable amount of information already is available in government files, scientific reports and other sources and the theme maps can be prepared primarily on the basis of this existing information.

The second way is by preparing land cover maps from air photos or satellite images. These land cover maps are especially useful where existing information is sparse, although the land cover maps can also be used to supplement or augment theme maps of existing information in cases where this seems desirable or important. Both the theme and land cover mapping approaches are usually supplemented by field studies 'on the ground'.

The ABC method has been used in many situations in Canada and in other countries such as Java and Poland. In this work, the method has been applied at various levels of detail and complexity. In some cases the method has served as a broad framework for organizing existing information of various kinds and inferring relationships among the different kinds of information. This was true in the case of the Grand River, Ontario, where the method was employed to organize, assess, and interpret geologic, biologic, archaeological, land use, and other information to determine whether the

Figure 5.

THE ABC RESOURCE SURVEY METHOD

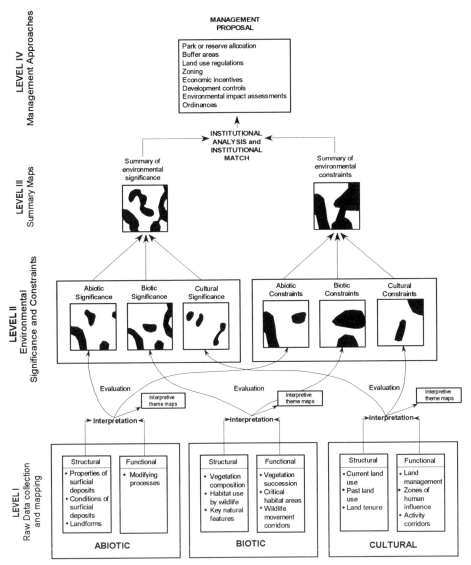

Adapted from: Grigoriew et al., 1985

Grand River had the outstanding, natural, cultural, and recreational qualities to merit designation as a Canadian Heritage River (Nelson and O'Neill 1989; Skibicki 1992).

In the Segara Anakan area, south central Java, the ABC was used as a framework for organizing and assessing existing documentary information in order to understand the kind, rate and distribution of rice cultivation, settlement, and other land use changes and their effects on mangrove as well as on sedimentation and other environmental processes (Nelson et al. 1992). The work with existing information was supplemented by analysis of land cover and land use changes from satellite images for 1978 and 1987. Recently this Segara Anakan work has been extended to include field studies and participatory research with residents in the area.

Similar use was made of existing documents as well as new information from historic aerial photographs and field studies in work on coastal planning for the Saugeen Valley Conservation Authority, Lake Huron, Ontario (Lawrence and Nelson 1992). These studies were conducted in close consultation with professionals and government agency personnel through technical meetings and with citizens through open houses and public meetings. The analysis and interpretation of 1950s, 1960s, 1970s, and 1990s aerial photographs revealed some surprises, notably fragmentation of forest cover to a degree and extent unknown or unappreciated by residents, government planners, and others (Figures 6, 7). As a result, forest fragmentation became an issue along with water quality, shore erosion, and other concerns in coastal planning.

Two recent applications of the ABC resource survey method to national park issues are of special interest. The first, on Pukaskwa National Park on the north shore of Lake Superior, involved detailed land use mapping which showed many land uses in what most people would view as a wilderness or near-wilderness area (Skibicki 1995). Included here were various kinds of mining sites, native hunting, sport fishing, and other recreational and tourism activities as well as government conservation and management programs (Figure 8).

An additional interesting aspect of the Pukaskwa National Park study was the attempt to develop methods of mapping a greater park ecosystem boundary around the national park. A major reason for doing this was to indicate not only to park people, but also land users in the surrounding region, what the linkages between the park and the region were - and so where common interest and the need to co-operate was to be found.

Two boundary indicators were selected and mapped (Figure 9). These were, first, the watersheds around the park and, second, the migratory range of caribou in and around the park. The area covered by the two indicators differed, making it clear that the greater park ecosystem is a complex thing, more achievable ideally than realistically. Nevertheless, the results of the exercise were valuable in providing local people, visitors, and planners with information which showed how various activities located some distance from the park affected its well-being. The results also showed how far outside park boundaries the effects of caribou and other national park conservation programs were felt.

Another relevant part of the Pukaskwa National Park work was mapping and analysis of the coastal zone, notably the near offshore waters (Skibicki 1996a). Coastal areas have long been used for canoeing, fishing, hiking, camping, and other purposes by visitors and local people. The offshore focus is especially interesting because water uses are seldom mapped. The Pukaskwa National Park land and water use research is now being used as a basis for discussion and possible collaborative planning among park officials and surrounding land and resource users.

Land use change was mapped and analyzed for a comprehensive survey and assessment of information for the preparation of an Environment Folio for the Long Point World Biosphere Reserve on Lake Erie (Nelson and Wilcox 1996). This project developed in close collaboration with local residents and others concerned about the ecological health and sustainable development of Long Point. This is a system of wetlands, beaches, and dunes jutting 40 km into Lake Erie. The site is one of Canada's outstanding marsh and wetland ecosystems and a major staging area for migratory waterfowl and passerine birds. Its outstanding qualities merited designation as a RAMSAR Site and World Biosphere Reserve by UNECSO in 1986.

Figure 6. Saugeen Coast, Land Use 1954

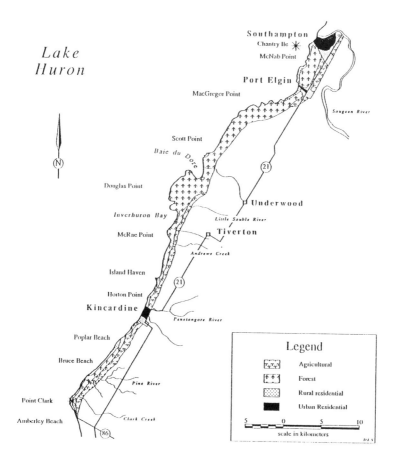

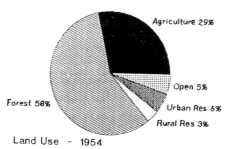

Figure 7. Saugeen Coast, Land Use 1990

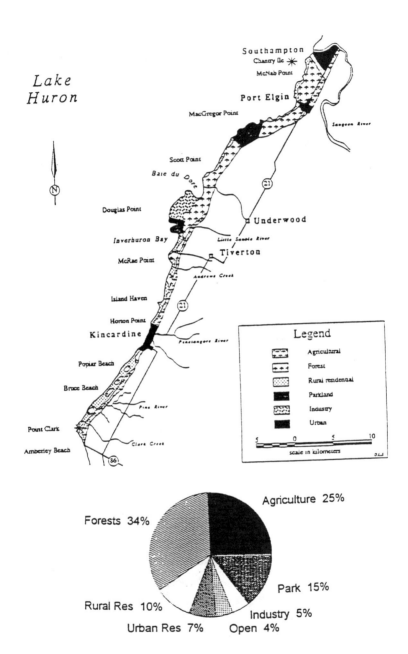

Figure 8. Proposed Greater Pukaskwa National Park Ecosystem

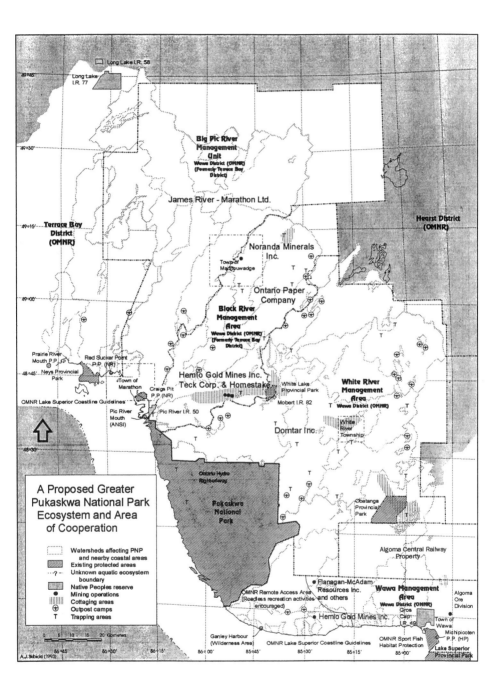

Figure 9. Biotic Process Map for Proposed Greater Pukaskwa National Park Ecosystem

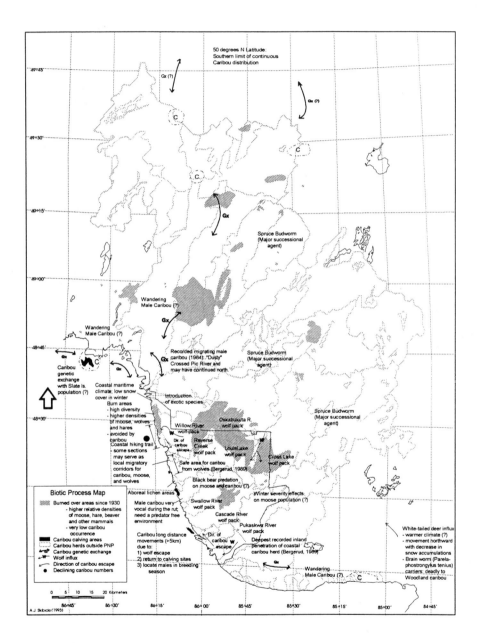

The Environmental Folio idea developed because residents and other members of the Biosphere Committee told us that while much information on Long Point was listed in bibliographies or published in scientific and scholarly papers and accounts, it was generally not available to them for several reasons. They had no physical access to the documents which were often quite technical and difficult to understand. They also dealt with an array of topics and disciplines in an unco-ordinated manner. The idea then developed of reviewing the literature with a team of students and mapping, analyzing, and assessing the results in the form of a Folio written and prepared in such a way as to be accessible to citizens as well as various kinds of professionals.

The folio consists of sixteen chapters on landforms, weather and climate, archaeology and culture history, historical economies, the current economy, forests, waterfowl, birds, and other topics. These topics were selected on the basis of professional judgement and the views of the Biosphere Committee. One chapter of the Folio focuses on land use and land cover changes (Lawrence et al. 1996). The work is based on an analysis of historic aerial photography and satellite images.

The land use analysis revealed that development was contributing not only to wetland losses but also to flooding, erosion, and hazard damages as well. The responses to the hazards have been largely engineering in nature, consisting of breakwalls, rip-rap, fill, and other measures to control lake levels, floods and erosion (Lawrence and Nelson 1996). These structures tend to break down over the years. The structures also interfere with normal erosion and transport along the shore and rob adajacent down-current areas of sediment, making them more vulnerable to future erosion, flooding and economic losses. At Long Point these effects are occurring near protected areas such as a National Wildlife Conservation Areas.

8 Planning, Management and Decision-Making

This brings us to our final topic: how we plan, manage, and make decisions about land use in and around national parks and protected areas. This is a huge issue which can only be discussed here in a preliminary way (MacFarlane and MacCauley 1984). It is also a topic on which relatively little research has been done, even less than on land use change itself.

At the outset it is necessary to provide some information on the meaning of terms such as planning, management, and decision-making. We are using the term, planning, in the general sense of thinking about and preparing for the future (Gertler 1972; Etzioni 1967; Friedmann 1973; Nelson 1991). Numerous different kinds of planning have, however, been recognized. Some of the major ones are shown on Table 1. These kinds differ quite strongly in character although all are applicable in one way or another to national parks and protected areas and related land use changes. Management, in the general sense, refers to the controlling or direction of activities. The term implies care, responsibility, and accountability. Once again, however, there are different kinds of management (Table 2) which can be used in various ways in dealing with land use changes in and around parks and protected areas.

Decision-making is used in the general sense of how we make choices as groups, communities or societies (Nelson and Serafin 1996). The term, therefore, includes the various kinds of planning and management noted previously. It also refers to the basic ways in which we make choices, for example by dictatorial or military methods, by oligarchy, by consensus, or by voting and other democratic means. Decision-making also has been used to refer to broad styles of decision-making - that is, managerial or corporate or participatory and civic.

The question is how such theory and thinking applies to planning, managing, and deciding upon land use changes in and around national parks and protected areas? In the past, stress has been placed on rational or synoptic planning, and on corporate management, in national parks and protected areas around the world. In North American these approaches took hold in the 1960's and 1970's. This was a time of great growth in demand for recreation, for environmental protection, and for parks and protected areas. National, provincial, and state systems plans were developed in countries such as Canada and the U.S. and bureaucracies were created to manage protected areas. Planning and management were concentrated within the boundaries of the protected areas, an approach that was later referred to by some critics as a fortress mentality. Management, resource conservation and visitor service plans were to be developed for each park and protected area, with the idea that they would be followed quite precisely in governing the areas in question.

Table 1. On Planning Ideas or Methods

Rational or Synoptic Planning	Thinking in terms of goals and objectives, and the use of scientific and related knowledge to attain them.
Incremental Planning	Responding to challenges and opportunities on a step by step basis without any necessary overall goals.
Mixed Scanning	Responding to challenges and opportunities on a step by step basis while guided by longer term goals and objectives.
Transactive Planning	Preparing for the future through regular communication and transactions with other parties.
Advocacy Planning	Planning intended to promote the aims and welfare of particular groups.

Table 2. Types of Management

Corporate Management	Controlling or directing an agency or group in accordance with a set of goals and objectives set by law, policy, and/or a Board.
Shared, Joint, or Co-management	Sharing of powers and responsibilities to varying degrees and in various ways, for example by legal agreement and memoranda of understanding.
Adaptive Management	Deciding on action through research and experiment, monitoring and assessment and adjusting to the results as deemed necessary.

Almost from the outset however, this command and control system had to be modified because of the need to take account of the reactions, opinions, and ideas of users and citizens. Public hearings and reviews were held on plans and they were often changed. Advisory scientific, and technical committees and other means were used to receive information and advice from people concerned about parks and protected areas.

Challenges began to develop in the areas around the parks and protected areas as well. Here park and protected area policies and practices caused difficulties for surrounding land users and vice versa. For example, elk or bear strayed out of the protected areas onto land set aside for grazing by domestic stock. Or lumber companies or hunters conducted their external activities up to or even beyond the boundary of parks and protected areas. These challenges arose not only between protected area managers and surrounding private land owners, but also among government agencies and citizens groups. Various consultative procedures were developed to address these challenges.

Since the 1960's and 1970's the foregoing challenges to parks and protected area management have continued to grow for many reasons. The number of uses, and users, and the associated controversy and conflict, have intensified in parks and protected areas as well as the lands and waters around them. One result has been the greater use of citizen participation, environmental impact assessment, mediation, and other conflict resolution methods within and outside the parks and protected areas (Nelson and Serafin 1996; Ross and Saunders 1992).

In North America, two very important challenges to the concept of corporate management, have been the demands and ideas of native or local people and the development of new scientific theory, including the landscape ecology and conservation biology referred to earlier in this paper. Native and local people have had a very basic effect by challenging two underpinnings of corporate management: first, the idea that all the land in national parks and protected areas should be publicly owned and controlled; and, second, that wildlife should not be hunted and harvested in national parks.. The scientific developments have spread interest from a focus on the parks and protected areas themselves to their role in the surrounding lands and waters -- or ecosystems. The result in the last few years, has been accelerating interest in: co-operative approaches to management; greater citizen participation in decisions; and private stewardship (Roseland et al. 1996).

Current thinking is, however, quite messy, especially from the standpoint of the managers of national parks and protected areas. These officials have been given the responsibility by law and government policy to direct land use within national parks and protected areas so as to meet certain broad social goals such as ecological integrity and visitor use which frequently are in conflict. In this context Round Tables and other committees, consisting mainly of representatives of user groups, have been established to develop vision statements or to advise on policy and practice within parks and protected areas. This is being done without much clear thought about how the results will be used by managers.

Ecosystem thinking and increasing human competition over use and conservation of resources and environments, are pushing us in the direction of co-operation, co-ordination, and collaboration over ever larger areas (Madgwick and O'Riordan 1995). Yet the number of users and interests tends to increase with distance from national parks and protected areas, increasing the difficulties of working together.

Conceptual frameworks or models have been advanced in the past to help us think about how to link the land use, conservation, planning, management, and decision-making arrangements involved in parks and protected areas. One familiar framework is associated with UNESCO Biosphere Reserves. Here land use, conservation, planning, management, and decision-making are envisioned as coming together on the ground in the form of a core, a surrounding buffer zone, and a zone of co-operation. The level of conservation and the degree of control increase toward the core. A comparable line of thinking is shown in Table 3. If the land management classes here were placed on the ground, then the preservation areas would more or less correspond to the core, the protected and multiple use areas to the buffer, and the multiple use and more highly exploited areas to the zone of co-operation. The land use spectrum (Table 3) is also valuable in indicating the kinds of land uses and the various stakeholders and institutions that apply to the core, the buffer, and the zone of co-operation.

Land use zoning in national parks follows similar principles. Zoning has been used for many years in North America but, largely because of the lack of attention to land use changes and their effects, these zoning systems have not been evaluated publically for their effectiveness. Applications of nodes and corridors thinking, or landscape ecology, to parks, protected areas and surrounding lands, is going to require careful thought about how they will be planned, managed, and decided upon 'on the ground'. The demands here are more precise and detailed than in the case of the general Biosphere and other zonal frameworks described previously.

9 The Georgian Bay Islands National Park Ecosystem Plan

We are working on challenges like the foregoing now, in the context of preparing an Ecosystem Conservation Plan for Georgian Bay Islands National Park (GBINP) and surrounding lands and waters, in the southeastern Georgian Bay area, Ontario (Dakin 1996; Skibicki 1996b; Skibicki and Nelson 1997). The national park consists of about 60 islands scattered over an area of several hundred km^2. The major island, Beausoleil, is vital to the protection of rare and threatened species and to the maintenance of biodiversity in the entire park and surrounding area. It is especially noteworthy as the habitat for snakes and amphibians. This habitat extends off the island however, onto adjacent private and Ontario government land. Snakes migrate seasonally from Beausoleil Island across intervening channels to neighbouring lands. Beausoleil Island also has high water quality which is important for waterfowl, fish, marshes, and other valued species and communities. The nearby channels and near shore waters are, however, stressed by high boat use and also are susceptible to changes in water chemistry as a result of pollution carried into the area from near-shore or more distant sources in the watersheds around the park.

Table 3. The Land Use Spectrum (Bastedo et al., 1984; Nelson, 1987)

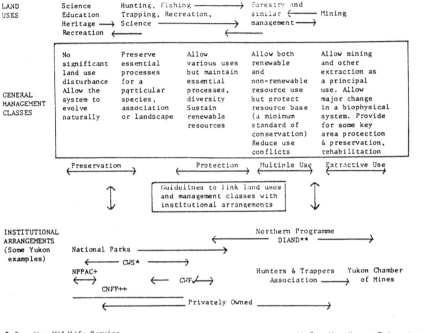

To handle the foregoing and related challenges, we are recommending that three areas be recognized by parks' managers and other users in the Georgian Bay Islands National Park situation (Figure 10). The first is the core. This focuses on Beausoleil Island and its nearshore waters. The second is a transition or near-core area which involves the adjacent waters and some key reptile and other habitat on neighbouring lands. The third is an area of co-operation and communication - the civic area. It encompasses the watersheds whose drainage most heavily affects or is likely to affect the core and near-core areas around Beausoleil Island.

Decisions about the core area should, in our view, be made by national park managers in consultation with neighbours and user groups. Parks Canada is responsible for the ecological health of Beausoleil by law and policy but should consult with users and neighbours for information and for political reasons. This consultation would be facilitated by the creation of a collaborative or Consultative Committee which would have as its main responsibility the land and water uses and environmental conditions in the near core areas. This committee would consist of key government and private stakeholders in the near-core area. Communication would be promoted in the zone of co-operation and communication - which encompasses the entire greater park ecosystem - by the creation of a Greater Park Ecosystem Forum. This would be a relatively open body, attended by people from an array of agencies, bureaus, and groups, as well as individuals involved in every day life in this area.

Figure 10. Spatial Framework for the Georgian Bay National Park Ecosystem Conservation Plan

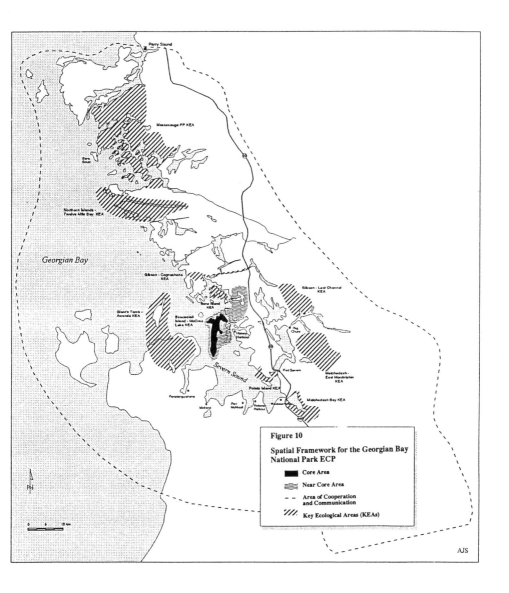

Table 4. Essential Processes in the Civics Approach (Nelson and Serafin, 1996).

Understanding	Broadly informing; comprehensive and pluralist; selective in terms of significance, assessment and action; focusing on preparedness for decision.
Communicating	Understanding and using varied means and media; personal and group communication skills; technical understanding and skills; inter-group or cross-cultural understanding and skills.
Assessing	Understanding of and ability to evaluate and select on the basis of principles and standards; pluralist in orientation; awareness of various kinds of social, economic and environmental assessment; understanding of trade-offs; importance of understanding and assessing institutions both as resources and as obstacles to desired change.
Planning or Visioning	Ability to think systematically and interactively about time and change; historical as well as a futuristic perspective; understanding of time in both natural (geologic, biologic), and human (historic) senses; a human ecological perspective.
Implementing	Understanding how to decide and act; ideas and models of cooperation and coordination; integrating the technical and socio-economic, the scientific and the humanistic; understanding and use of bridging institutions, of demonstration, of a research and experimental approach, of mixed scanning and transactive planning (Etzioni, 1967; Friedmann, 1973).
Monitoring	Generally following or tracking issues and current events; understanding and use of auditing and follow-up procedures as part of all aspects of civics; understanding of different kinds of monitoring and the pluralist nature of monitoring; regular, periodic, and technical monitoring and assessment.
Adapting	Understanding that continuous adjustments to turbulent and changing circumstances are part of the civics model; objectives and activities frequently change among individuals, groups and nations in a dynamic world; capacity to foresee and adapt; evolutionary, interactive, competitive and accommodating; tolerance for ambiguity.

The Forum would also include people from the core and near-core areas. The Forum would promote research, monitoring, assessment, communication, and other activities. It would meet at least annually to receive reports on changes, to secure ideas, and to discuss State of the Environment and other reports. The Forum would promote a sense of civics. In this respect the underlying idea would be a committment to stewardship - to care for environment and concern about sustainable development by public and private agencies and groups. In this context it is important that those involved in and responsible for the core, the near-core, and outlying lands and waters in the greater park ecosystem, be aware of the major processes that are considered to be essential to planning and managing in a civic context (Nelson and Serafin 1996; Andrey and Nelson 1993). Our view of these is shown in Table 4. All of the processes shown in Table 4 have to be thought about to conduct efficient, effective, and sensitive planning and management in a civic society. All of these processes rely to a considerable extent on the availability and effective use of scientific, scholarly and local knowledge and study. We need to learn more about how to combine these various ways of learning. In this respect we are especially challenged by the 'surprises' - and opportunities - that gradual or chaotic changes open up for study and decision-making in parks and protected areas and the lands and waters around them (Holling and Bocking 1990).

10 References

Andrey, J. and J.G. Nelson. 1993. *Public Issues: A Geographical Perspective*. University of Waterloo, Department of Geography, Publication Series, No. 41. Waterloo, Ontario.

Bastedo, J.D., J.G. Nelson and J.B. Theberge. 1984. An Ecological Approach to Resource Survey and Planning for Environmentally Significant Areas: The ABC Method. *Environmental Management* 8: 125-135.

Boyd, J.M. and I.L. Boyd. 1990. The Hebrides: A Natural History. Collins, London, UK.

Dakin, S. 1996. *Georgian Bay Islands National Park Ecosystem Conservation Plan: Communications Strategy report*. Unpublished report submitted to Parks Canada, Cornwall, Ontario.

Etzioni, A. 1967. Mixed Scanning: A Third Approach to Decision-making. *Public Administration Review*. 27: 385-392.

Forman, R.T.T. and M. Godron. 1986. *Landscape Ecology*. Wiley, New York and Toronto.

Friedmann, J. 1973. *Retracking America: A Theory of Transactive Planning*. Anchor Press, Garden City, NY.

Gertler, L.O. 1972. *Regional Planning in Canada: A Planners Testament*. Harvest House, Montreal, Quebec.

Grigoriew, P., J.B. Theberge and J.G. Nelson. 1985. *Park Boundary Delineation Manual: The ABC Resource Survey Approach*. Occasional Paper No. 4. University of Waterloo, Heritage Resources Centre, Waterloo, Ontario.

Holling, C.S. and S. Bocking. 1990. Surprise and Opportunity: In Evolution, in Ecosystems, in Society. In C. Mungall and D.J. McLaren (eds.) *Planet Under Stress*. Oxford University Press, New York, NY: 285-300.

IUCN. 1980. *World Conservation Strategy: Living Resource Conservation for Sustainable Development*. IUCN, Gland, Switzerland.

Lawrence, P.L. and J.G. Nelson. 1992. *Preparing for a Shoreline Management Plan for the Saugeen Valley Conservation Authority*. University of Waterloo, Heritage Resources Centre, Waterloo, Ontario, Waterloo, Ontario.

Lawrence, P.L., Beazley, K. and C.L. Yeung. 1996. Chapter 12: Analysis of Land Cover Change in the Long Point Area.In: J.G. Nelson and K. Wilcox (eds.) *Long Point Environmental Folio*. University of Waterloo, Heritage Resources Centre,Waterloo, Ontario.

Lawrence, P.L. and J.G. Nelson. 1996. Chapter 14: Shoreline Flooding and Erosion Hazards in the Long Point Area.In: J.G. Nelson and K. Wilcox (eds.) *Long Point Environmental Folio*. University of Waterloo, Heritage Resources Centre,Waterloo, Ontario.

MacFarlane, C.B. and R.N. MacCauley. 1984. *Land Use Planning: Practice, Procedure and Policy*. Butterworths, Scarborough, Ontario.

Madgwick, J. and T. O'Riordan. 1995. Assessing Environmental Management in the Norfolk Broads, U.K. *Environments* 23(1): 44-51

McKibben, B. 1990. *The End of Nature*. Anchor Books, New York, NY.

Meyer, W.B. and B.L. Turner II (eds.) 1994. *Changes in Land Use and Land Cover: A Global Perspective*. Cambridge University Press, New York, NY.

Nash, R. 1967. *Wilderness and the American Mind*. Yale University Press, New Haven.

Nelson, J.G. 1987. National Parks and Protected Areas, National Conservation Strategies and Sustainable Development. *Geoforum* 18(3): 291-319.

Nelson, J.G. 1991. Research in Human Ecology and Planning: An Interactive, Adaptive Approach. *The Canadian Geographer*. 35(2): 114-127.

Nelson, J.G. 1994. The Spread of Ecotourism: Some Planning Implications. *Environmental Conservation* 21(3): 248-255.

Nelson, J.G. 1995. Natural and Cultural Heritage Planning, Protection and Interpretation: From Ideology to Practice, a Civics Approach. In: J. Marsh and J. Fialkowski (eds.) *Linking Cultural and Natural Heritage*. Proceedings of a conference at Trent University, Peterborough, Ontario, June 11-13, 1992. The Frost Centre for Canadian Heritage Development Studies, Trent University, Peterborough, Ontario: 33-43.

Nelson, J.G., R. Butler, and G. Wall (eds.). 1993. *Tourism and Sustainable Development: Monitoring, Planning, Managing*. University of Waterloo, Department of Geography, Publication Series, No.37, Waterloo, Ontario.

Nelson, J.G., P. Grigoriew, PG.R. Smith and J.B. Theberge. 1988. The ABC Resource Survey Method, the ESA Concept and Comprehensive Land Use Planning and Management. In M.R. Moss (ed.) *Landscape Ecology and Management*. Proceedings of the First Symposium of the Canadian Society for Landscape Ecology and Management. University of Guelph, Polyscience Publications Inc, Toronto, Ontario: 143-175.

Nelson, J.G., E. LeDrew, Dulbahri, J. Harris and C. Olive. 1992. *Land Use Change and Sustainable Development in the Segara Anakan Area of Java, Indonesia*. A Joint Publication of the Earth Observation Laboratory of the Institute for Space and Terrestrial Science and the Heritage Resources Centre, University of Waterloo, Waterloo, Ontario.

Nelson, J.G. and P. O'Neil. 1989. *The Grand as a Canadian Heritage River*. University of Waterloo, Heritage Resources Centre, Waterloo, Ontario.

Nelson, J.G. and R. Serafin. 1992. Assessing Biodiversity: A Human Ecological Approach. *Ambio*. 21(3): 212-218.

Nelson, J.G. and R. Serafin (eds.). 1995. Learning from Experience: Post Hoc Assessment and Environmental Planning, Management and Decision-making. Special Issue. *Environments* 23(1).

Nelson, J.G. and K.L. Wilcox (eds.) 1996. Long Point Environmental Folio. University of Waterloo, Heritage Resources Centre, Waterloo, Ontario.

Nelson, J.G. and R. Serafin. 1996. Environment and Resource Planning and Decision-making in Canada: A Human Ecological and Civics Approach. In R. Vogelsang (ed.) *Canada in Transition: Results of Environmental and Geographical Research*. Universitatsverlag Dr N Brodsmeyer; Bockum: 1-25.

Rackham, O. 1986. *The History of the Countryside*. London and Melbourne: J.M. Dent and Sons Ltd.

Roseland, M., D.M. Duffy and T.I. Gunton (eds.) 1996. Shared Decision-making and Natural Resource Planning: Canadian Insights. Special Issue. Environments 23(2).

Ross, M. and T. O. Saunders (eds.). 1992. *Growing Demands on a Shrinking Heritage: Managing Resource Use Conflicts*. Canadian Institute of Resource Law, Calgary, Alberta.

Skibicki, A.J. 1992. *Planning for Heritage Resources in a Changing Landscape: Sustainable Development and Conservation in the Grand River Forests Area of North and South Dumfries, Ontario*. M.A. Thesis in Urban and Regional Planning. University of Waterloo. Waterloo, Ontario.

Skibicki, A.J. 1995. *Preliminary Boundary Analysis of the Greater Pukaskwa National Park Ecosystem using the ABC Resource Survey Approach*. Occasional Paper No.6. Department of Canadian Heritage, Parks Canada, Ottawa, Ontario.

Skibicki, A.J. 1996a. *The Pukaskwa National Park Greater Ecosystem: The Lake Superior Coastal Zone*. Report submitted to Parks Canada. University of Waterloo, Heritage Resources Centre, Waterloo, Ontario.

Skibicki, A.J. 1996b. *Georgian Bay Islands National Park Ecosystem Conservation Plan: ABC resource survey of the Georgian Bay Islands National Park Greater Park Ecosystem*. Report submitted to Parks Canada. University of Waterloo, Heritage Resources Centre, Waterloo, Ontario.

Skibicki, A.J. and Nelson, J.G. 1997. *An Ecosystem Conservation Plan for Georgian Bay Islands National Park: Planning for Georgian Bay Islands National Park and Nature Conservation in the Surrounding Region*. Report submitted to Parks Canada. University of Waterloo, Heritage Resources Centre, Waterloo, Ontario.

Tebeau, C. 1986. *Man in the Everglades: 200 Years of Human History in the Everglades National Park*. University of Miami Press, Coral Gables, Florida.

Tubbs, C. 1968. The New Forest: *An Ecological History*. David and Charles Ltd, Newton Abbot, Devon.

Turner, B.L., W.C. Clark, R.W. Kater, J.F. Richards, J. T. Matthews and W.B. Meyer (eds.) 1990. *The Earth as Transformed by Human Action*. The Cambridge University Press, New York, NY.

Woodley, S. 1996. A Scheme for Ecological Monitoring in National Parks and Protected Areas. *Environments* 23(3): 50-73.

World Commission on Environment and Development. 1987. *Our Common Future*. Oxford University Press. Oxford, UK.

A Clash of Values: Planning to Protect The Niagara Escarpment in Ontario, Canada

Anne Varangu[1]

[1] School of Urban and Regional Planning
University of Waterloo, Waterloo, Ontario, Canada N2L 3G1

Abstract. The Niagara Escarpment Plan in Ontario, Canada was approved in 1985. A clash of values continues to characterize efforts to protect the Escarpment. Regulatory controls must be accompanied by provisions which encourage voluntary compliance. Planning needs to address values in order to achieve objectives in protected areas. Values cannot be addressed without public participation in an interactive planning process. Our experiences in planning for protected areas such as the Niagara Escarpment will help us to plan in areas that are not protected.

Keywords. Agriculture, economy, environment, heritage, parks, preservation, protection, biosphere, private property, urbanization, sustainability

1 Introduction

Canada's first large-scale environmental plan was enacted in 1973 with the passage of the Niagara Escarpment Planning and Development Act (NEPDA). The Act provided for the establishment of the Niagara Escarpment Commission (NEC) which was given the responsibility of creating the Niagara Escarpment Plan (NEP). This plan was approved on June 12, 1985 and revised as the result of a review initiated in 1990. The purpose of the Niagara Escarpment Plan is to provide for the maintenance of the Niagara Escarpment and land in its vicinity substantially as a continuous natural environment, and to ensure only such development occurs as is compatible with that natural environment (Ontario Ministry of Environment and Energy 1994). The Niagara Escarpment was designated a World Biosphere Reserve by UNESCO in 1990.

Stretching 725 km, and including more than 100 parks, the Niagara Escarpment is the predominant land form in southern Ontario. It has shaped the lives of many individuals without people necessarily realizing that the part of the Niagara Escarpment with which they, personally, have a relationship, is also part of the larger landform which runs through Ontario but starts and ends in the United States. (Figure 1)

Our individual relationships to the Niagara Escarpment are important in developing the value we see in the Escarpment. For some of us, the Escarpment is 'home'. For others it is a convenient source of aggregates, a recreation area, a panoramic vista, a source of research material, a source of drinking water, a tourist destination, a place to grow grapes or let cattle roam.

Some of our values are related directly to the Niagara Escarpment while others are socially constructed independently of this specific issue. More general concerns for land ethics, the future of the biosphere, and the effect of our human activities on our environment all provide a perspective from which to view the Niagara Escarpment. Values challenge values, notably:

- a faith that science, technology and progress allow us to conquer nature and that we no longer need to nurture our natural environment in order to survive;
- a belief in the existence and ability of the free market economy to eventually right all wrongs; and
- a belief that individual property owners have rights which supersede those of society in general.

How can or should land use planning accommodate changing values? In the relatively short planning history of the Niagara Escarpment, values have continued to clash when more general values take on specific meaning with respect to the Escarpment. The Niagara Escarpment, for some, has become a battleground.

Rather than simply hoping that value-based clashes will disappear, both planners and members of the public should be actively encouraged to engage their values and reflect on the context in which values develop. The experience we gain from planning exercises in protected areas such as the Niagara Escarpment will guide us when we encounter these same issues in planning for areas that are not 'protected.'

Figure 1 Niagara Escarpment and related Geological Features, Southern Ontario

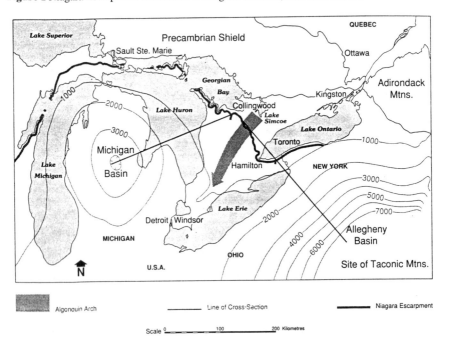

2 Summary of the Planning Process to Date

Although land associated with the Niagara Escarpment is owned by a mix of federal, provincial, municipal, and private owners, the majority of land is owned privately. Responding to demands that the Niagara Escarpment be protected from being totally carved up and developed, the provincial government commissioned a study. In 1968 this study, known as the Gertler Report (1968), recommended that the Escarpment be preserved, that a park system be established, and that the aggregate extraction industry be subjected to controls. A Task Force was given the job of deciding how to implement these recommendations (Ontario Ministry of Treasury Economics and Intergovernmental Affairs 1972). Its 1972 report led to the passage of the NEPDA in 1973 by the provincial government (Ontario Government 1973). The then Provincial Premier William Davis commented:

'Traditionally, ownership of a plot of land has been an individual's most cherished mark of independence, and there was a time when the owner enjoyed the privilege of putting his land to whatever use he chose. In this country, however, zoning restrictions began to provide a check against indiscriminate or incompatible use of land - though in certain areas there is evidence that zoning bylaws alone cannot preserve for this and future generations our landscape as we would like it to be and as we would wish to leave it for our children - all of us - with the co-operation of the private sector - must agree upon some basic principles governing the future use of our land'. (NEC 1981, p.13-14 Reprinted with permission.)

Response by private property owners was mixed. Zoning was one thing; a land use planning act such as the NEPDA was perceived by some as much more threatening; as what they perceived to be an infringement of their rights. Others were very supportive. A still active non-governmental agency (NGO), the Coalition on the Niagara Escarpment (CONE), was formed in 1978 to give major environmental groups and supporters a stronger voice with which to lobby for Escarpment protection.

From the responses to the Proposed Plan during the commenting period (NEC 1981), it was clear that the public in general did not have much understanding about the Escarpment. Some municipal governments applauded the Plan while others insisted that local governments, rather than a single agency such as the NEC, should be given responsibility for implementation of the Plan. Some municipalities and private individuals, then as today, insisted that the Plan should only govern the obvious physical features of the Niagara Escarpment. As a result of appeals, the total area first proposed to be included in the Plan was slashed by 63% by the time that the Plan was finally approved, dividing the Escarpment into a Plan Area and a peripheral Planning Area (Figure 2).

The limitations set for development within the Niagara Escarpment Plan Area seemingly will be met in the next few years. Residents and politicians will be forced to confront the immediate application and impacts of a heretofore abstract concept, limits to growth.

3 Features of the Niagara Escarpment

The Escarpment is composed of limestones, dolostones, sandstones and shales. The softer shales have eroded and retreated relatively rapidly with the harder dolostones and limestones acting as cap rocks for the formation of the steep cliffs and slopes of the Escarpment. The rugged topography has limited agricultural and other settlement and helped protect relatively large and extensive forests, wildlife habitats, waterfalls and scenery which are set amidst some of the most developed lands in Ontario. More details on the geology, biology and other attributes of the Escarpment are given in Appendix 1.

3.1 Geological

The Niagara Escarpment marks changes over time to the area in which we now live. We are still interpreting the stories it tells. It is living history; our heritage and the history of the place in which we now live.

'The Niagara Escarpment consists of Ordovician and Silurian rocks formed from sediments deposited in ancient seas between 445 and 420 million years ago. It is also known that between 23,000 and 12,000 years ago the escarpment was covered with 2-3 km of ice for the last time...Further evidence indicates that the Niagara Escarpment came into existence during the long preglacial interval and that continuous erosion must have been the cause of its formation. When the land that is now southern Ontario emerged from the seas of the Paleozoic Era at least 245 million or more years ago, drainage networks developed and began the task of eroding the land by removing immense quantities of rocky materials...Because of erosion the escarpment may have migrated southwestward over much of southern Ontario'
(quoted in Tovell 1992, p.75 Reprinted with permission).

With the ridge itself over 100 metres high in some locations, many waterfalls are associated with the Niagara Escarpment (Figure 1). Internationally, Niagara Falls is probably the best known Escarpment feature. In the 19th century, examination of rocks and erosion patterns in the Niagara

Gorge helped geologists to begin to understand the geological time clock associated with the Niagara Escarpment (Tovell 1992). Geological features are not just of interest to geologists. Kettles, caves, and moraines associated with the Escarpment are all familiar landforms to those who live near them. The kettle may be a favorite swimming hole, the caves a place to play hide and seek, and the moraine a gently sloping field for crops.

3.2 Forests, Flora and Fauna

Forests now protected by the Plan remain threatened by illegal logging. Before European settlement, much of the Niagara Escarpment was forested. The area described in the following quotation from 1838 is now predominantly agricultural or residential.

'No one who has a single atom of imagination can travel through the forest roads of Canada without being strongly impressed and excited. The seemingly interminable line of trees before you; the boundless wilderness around; the mysterious depths and multitudinous foliage, where foot of man hath never penetrated, - and which partial gleams of light of the noontide sun, now seen, now lost, lit with a magical beauty as the wondrous splendor and novelty of the flowers'. (Mrs. Jameson, writing about the Brantford-Hamilton area in 1838, quoted in Tovell 1992, p. 148 Reprinted with permission.)

The Escarpment provides habitat and corridors for wildlife. Deer, coyotes, foxes and other animals attempt to cohabit with humans. Because the Escarpment has been protected it is also still home to many species of plant life, such as unusual ferns, orchids, and, in the southern part, Carolinian species. On the cliff face itself, living, dwarf white cedar trees have been discovered to have survived for over 1,000 years despite an inhospitable environment (See Appendix 1).

3.3 Hydrology/Groundwater

The Escarpment is also an hydrological feature. The headwaters of several major rivers are located upon it (Tovell 1992, p. 17). Many people still understand the 'escarpment' in a narrow sense as meaning only the cliff face and lands immediately adjacent. From a panoramic perspective, it may be easier to understand why lands below the Escarpment should be protected. But as the source of our water, there is also an urgent reason for protecting lands on top of the Escarpment.

'The headwater zones of the major escarpment watercourses extend well away from the escarpment into the escarpment upland. In these areas, water from rain and melting snow is retained in swamps and ponds...and in porous sediments and soils as (groundwater). The waters leak out from these reservoirs as small trickles and springs that gather into rivulets and then combine to form streams, creeks, larger creeks and rivers' (Tovell 1992, p. 16-17 Reprinted with permission).

Partly for this reason, the practice of landfilling in abandoned quarries within the Plan area was ended in 1994. But because the Plan Area is quite narrow in some locations, the Niagara Escarpment remains threatened by leachate from landfills located above the ridge.

4 Issues

The Niagara Escarpment is as much a cultural phenomenon as it is geological, biological, and hydrological. We declare the value of the Niagara Escarpment through the meaning we create for it. The features of the Niagara Escarpment are important to us because we interpret these features through the lens of our value system.

Figure 2. Preliminary Niagara Escarpment Planning Area and Final Approved Plan Area.

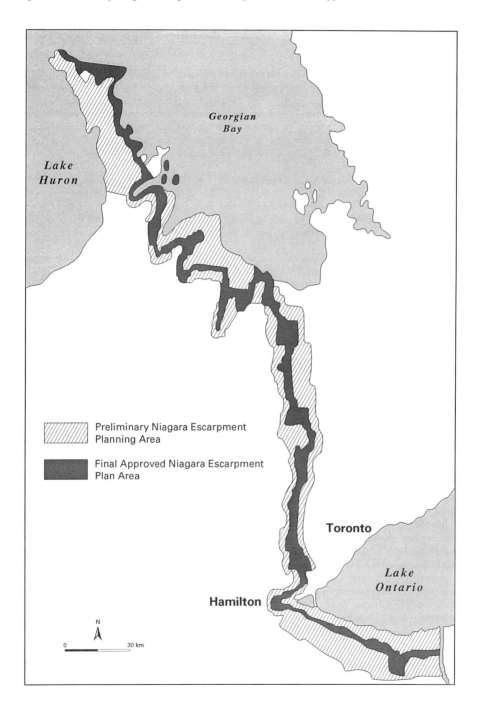

4.1 Preservation of what?

Two hundred years ago the European settlers cleared land for agriculture and for safety. We are still living with the legacy created by humans who altered the landscape for their own purposes at one point in time.

'Such changes in the biosphere had an impact on the physical environment. Permanent streams became intermittent and disappeared; wells and ponds dried up; erosion by wind and water and flooding altered the lands. In more recent years, the Niagara Escarpment has been subjected to developments that lacked understanding of the environment. At the southern end especially, housing developments line the edge of the escarpment, while one quarry after another has defaced the landform's natural beauty. In some sections development has invaded the headwater zone of major escarpment rivers' (Tovell 1992, p. 148 Reprinted with permission).

What is the 'natural' state of an evolving landform? Is our goal to recreate what is 'natural'? 'Natural' at what point in time? Should we aspire to return to forest cover and the character of the Escarpment prior to settlement and development by Europeans? If this is our choice we must be able to state our reasons convincingly. If people no longer see themselves as dependent upon their environment, if they believe that technology and devices manufactured by humans can be an adequate substitute for nature, then they will see no need to return to the past. They will dismiss as trivial nostalgia any plans which include objectives such as 'natural'. Alienation from our natural environment permits people to create all kinds of arguments against preservation. There may also be considerable confusion between what we mean by 'natural' and what we mean by 'scenic'.

4.2 Panorama

What is 'scenic'? A traveller recorded the following in the pre-developed Ontario of 1796:

'I saw very grand rocks in going towards the Mountain and passed three water falls, the first sombre and beautiful from the water falling from various directions over dark, mossy rocks. The second was pretty from the fine scenery of tall trees, thro' which it shone - the third, just below an old saw mill, falls smoothly for some feet, and is a bright copper colour, having passed through swamps; it then rushes into white foam over regular ledges of rocks spreading like a bell, and the difference in colour is a fine contrast. The course of this river is a series of falls over wild rocks, the perpendicular banks on each side very high, covered from top to bottom with hemlock, pines, cedars and all forest trees of an immense height' (Mrs. Simcoe's entry in her diary for June 9, 1796. Quoted in Tovell 1992, p. 15 Reprinted with permission).

In the area described above there are now homes, businesses, back yards and a greatly reduced flow of water. Should this area have been preserved only because it was scenic? Is it still 'scenic'? Are the old European town centres scenic? If yes, then the panoramic objectives for natural areas should include modifiers which explain why one 'scenic' area is more significant than another and elaborate the criteria by which that decision is made.

E.G. Zube and D.G. Pitt argued in 1981 that what we mean by 'scenic' is 'strongly influenced by the cultural traditions and the range of opportunities available where one lives' (p.85). If, as the studies they cite suggest, 'perceptual responses are conditioned by learning and learning varies with culture and environment' (p.85) then the 'management problem may also be compounded when landscapes attract visitors from diverse cultures and when host and visitor perceptions differ ' (p.86).

4.3 Urbanization and Residential Development

Seven million people in the United States and Canada live within 100 km of the Niagara Escarpment. About 120,000 people live within the Niagara Escarpment Plan Area, the area designated as a World

Biosphere Reserve (NEC, 1995). Despite the existence of the Niagara Escarpment Plan, the pressure to develop new residential areas is intense in the Niagara to Hamilton portion of the corridor. Traditionally, farmers have been lured by developers to sell prime tender fruit lands in the Niagara area for residential development at higher prices than they thought they could ever earn farming. Recently a rare statement refuting this view was made by a grape grower and neighbouring landowner to a proposed residential development. He argued that the Escarpment's 'bench', a plateau along parts of the Niagara Escarpment in the Niagara to Hamilton corridor, would actually bring more money to the economy if it were planted in grapes than if it were developed into high-end estate housing (see Appendix 2).

4.4 Non-Renewable Resources and the Niagara Escarpment

Sustainability may be a widely touted criterion for assessing our plans, but this concept is as much a source of debate as it is a solution. Not everyone agrees on the speed, the specific course of action, nor even the need to ensure that our activities are sustainable. The core questions for sustainability revolve around establishing criteria for the use of non-renewable resources, linking environmental with social and economic issues. We are accustomed to solving social problems by using non-renewable resources. Issues which affect the Niagara Escarpment are the same issues which extend beyond the Escarpment area but have an added variable in that by depleting a particular natural resource we also destroy the Niagara Escarpment.

4.4.1 Land

Whether farmers can make ends meet by farming is a social problem in the broad sense and an economic problem in a narrower sense. As indicated above, the drive to develop land is so strong that even with the Niagara Escarpment Plan in place there are still development proposals for subdivisions. In developing land for housing, we should be aware that we are using a non-renewable resource to solve social problems such as urban sprawl and its accompanying tendency to encourage speculation on land values.

4.4.2 Aggregates

Pressure from urbanization creates a heavy demand for a convenient source of aggregate materials. Worries that the Escarpment was being destroyed by quarrying were a driving force for demands that the Niagara Escarpment be protected. From an economic perspective quarrying continues to be a major issue. Conservationists and aggregate producers take diametrically opposed positions about the need to quarry the Niagara Escarpment and about whether the Niagara Escarpment should be treated differently from other areas.

4.4.3 Water

Development creates pressures on headwaters located on the Niagara Escarpment. Water is generally not thought of as a non-renewable resource by the general population except in crisis situations where the relationship of humans to water is forcefully brought to mind. In urban areas where people no longer rely on well water or streams for their everyday water supply, our relationship to groundwater disappears and is replaced by a reliance on systematic and engineered 'solutions'. With the disappearance of this personal relationship, the value we place on protecting our water supply decreases. The result is that instead of noticing the gradual disappearance of quality ground water we generally do not realize that we have a problem until faced with an acute situation.

4.5 Regulation vs Persuasion/Public vs Private

Education is critical. Just as our values about what is scenic are based on learning and conditioning provided by the experience of our culture, so too other attitudes are functions of the social construction of value. The question of how much can be accomplished by regulatory control is cross-linked to attitudes about the individual's relationship to society. As elsewhere in Ontario, much of the conflict in values on the Niagara Escarpment is created as a result of proposed changes in land use. People still argue about compensation and development rights which they see as having been unjustifiably restricted when the Niagara Escarpment Plan was created.

5 Incorporating Values into the Planning Process

Planners and plans need to incorporate values. Values cannot be addressed without public participation in an interactive planning process. This type of planning process is substantially different from the planning process applied to the Niagara Escarpment.

In a recent work C. Paine and J. R. Taylor (1995) favour the term, cultural landscapes, since the word cultural suggests an 'integration of both natural and manmade resources' (p.5) and 'also suggests a wider scope for interpretation of associated human values. Cultural landscapes may embody not only aesthetic and associative dimensions, but also utilitarian, social, symbolic, economic, scientific, or other values as well.' (p.6)

Understanding aesthetic or perceptual perspectives potentially requires a high degree of personal involvement on the part of the landscape assessor. The interpretive dimensions of meaning and place can only be determined through historic research and local informant input. The values inherent in the cultural landscape, such as knowledge of place and identity, are best determined by local residents (Paine and Taylor 1995, p. 8-9 Reprinted with permission). This will require a different type of ongoing planning process. The schematic for the model proposed by Paine and Taylor (1995) for the Niagara Escarpment is shown in Figure 3.

Figure 3. Proposed Cultural Landscape Assessment Model

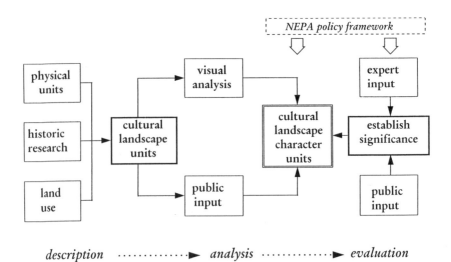

6 Current Planning Challenges

6.1 Values

Value based clashes characterize the planning process for the Niagara Escarpment. How can the scope and degree of public participation in the planning process be increased without threatening the integrity of plans to protect natural areas? Which values are culture-specific and which are cross-cultural? How can dialogue and reflection have an impact on the elasticity of values? How might incorporating values into the planning process affect the end result?

6.2 Niagara Escarpment as a Unit

People identify with the Niagara Escarpment through the individual relationships they create with parts of the escarpment. There seems to be difficulty in conceiving of the Niagara Escarpment as a larger unit, a 'whole'. How can we generate new meaning for natural areas? How can we encourage people to see the small part with which they may be familiar as a part of a larger 'whole', in both a physical and cultural sense.

6.3 Resource Use

The existence of a Niagara Escarpment Plan has not precluded competing demands driven by economic and development interests. This situation is particularly aggravated because of the non-renewable resources potentially available on the Niagara Escarpment. Although the Niagara Escarpment Commission has been given the responsibility of interpreting the Plan, this agency does not have complete control over the Niagara Escarpment. Responsibility for the Niagara Escarpment is now shared by the NEC, the Ontario Ministry of Natural Resources, the Ontario Heritage Foundation, and the Canadian Parks Service.

Will the consolidation of responsibilities for natural areas minimize the opportunity to play off one agency or department against another and better serve to protect natural areas? What benefits might accrue from 'one-window' control over the Niagara Escarpment? This might be achieved by transferring responsibility for the Niagara Escarpment to a single agency, as recommended by UNESCO (UNESCO 1974, p.16, p.24) and the Niagara Escarpment Program Evaluation Final Report (Cresap 1988).

6.4 Nature Protection

Simply changing the rules for certain natural areas seems to be an inadequate way to achieve protection and insufficient to prepare our societies to address questions about conservation and protection in areas that are not subject to regulatory control. What factors jointly affect the management, the perception, and the success of nature protection in protected areas and in unprotected areas? How can these factors be addressed in a way that does not threaten the integrity of protected areas while at the same time becoming part of the approach by which the wider society addresses unprotected areas?

7 Conclusion

Nearly twenty years ago, a Statistics Canada report argued for the need to understand the impact of our activities on our environment.

'That human activities can drastically alter the suitability of landscapes for human existence is a fact that has been recorded since ancient times. Today the scope and scale of these transformations has increased markedly. As the amount of wilderness area decreases at an increasingly rapid rate, man no longer has the ready option to become a fugitive species, to escape to a new continent, as in centuries past. Having run out of new parts of the planet to colonize, we are now forced to turn our attention to maintaining the quality of the present environment to ensure the survival of future generations' (Statistics Canada 1978, p. 7 Reprinted with permission).

Unless we use the opportunity provided by areas like the Niagara Escarpment to understand the human impact on our environment and to explore ways to change our habits outside protected areas, protected areas themselves will continue to be subjected to great pressures for development. Plans for protected areas cannot be considered static. With a view to educating and gaining support from members of the public, plans must continually be interpreted in an attempt to answer concerns and maintain the integrity of the plan. The Plan for the Niagara Escarpment, like the geological feature itself, is an interpretive work in progress. The challenge is to answer concerns without losing sight of the overall goal of protection.

8 References

deBoer, G.. 1993. *Understanding the Process of Evaluating Candidate Biosphere Reserves* (unpublished paper). University of Guelph, Guelph, Ontario.

Cresap Management Consultants. 1988. *Niagara Escarpment Program Evaluation, Final Report* Toronto, Ontario.

Gertler, L.O. 1968. *Niagara Escarpment Study, Conservation and Recreation Report.* Niagara Escarpment Study Group, Regional Development Branch, Treasury Branch, Province of Ontario, Toronto, Ontario,

Lindgren, R. D. (no date). *Submissions of the Coalition on the Niagara Escarpment (CONE) Regarding Proposed Revisions to the Niagara Escarpment Plan.* (Response to the first Five Year Review) Canadian Environmental Law Association, Toronto, Ontario.

Nelson, J.G., and L. Shultz. 1991. *Sharing Responses to the Niagara Escarpment Plan Review.* Heritage Resources Centre, University of Waterloo, Waterloo, Ontario.

Niagara Escarpment Commission. April 1976. *Pits and Quarry Development within the Niagara Escarpment Planning Area.* Georgetown, Ontario.

Niagara Escarpment Commission. 1977. *Land Use Land Fragmentation.* Georgetown, Ontario.

Niagara Escarpment Commission. 1977. *Report of the Bruce Trail Study Group.* Georgetown, Ontario.

Niagara Escarpment Commission. 1977. *Preliminary Proposals.* Georgetown, Ontario.

Niagara Escarpment Commission. 1977. *Preliminary Proposals: Background Data.* Georgetown, Ontario.

Niagara Escarpment Commission. 1978. *Niagara Escarpment News,* Georgetown, Ontario.

Niagara Escarpment Commission. 1979. *The Proposed Plan for the Niagara Escarpment.* Georgetown, Ontario.

Niagara Escarpment Commission. 1981. *Information Update Report on the Proposed Plan for the NiagaraEscarpment.* Georgetown, Ontario.

Niagara Escarpment Commission. 1990. *Five Year Review Niagara Escarpment Plan Policy Paper 13: The Niagara Escarpment Parks System.* Georgetown, Ontario.

Niagara Escarpment Commission. 1990. *Five Year Review Niagara Escarpment Plan Policy Paper 11: Mineral Resources Extraction.* Georgetown, Ontario.

Niagara Escarpment Commission. 1991. *Proposed Revisions to the Niagara Escarpment Plan.* Georgetown, Ontario.

Niagara Escarpment Commission. 1993. *Renewing the Vision: Response to the Hearing Officers' Report.*Georgetown, Ontario.

Niagara Escarpment Commission. 1995. *Ontario's Niagara Escarpment (Ontario, Canada): Implementing theBiosphere Reserve Concept in a Highly Developed Region.* Georgetown, Ontario.

Ontario Department of Treasury and Economics, Niagara Escarpment Study Group. 1972. *The Niagara Escarpment Study. Fruit Belt Report.* Ontario Department of Treasury and Economics, Niagara Escarpment Study Group. Toronto, Ontario.

Ontario Government, 1973. *Bill 129 -- An Act to Provide for Planning and Development of the Niagara Escarpment and its Vicinity,* Amendments -- Bill 86 1974; Bill 135, 1975; and Bill 9, 1976. The Government of Ontario. Toronto, Ontario.

Ontario Government. Revised Statues of Ontario, 1990. Chapter N.2. Jan. 1992. *Niagara Escarpment Planning and Development Act.* The Government of Ontario. Toronto, Ontario. Toronto, Ontario.

Ontario Ministry of Culture and Recreation. 1976, *Historical Resources in the Niagara Escarpment Planning Area.* (+maps). Ontario Ministry of Culture and Recreation. Toronto, Ontario.

Ontario Ministry of Culture and Recreation. 1976. *Niagara Escarpment Planning Area. Archaeological Resources.* (+maps). Ontario Ministry of Culture and Recreation. Toronto, Ontario.

Ontario Ministry of Environment. 1993. *Niagara Escarpment Plan Review Report.* Ontario Ministry of the Environment and Energy, Toronto, Ontario.

Ontario Ministry of Environment and Energy. 1994. *The Niagara Escarpment Plan.* Ecosystem Planning Series, Ministry of the Environment and Energy, Toronto, Ontario.

Ontario Ministry of Environment and Niagara Escarpment Commission. 1995. *Leading Edge '94 Conference Proceedings.* Ontario Ministry of the Environment and Energy. Toronto, Ontario.

Ontario Ministry of Environment and Niagara Escarpment Commission. 1996. *Leading Edge '95 Conference Proceedings.* Ecosystem Planning Series, Ministry of the Environment and Energy. Toronto, Ontario.

Ontario Ministry of Natural Resources. 1976. *Hazard Lands in the Niagara Escarpment Planning Area* (+maps). Ontario Ministry of Culture and Recreation. Toronto, Ontario.

Ontario Ministry of Natural Resources. 1976. *Nature Reserve Candidates in the Niagara Escarpment Planning Area.* (+maps). Ontario Ministry of Culture and Recreation. Toronto, Ontario.

Ontario Ministry of Natural Resources. 1976. *A Policy for Mineral Aggregate Resource Management in Ontario.Report of the Ontario Mineral Aggregate Working Party.* Ontario Ministry of Natural Resources. Toronto, Ontario.

Ontario Ministry of Natural Resources. 1977. *Land Acquisition Program of the Ministry of Natural Resources and the Conservation Authorities in the Niagara Escarpment Planning Area.* (+maps) Ontario Ministry of Natural Resources. Toronto, Ontario.

Ontario Ministry of Natural Resources. 1977. *Ministry of Natural Resources in the Niagara Escarpment Planning Area: A Summary Report.* (+maps). Ontario Ministry of Natural Resources. Toronto, Ontario.

Ontario Ministry of Treasury, Economics, and Intergovernmental Affairs. 1972. *To Save The Escarpment.* Report of the Niagara Escarpment Task Force. Ontario Ministry of Treasury, Economics, and Intergovernmental Affairs. Toronto, Ontario.

Ontario Ministry of Treasury, Economics, and Intergovernmental Affairs, 1973a. *Regional Planning in Ontario:The Niagara Escarpment.* (Government Policy for Niagara Escarpment). Ontario Ministry of Treasury, Economics and Intergovernmental Affairs. Toronto, Ontario.

Ontario Ministry of Treasury, Economics, and Intergovernmental Affairs. 1973b. *Development Planning in Ontario: The Niagara Escarpment.* (Government Policy for Niagara Escarpment). Ministry of Treasury, Economics and Intergovernmental Affairs. Toronto, Ontario.

Ontario Ministry of Treasury, Economics and Intergovernmental Affairs. July 1978. *The Parkway Belt West Plan.* Parkway Belt Group. Parkway Belt Section, Ministry of Treasury, Economics and Intergovernmental Affairs. Toronto, Ontario.

Paine, C. and J. R. Taylor. 1995. *Cultural Landscape Assessment: A Comparison of Current Methods and Their Potential for Application Within The Niagara Escarpment.* Landscape Research Group School of Landscape Architecture, University of Guelph, Guelph, Ontario.

Parkway Consultants for Tri-County Committee. 1968. *Niagara Escarpment Scenic Drive Feasibility Study.* Toronto, Ontario.

Plaunt, M. 1978. *The Decision Making Process that Led Up to the Passing of the Niagara Escarpment Planning land Development Act.* Faculty of Administrative Studies (unpublished paper). York University. Toronto, Ontario.

Reid, I. 1977. *Land In Demand: The Niagara Escarpment.* The Book Society of Canada Ltd. Agincourt, Ontario.

Richards, R.N. 1991. *Abuses and Improprieties of the Niagara Escarpment Commission.* Medric Ltd. Willowdale, Ontario.

Soil Conservation Society of America. 1974. *Land Use: Persuasion or Regulation? Proceedings of the 29th Annual Meeting at Syracuse, New York.* Ankeny, Iowa.

St. Catharines Standard, J. Meyers. February 6, 1996. *NEC, Developers Take First Shots.* St. Catharines Standard, St. Catharines, Ontario.

Statistics Canada. 1978. *Human Activity and the Environment.* Ministry of Industry, Trade and Commerce, Ottawa, Canada.

Tovell, W. M. 1992. *Guide to the Geology of the Niagara Escarpment.* Niagara Escarpment Commission, Georgetown, Ontario.

UNESCO. 1974. *Task Force on Criteria & Guidelines for the Choice and Establishment of Biosphere Reserves- Final Report.* UNESCO Programme on Man and the Biosphere Report Series 22. Paris, France.

Zube, E.H. and D.G. Pitt. 1981. Cross-Cultural Perceptions of Scenic and Heritage Landscapes. *Landscape Planning,* 8 (1981) 69-87.

Appendix 1. The Niagara Escarpment Biosphere Reserve Area of the Biosphere Reserve*

Niagara Escarpment Plan Area	183,694 ha
Bruce Peninsula National Park (portion outside Plan Area)	5,684 ha
Fathom Five National Marine Park (portion outside Plan Area)	1,275 ha
TOTAL	190,654 ha

CORE AREA

Escarpment Natural Area (includes portions of Bruce Peninsula National Park and Fathom Five National Marine Park that are Escarpment Natural)	48,403 ha
Bruce Peninsula National Marine Park (minus that which is Escarpment Natural)	13,584 ha
Fathom Five National Marine Park (portion outside Plan Area + portion that is Escarpment Protection (land base), the remainder being Escarpment Natural)	1,415 ha
TOTAL	63,402 ha

BUFFER AREA

Escarpment Protection Area (minus) those parts of Bruce Peninsula National Park & Fathom Five National Marine Park that are Escarpment Protection)	65,193 ha
Escarpment Rural Area (minus those parts of Bruce Peninsula National Park that are Escarpment Rural)	48,095 ha
TOTAL	113,288 ha

TRANSITION AREA (ZONE OF COOPERATION)

Urban Area	3,511 ha
Escarpment Recreation Area	7,539 ha
Mineral Resource Extraction Area	2,914 ha
TOTAL	13,964 ha

*Figures presented here reflect updated information since designation of the biosphere reserve in 1990.

Elevation
Lowest point: 98 metres asl (excluding Fathom Five National Marine Park)
Highest point: 532 metres asl

Biogeographic Region - Latitudinal trends are well displayed on the north-south trending Escarpment:

Deciduous Forest Region
At the south end, common vegetation communities include rich slope forests of Sugar Maple and Black Maple mixed with Tulip-tree and Red Elm. Drier slopes have oak-hickory forests of Red, White, Black and Chinquapin Oaks mixed with Bitternut, Shagbark, and Pignut Hickories - all trees characteristic of the eastern deciduous forest region. Understorey species are largely southern ones such as Flowering Dogwood, Running Strawberry bush and Yellow Mandarin.

Great Lakes-St. Lawrence Forest Region
All of the Ontario Niagara Escarpment north of the Niagara Peninsula occurs within this forest region. At its north end on the Bruce Peninsula, there are fire-successional forests of White Birch, Trembling Aspen and Eastern White Cedar, with boreal species such as Balsam Fir and White Spruce. Jack Pine occurs at the southern limit of its range. Understorey species include northern ones such as Striped Coralroot and Bearberry and such circumboreal species as Hudson's Club-rush. There is also a concentration of western species of flora and fauna in the northern half of the Escarpment. e.g., eight plant species have disjunct populations, including Trail-plant, Menzie's Rattlesnake-plantain, Holly Fern and Alaskan Orchid.

Climate
Maximum average temperature, warmest month. 29.9 C
Minimum average temperature, coldest month: -7. C
Mean annual precipitation 818.5 mm (13% as snow)

Geology
Escarpment formation: Ordovician and Silurian Periods (420-445 million years Before Present)
Sedimentary rocks of the Niagara Escarpment: limestones, dolostones, shales, sandstones

Biology

Flora
Over 1500 species of vascular plants (including 40% of Ontario's rare flora):

In the south: Cucumber-tree, Paw-paw, Green Dragon, Tuckahoe, American Columbo
In the north: Rand's Goldenrod and Roundleaf Ragwort.
The threatened American Ginseng occurs in rich Sugar Maple forests along much of the Escarpment.

Significant species endemic to the Great Lakes occur on the Bruce Peninsula portion of the Escarpment, including Lakeside Daisy, Dwarf Lake Iris, Hill's Thistle, Provancher's Philadelphia Fleabane and Ohio Goldenrod.

Ferns: 50 species recorded, including Wall-rue, an Appalachian species rare in Canada. Most of the world population of the North American subspecies of Hart's tongue fern occurs along the Escarpment.

Orchids: 37 species recorded in the northern parts of the Escarpment, including Calypso Orchid, Ram's-head Lady-slipper and Alaska Rein Orchid.

Oldest trees in eastern North America (1000 years): Eastern White Cedar

Fauna
Over 300 bird species (of which 200 species have shown evidence of breeding in the Niagara Escarpment). Of the breeding species, 25 are considered nationally or provincially endangered, threatened or vulnerable, including Bald Eagle, Red-shouldered hawk, Black Tern, Louisiana Waterthrush and Hooded Warbler.
55 mammal species and 34 species of reptiles and amphibians have been recorded. Rare species include the endangered North Dusky Salamander, the threatened Eastern Massasauga Rattlesnake, the vulnerable Southern Flying-squirrel and the rare Eastern Pipistrelle.

Adapted with permission from the Niagara Escarpment Commission. 1995. *Ontario's Niagara Escarpment (Ontario, Canada): Implementing the Biosphere Reserve Concept in a Highly Developed Region.*

Appendix 2

The economic argument against using land for residential development: In February 1996, M. Kriluck testified at hearings into the proposed Twenty Valley Estates in Jordan, near Balls Falls. He calculated: One acre of vinifera grapes produces approximately 4000 bottles of wine. At the low end price of $13.00 CAN. for a bottle of wine, approximately $6.00 CAN goes back to federal and provincial governments in the form of taxes, duties, and markups. At $6 CAN per bottle, the revenue going back to society per acre would be $24,000 CAN. On a 30 acre property this would be $720,000 CAN per year. In contrast, municipal taxes on the same 30 acres would bring in $176,000 CAN per year. It is better for society to grow grapes than put scarce vinifera grape growing land out of production. References: NEC Hearing Notes; St. Catharines Standard; personal conversation July 11, 1996.

The Economic Pitfalls and Barriers of the Sustainable Tourism Concept in the Case of National Parks

Jan van der Straaten[1]

[1] Department of Leisure Studies and European Centre for Nature Conservation
Tilburg University, P.O. Box 1352, 5004 BJ Tilburg, The Netherlands

Abstract. It is often argued that national parks and other aspects of nature have general economic value. In saying this, the conclusion is drawn that this strong argument (economic value) can give an extra impetus to the protection possibilities of these parks. Sustainable tourism can be seen as an instrument to demonstrate this economic value. It cannot be denied that these statements are true at least in principle. The crucial question, however, is just how strong is the theoretical basis for such statements? To make clear why this is so a short overview is given of the possibilities and impossibilities of giving an economic value or a price to national parks. It can be concluded that nature and the environment are not well defined in economics; additionally the theoretical basis of value in economics is rather weak. This implies that the theoretical basis of the sustainable tourism concept cannot be stronger than the weak economic foundation. Does this mean that we do not have economic instruments to strengthen the position of national parks in the societal debate? This is not true, as can be seen from certain examples which are given. The general conclusion is that we have certain economic instruments to support the concept of sustainable tourism in national parks, but we have to be aware of the theoretical limitations of these concepts.

Keywords. Sustainable tourism, economic value, national parks and protected areas

1 Introduction

With respect to tourism in national parks, two approaches are generally found. On the one hand, the opinion is articulated that national parks have been established to protect landscapes of outstanding beauty, including the organisms living in them. In the beginning, these conservation practices did explicitly include a certain level of recreational and tourist activities. Nature was protected to give people the opportunity to view and admire it. On the other hand, in a number of European countries the idea has recently been advanced that tourism could be a threat to nature, including nature in national parks. If the level of funding is not sufficient, as is currently often the case, tourism is then propagated as an instrument for solving these financial problems. Both approaches have in common that national parks are seen as special places not directly related to the economic problems of the countryside around them. In this paper, it is argued that these approaches cannot achieve real solutions. Here national parks are seen as part of the problem of rural development. Additionally, arguments are given why tourism, national parks and rural problems should be investigated in a coordinated fashion.

The central question in this paper is what strategy should be evolved to guarantee the long term development of national parks. In Section 2, a sketch is given of the regional economic context of national parks. This context is used to describe and analyze the options which can be chosen for the development of parks. In Sections 3 and 4, special attention is given to the internal and the external relationships of national parks. In Section 5, the value of the nature concept, as it has been developed in environmental economics, is discussed. This can lead to a better understanding of the challenges and limitations of the marketing concepts discussed in the previous sections.

2 A Definition of the Problem

National parks are located in regions in which original landscapes are still found due to rare combinations of special geomorphological, climatic, or botanic features and processes. In most cases traditional agriculture is significant, as modern agriculture tends to destroy most of these features. This implies that national parks are often located at the periphery of economic activities, where modern agriculture could not penetrate for one reason or another.

Traditional or low-impact agriculture is in many cases, a precondition for the survival of the national parks. However, historic and current economic development is heading in the other direction. Agriculture and rural economic development are often seen as not beneficial to national parks. It is often assumed that investments have to be realized in the region which will result in an increase in economic activities, jobs, and income. These ideas are based on a traditional Keynesian approach which completely overlooks the difficulties resulting from an uneven distribution of economic activities in space. These problems can only be understood when analyzing recent rural developments.

In the twentieth century, agriculture and related economic activities have reacted to market forces, resulting in an increase in demand due to a rise in population and purchasing power in urban European regions. Additionally, it should not be overlooked that, in many European countries, the experience of two wars resulted in the idea that countries should be able to meet the demand for feeding their own population as much as possible. New technologies made it possible to meet this increasing demand. These new technologies aimed at the intensification of agricultural techniques, implying that agricultural production factors increased their productivity dramatically in Europe. One of these production factors is labour. Intensification in labour productivity in agriculture resulted in a surplus of labour in the countryside. Subsequently, people migrated to nearby cities where industrialization created new jobs.

Depopulation of the countryside was the result. When, in the course of time, the number of inhabitants of some villages fell below a certain level, normal services could no longer be provided in these villages because of a decrease in purchasing power of the people still living there. Schools and shops were closed, which again lowered the standard of living in these villages, thus decreasing the attractiveness of the villages as residential areas and as work places.

Recent developments in Europe are moving in two different directions. On the one hand, there is still the intensification of agriculture in relatively favourable areas close to the cities. On the other hand, extensification and abandonment of land is a common practice in the relatively peripheral areas. In many cases, these processes of intensification and extensification take place in fairly close proximity. In the relatively fertile valleys, intensification occurs. Higher up on the plateaus and the mountain slopes extensification is the norm. Both practices are detrimental to nature. Extensification leads to higher levels of fertilizer and pesticides with all their subsequent negative effects on the quality of water and soil, resulting in a dramatic decrease in many protected species. Extensification, particularly in mountain and Mediterranean regions, leads to erosion, desertification, and to a decline of open space and related flora and fauna.

While the foregoing changes were underway, national parks were established in many European countries. Most of these parks were located where modern agriculture was not such a significant factor. This implies that agriculture in and around these parks was, from the agricultural point of view, not very well developed. This relates to the question of how to develop and maintain the national parks and what the function should be of the surrounding countryside. In this respect, what is the position of agriculture? It can be taken as a starting point that 'normal' rural development can only increase the problems in the long run. Development consisting of high levels of investment, as stimulated by the European Union for many years through the instruments of the Rural Fund, the Cohesion Fund, and other sources, can only increase the intensification of agriculture. This in turn, leads to increases in labour productivity which result in migration, a rise in environmental pollution, and threats to landscapes and nature. Hence, it can be concluded that the intensification of agriculture using traditional models of economic development, does not benefit the countryside as much as many people would prefer.

The disadvantages of such development are increasingly being recognized by politicians, particularly within the European Union. In recent European documents, such as the Fifth Action Programme, it is argued that the traditional development of the countryside should be stopped and that a sustainable development of society should result in limitations to the 'normal' economic development of regions. However, recognition of the problem is not the same as solving it. We are now at the beginning of a transition process, in which the European Union and national authorities intend to guide development so that it will be more beneficial to the countryside, landscape and nature. Local and regional authorities are often not aware of the rapid change in the approaches to regional development and nature protection. In any case, national park authorities should be aware of the changing situation regarding rural development. They are now in a position to become actors in the development of new approaches.

3 The Internal Approach

From the previous discussion, it can be concluded that the problems of national parks and the problems of the countryside, are the result of the same economic development. They are more or less two sides of the same coin. This implies that a solution can only be found if national parks and the countryside are investigated as a common problem. National parks can only survive in the long run if the intensification of agriculture in and around the national parks is stopped. On the other hand, further decline in agriculture could have detrimental effects on population, nature, the landscape, the national parks and the countryside. There are two options. The first is to conclude that regional agricultural and related development are no longer possible once the economic and human resources of the region fall below a certain threshold level. In that case, it might make sense to take this situation as a starting point and develop the region completely in the direction of national parks of a much bigger size than is common now. These new national parks can be compared with the American parks in which normal economic activities are no longer found. The advantage of this approach is that economic funds can be concentrated in those regions which have a more favourable economic level.

The more favourable regions should receive sufficient funds for sound development. How should this be done? From the previous discussion, one can conclude that making a separate plan for the national parks and another plan for the regions does not make sense. If the problems of national parks and the surrounding countryside are the same from the economic perspective, a comprehensive plan for a national park and the surrounding countryside should be the starting point. In most cases, the national parks have sufficient expertise at their disposal to make such a plan. They have the know-how and often the human resources needed for such work. It goes without saying that communication and involvement of the local and the regional communities is highly recommended in drawing up such plans.

One of the most striking aims of these plans is job creation, which is a precondition for every rural development outside the big national parks, as suggested previously. Among the various economic sectors, tourism is the only major alternative for job creation in rural areas where national parks are located. The advantage of rural tourism is that there is a dramatic increase in the demand for tourism located in areas with high landscape values. The quality of this type of tourism is not signified by five-star hotels with indoor and outdoor swimming pools. Rather the quality of the tourism infrastructure is found in the quality of nature in the region. When tourism in and around national parks is developed, the region should never have to compete with high-standard tourism areas along the coast and in other well-established tourist locations. It has to be stressed that there is an abundance of these types of tourist destinations. It makes no sense, from the marketing point of view, to compete with these regions.

Sustainable tourism, which economically benefits the regions and their inhabitants, is the only way to tackle the problems of the national parks and the surrounding regions. It needs to be reiterated that job creation is necessary for the long term survival of the national parks as well as the countryside.

The next question is, of course, how to maintain the quality of nature and the landscape in and around the national parks with growth of tourism? People responsible for the quality of national parks are often afraid that tourism development will decrease the quality of nature. It is often argued that people will always disturb nature; the two are incompatible. It has to be said, however, that many investigations have not made it clear what disturbances would inevitably occur in the case of tourism and recreational activities (see for example, Cocossis and Parpairis 1992). Of course, if vulnerable vegetation is found in a certain part of a national park it goes without saying that paths should not be constructed in these areas. Whenever it is known that visitors can cause damage, conservation measures have to be taken.

We should stress the point that the interactions of tourists with nature and the landscape can be made compatible with the limitations of the national park. It is the tourism infrastructure which takes the tourist to a certain place. The development of this infrastructure is often in the hands of the national park authorities. Mass tourism, for instance, can only be developed after establishing a certain level of infrastructure. National parks authorities do not usually promote mass tourism. They should ascertain what level of a given type of tourism is, generally speaking, compatible with the scope and limitations of the national park in question. After defining this level, they could support the infrastructure needed for that level. It is of the utmost importance that they investigate to what degree existing infrastructures such as barns and farm houses are still available in the region. By using this infrastructure, they can demonstrate that they are willing to promote the economic opportunities of the people living in the region.

Many national park authorities do not have the vaguest idea what to do with sustainable tourism. One thing is clear; sustainable tourism is consistent with the possibilities of the national park and the surrounding countryside. However, one should not forget that sustainable tourism is, from the marketing point of view, different from other forms of tourism. This implies that marketing should concentrate on the type of sustainable tourism that national park authorities intend to promote. It does not make any sense to argue that a certain type of tourism can be accepted in the region, and then wait until such tourists come. In that case, they will never come.

Marketing approaches recognize different types of tourists. Some groups are relevant for the development of sustainable tourism. An example is bird watchers. One should not overlook that they are concentrated in some European countries such as the Netherlands, Great Britain, Germany and Scandinavia. There are hardly any in Italy, Spain, Portugal and Greece. The promotion of birdwatching is possible by advertisement in the former countries. In that case, the questions are: What type of accommodation is 'normal' for this type of tourist? Do they need guided tours? Where can watch towers be built? Which part of the national park should be closed to them to protect breeding birds? The relevant information has to be communicated to birdwatchers. The easiest way is to contact relevant organizations such as the Royal Society for the Protection of Birds and the international organization, BirdLife International. National park authorities should stress the benefits of these visits for bird protection. But it is not only birds which are relevant in this respect. In Western countries, there are organizations investigating plants, reptiles, amphibians, and butterflies, among others. There are many possibilities; however, communication should always focus on the information which is relevant to appropriate organizations and their members.

In Western countries, aside from bird watchers, many other people are interested in nature and an interesting landscape. For instance, the total number of active birdwatchers in the Netherlands is approximately 100,000; but more than 800,000 people are members of Dutch Nature.

Public awareness, nature conservation, and sustainable tourism should be dealt with and evaluated as one entity. Only by communicating these ideas to the people around the national parks, can sufficient support be generated. It is of utmost importance to make it clear to people living around the park that the park authorities are willing to support their economic endeavours. Furthermore, it needs to be stressed that only certain types of tourism can be accepted in a national park. A large number of visitors to national parks creates a high level of support in society itself. A very good example of this strategy can be found in the national parks of the USA, where willingness to provide public goods and services is generally very low. This is not the case with national parks, however, which are supported by the government. This is only possible because of the very large number of visitors to and public support for national parks.

A crucial issue is the financial and administrative situation of national parks. When national parks are funded by national authorities, the income from sustainable tourism realized by the national park itself often has to be paid back to the government. In that case, it makes sense to investigate to what extent it is possible to create new administrative structures or foundations in which tourism activities can be realized without the intervention of the authorities. There are two financial instruments which benefit national parks that can be used in nearly all situations: an entrance fee and a charge for guided tours. There are many arguments for asking a high price for these services. The rationale behind this statement is given in Section 5.

4 The External Approach

In the previous section, it was argued that appropriate marketing is a prerequisite for achieving sustainable tourism in national parks. Of course, many national park authorities are not accustomed to marketing and related types of commercial activities, and in many cases there is a certain amount of hostility towards the whole idea. This is, of course, not surprising given that marketing is currently often used to draw tourists to any place where tour operators can make money, without regard to the effects of tourism on nature and the environment. Two comments have to be made here. Firstly, this attitude is changing rather rapidly, with an increasing number of tourists taking nature and the environment into consideration when planning their holidays. In everyday life, the effects of behaviour on nature and the environment are often given full attention in the public debate. People are becoming more aware of these issues during their holidays as well. Secondly, if it is the responsibility of national park authorities to promote sustainable tourism in and around national parks, it does not make any sense to fail to use the appropriate instruments to achieve that goal.

It should be noted that modern tourism is full of visual symbols (Goossens 1992). Sustainable tourism is no exception in this respect. This implies that leaflets with pictures, signposts in the national parks, brochures for the tour operators, among other things, should provide such symbols. Modern tourists, including many nature lovers, are not able to recognize the value of landscape, vegetation, birds, and wildlife without help. If they are informed that a certain region is a national park with rare animals and plants, they will 'know' that they have made the right choice. A striking example is the Abruzzi National Park in Italy which has large numbers of visitors who are absolutely convinced that this park is of outstanding quality since there are brown bears and wolves there. They will never see them, but they will tell their neighbours that they have been to a place with such outstanding nature that there are even brown bears and wolves there. This type of information has to be communicated to potential visitors.

5 The Value of Nature and the Environment in Environmental Economics

From the previous discussion, one can conclude that nature, landscape, and the environment are given, often implicitly, a high value in society. The European Union policy, for example, aims at sustainable development of the economy, implying that this will not come about in a normal market situation. Hence, public policy is needed. In this respect, tourists are willing to spend a lot of money to visit unspoiled landscapes and to observe rare animals and plants. From an economic point of view, one could argue that tourists are willing to spend scarce economic resources to get satisfaction from the observation of unspoiled nature. Presumably then, nature and unspoiled landscapes have a high economic value.

One cannot overlook the problem, however, that this economic scarcity is only reflected to a limited degree in the market. It is often suggested that many other economic goods and services provide a better reflection of economic scarcity, as they are sold and bought on a market. In this case, the market price can be seen as a reflection of economic scarcity. However, this is only partly true. An automobile, for instance, has a certain market price which can be seen as the economic value which is given to that product by the car owner. Cars, however, do have negative effects on the environment; these negative effects are, in fact, societal costs, and should, therefore, be subtracted from the market price in order to calculate 'true' economic value. This example demonstrates that even market prices do not reflect 'true' economic value. It is the market price of the car which gives the impression of economic value. Thus, market prices cannot be seen, in many cases, as a true reflection of economic scarcity. Economists have recognized such problems and have developed a theoretical framework to cope with them. This will be discussed in the following sections.

5.1 Traditional Approaches

To understand the significance of the economic value of nature, landscape and the environment, we need to pay attention to the theoretical value framework in traditional economics. In the traditional neoclassical framework predominant in Western countries, the value of a good can be measured in a market. Producers sell products in a market where consumers can buy these products. The equilibrium price of the market is generally seen as a reflection of the economic value of the product. On the one hand, the price is related to the costs the producers incur in the production process. Normally, these costs include the price of labour and capital and natural resources - when these are traded on a market - as well as all other delivery costs. On the other hand, the market price bears a strong relationship to the consumer's willingness to pay, as a result of the revealed preferences of consumers. In this framework, the economic value of a good is normally the same as the market price. In neoclassical theories and methodologies, the individual is a cornerstone of the framework. The idea is that all individual preferences can be summarized to define national or total welfare.

This economic system is thought to bring about an optimal allocation of production factors due to the ability of the market mechanism to steer production and consumption into a societal desirable direction. One should not overlook the fact that in this framework, economic value can, by definition, only be measured in a market. Outside the market, it is assumed, there is no relevant economic value.

As early as the nineteenth century Marshall (1890) was aware of some pitfalls in this thinking. He formulated a large number of assumptions which have to be fulfilled before this theoretical model of optimal allocation can function. One of these assumptions was the absence of significant external effects. These were the welfare effects on economic agents other than the current market parties. In Marshall's publications we find only positive external effects. When someone wants to allocate a new factory, for example, he or she will, in most cases, give priority to a location where many other services are available for which the investor does not need to pay. This will bring the newcomer positive effects which are external to his or her decision. Negative external effects cannot be found in Marshall's publications.

Pigou (1920) was the first economist to pay attention to negative external effects. He argued that optimal allocation of production factors is no longer apparent if substantial negative external effects are the result of production processes. Authorities should, in this view, investigate the damage to people outside the market, caused by polluting industries. This damage has to be given a monetary value. The monetary value of the damage is seen as that segment of the production costs that is shifted to other economic actors. In this view, there is a dislocation of costs, resulting in a lower level of costs to the polluting industry. The products are too cheap, as the production of these products is not confronted with all relevant costs. The dislocation can only be corrected by the government imposing a levy on the polluting industry. This levy can be seen as an attempt by the authorities to return that part of the production costs that has been shifted to others, to the polluting industry. If this is done the dislocation of costs is neutralized by the environmental policy. Pigou published his arguments as early as 1920, when environmental disruption was not as relevant. Most economists were thus of the opinion that the Pigovian approach could be valued as an elegant solution to a difficult theoretical problem which hardly bore any relationship to normal life.

When environmental pollution became a 'normal' phenomenon in the fifties and sixties, discussion among economists to correct a dislocation of production factors resulting from environmental pollution arose again. Coase (1960) argued that the Pigovian approach need not lead to an optimal solution. If purification costs are high a more cost-effective solution could result from negotiations between polluters and consumers suffering from pollution. We will not discuss the differences between Coase and Pigou here as they are not relevant to our approach. What does require discussion, however, is the fact that Coase used the same assumption as Pigou. In both approaches the polluter and the victim are defined and there is a clear dose-effect relationship. In addition, in this approach the authorities can put a specific price on the societal costs of environmental pollution. However, the Coasian approach was not given serious attention.

In the course of the seventies and the eighties, Western countries were confronted with a growing level of environmental awareness among their citizens, resulting in the implementation of many environmental laws. In most cases, these laws do not reflect the introduction of Pigovian taxes. In the laws, a certain level of pollution is defined as acceptable and emissions have to be brought below that level. Permits and control were the normal instruments. In our context, the relevant question is why the authorities adopted an approach that is not Pigovian. Perhaps we can answer this question when we investigate the pitfalls and barriers which can be found in the theoretical Pigovian neoclassical system.

5.2 Pitfalls and Barriers in the Theoretical Framework

Among the most significant shortcomings of the Pigovian approach is first, that in the model it is assumed that only a limited number of polluters are involved, and second, it is known exactly who the consumers are. Of course, this is far from current reality, since, for example, in the case of cars, there are many polluters and whole societies suffering from pollution. In such a situation, it is completely impossible to shift the environmental costs from consumers on to the polluting industries. This is complicated by the fact that in the Pigovian model, a national authority is the entity par excellence to handle the environmental problem. On the one hand, in modern production processes, many emissions are transboundary, which means that only an international authority is able to deal with these types of environmental problems. On the other hand, in most cases countries or nation-states have ultimate jurisdiction, and are reluctant to surrender it, which makes the international environmental problems difficult to solve.

National authorities are in a difficult position if they try to implement environmental policies with strict norms. In the Pigovian approach, the government is seen as an objective economic agent not involved in the controversies between polluters and consumers. The government has the power, in this approach, to implement an environmental policy with strict norms when it is necessary to restore an optimal allocation of production factors. In modern societies, however, authorities are held responsible for the results of the economic process. When, for example, there is a high level of unemployment, it is the government which is blamed. When the implementation of environmental policies is accompanied by detrimental effects on employment or the international competitive position of national industries, the authorities also get the blame. This implies that in modern societies, the government cannot be seen as an objective economic actor able to implement a strict environmental policy that conflicts with leading economic interests.

In the Pigovian approach, it is assumed that environmental damage can be translated into monetary value. However, there are often no markets for nature and the environment. Indeed, crude oil, natural gas, and iron ore are traded on a market and therefore have a price. But this is not the case with the hole in the ozone layer, the greenhouse effect, the pollution of rivers and oceans, the decrease of biodiversity, and the damage done by acid rain. In the neoclassical approach, there is no economic value outside the market. This implies that nature and the environment, when they are not sold and bought on a market, do not have an economic value. How is it possible then, to shift the environmental costs back to the polluting industries? The Polluter Pays Principle is generally seen as a reflection of the Pigovian approach. But how can the polluter pay when we do not know how high the environmental costs are?

There are also many complications in the ecosystem itself. First, there is the problem of the relationship between emissions and deposits. Which chimneys in the Netherlands, for example, are responsible for the dying off of German forests due to acid rain? Of course, it is possible to claim that Dutch refineries are responsible for acid rain in Germany, but this does not mean that we are able to calculate all dose-effect relations in Dutch industry. However, we should not overlook the fact that these dose-effect relations are a prerequisite in Pigovian approaches. Another problem is the thresholds which are often at work in ecosystems. Pollution will often not have a really detrimental effect at the beginning of the pollution, as the ecosystem is able to accommodate this pollution. However, ecosystems can do this only up to a certain level. As soon as this level is passed, the ecosystem may be substantially damaged. In this situation, the question arises as to which polluting industries have the right to emit before the threshold is passed. Finally, in an ecosystem synergetic effects are at work. The combined effect of a number of pollutants is often more significant than would be expected on the basis of the sum of all the damage done by these polluting factors separately.

We may conclude that the introduction of a Pigovian tax or any other type of economic instrument based on neoclassical approaches is very complicated. Many economists are aware of the shortcomings of an approach using these types of assumptions. This is why Baumol and Oates (1988) introduced a different approach. They argued that in cases where environmental costs cannot be calculated, a levy should be introduced aimed at reducing the level of pollution to a certain degree. This level of pollution has to be established outside the realm of economic theory. It has to be based on political decisions. This approach was elaborated and given the name 'critical loads'.

5.3 The Practice of Critical Loads in Environmental Policies

The theoretical framework discussed in the previous section is seldom seen in current environmental policies. In most Western countries environmental awareness increased in the course of the 1960s, resulting in the opinion 'that something has to be done'. Generally speaking, authorities did not pay attention to the theoretical framework of environmental policies. They tackled the problem by introducing environmental legislation in which permits and norms were the general instruments. Economic instruments were virtually non-existent in these types of environmental legislation. It was said in these laws that environmental pollution and damage should be decreased. Permits were seen as the adequate instruments for realizing this. However, permits can only be effective when certain norms are introduced.

In most cases, these norms were not related to critical loads, which reflect some idea of the relationship between the level of emissions and the level of deposits. In reality, these relationships are almost unknown. Environmental legislation used a different point of departure. The levels of the norms increasingly became the result of a bargaining process between the polluting industries and the national authorities.

This put the authorities in a difficult position. Generally speaking, the technical knowledge possessed by the Ministry of the Environment about pollution resulting from certain industrial processes is relatively limited. It is the polluting industries themselves that are knowledgeable about these processes. So their bargaining position is considerably better than that of the government. The result is a strong tendency for the norms that are implemented to be weaker than the government originally intended.

However, it is not only a lack of technical knowledge which puts the government into a difficult position. In the Pigovian approach, in which costs and benefits of environmental measures are known, there is a strong theoretical basis for the implementation of a levy. The government is assumed to be able to demonstrate that, in a given case, a levy would be desirable based on established economic theories. However, as was argued previously, governments are not able to make these types of calculations. If the government intends to introduce strict norms which are used to calculate the maximum level of the emissions or deposits, the authorities are not able to 'prove' what the economic advantage is. Indeed, they often are not able to suggest the correct environmental improvements, as clear relationships between emissions, deposits and monetary values are not known.

This creates a situation in which the introduction of a strict norm could be defended by arguing that this is a good thing for the environment, as it will lower the level of pollution. The government is however, not able to calculate the economic advantage of this introduction. Yet polluting industries often do know what the extra costs of production are after the introduction of the strict norms.

Polluting industries are very well informed about the shortcomings of this approach. They focus on the heavy economic burden they will have to shoulder as a result of the introduction of strict norms. Generally speaking, they will not argue against a sound environmental policy as such, but they will demonstrate the rising costs for their industries resulting from such a policy. It is this mechanism which, in many cases, has hindered the introduction of strict norms. We may conclude that the legislative framework provides promising possibilities for implementing a sound environmental policy. However, for some of the reasons just discussed, the Western European countries have not been able to reduce the level of pollution to an acceptable level in the last twenty years.

From this point of view there are no differences between economic instruments on the one hand, and command and control approaches with permits, on the other. In both cases, a certain level of pollution has to be defined which can be achieved either by strict norms in the permits or by the maximum level used in implementing economic instruments.

There are two other mechanisms which exacerbate this situation. The first is the mechanism of economic growth. In certain cases, such as the installment of catalysts in automobiles, European Union countries have been able to introduce strict norms. These catalysts reduce the level of pollution by 90 percent. The catalysts are installed in new cars. People buy a new car about once every ten years. Thus, after a period of ten years all cars will have a catalyst. At that point the level of emissions by cars will have been reduced by 90 percent. However, the increase in the mileage of cars and the number of cars is so high that this will neutralize the effect of the catalyst. The result is that ten years after the introduction of catalysts, the emissions of acidifying substances will not have been reduced at all. It is the mechanism of economic growth which neutralizes the beneficial effects of the introduction of strict norms.

This mechanism of economic growth often leads to a situation in which the pollution level does not decrease in the long run. There are societal reasons why there is pressure to increase the number of cars and their mileage. When lower income groups achieve a higher level of income due to economic growth, there is a strong tendency to buy a car as soon as they are able to do so. Hence, the distribution of income and human desires increase pollution, even after the introduction of strict norms.

The second mechanism is the increasing effect of international environmental problems, as in the case of transboundary pollution. Global effects, such as the greenhouse effect and acid rain, cannot be neutralized by the actions of one country. Yet, the introduction of strict norms will benefit the environment of other countries as well. Hence, there is a strong need for international cooperation in order to reduce transboundary pollution.

5.4 The Value of Nature Revisited

The shortcomings of the traditional approaches in neoclassical economics have been recognized by many groups in society. The concept of sustainable development, introduced in the World Conservation Strategy of the International Union for the Conservation of Nature (1981) and later elaborated by the World Commission on Environment and Development (1987), can be seen as an attempt to overcome these shortcomings. In many publications, the concept of sustainable development, or sustainable tourism, is welcomed as a panacea to solve problems with environmental policies. This is a very optimistic and perhaps naive opinion. As we have seen in the previous sections, the crucial problem in environmental economics and policies is to define the economically correct level of pollution or environmental disturbance. The concept of sustainable tourism does not help us with this as there remains the problem of defining the acceptable level of tourism and by what means. Three potentially important approaches have been developed to address these problems:

a) the contingent valuation method where a pseudo market is created by asking people what they are willing to pay for a certain environmental issue or a certain type of nature;
b) the hedonic pricing method where the differences between the market values of the same types of houses located in different natural and environmental settings, are taken as a proxy for the value of nature and the environment (Hanley and Spash, 1993);
c) the travel cost method where the travel costs people are willing to incur for visiting a natural area are taken as a proxy of the value of the particular nature area.

The contingent valuation method in particular has received special attention recently. It is seen by many environmental economists as a relatively good method for securing relevant information on the value of environmental and natural assets. This paper cannot cover all the pitfalls and barriers of this method. A good overview would include the contributions of Navrud (1992), Hanley and Spash

(1993), and Hoevenagel (1994). One of the most striking problems of the method concerns the information which is available to consumers. In situations where the environmental situation is rather complicated as is the case with the greenhouse effect, nuclear waste, acid rain, and tropical rain forests, people are, generally speaking, insufficiently informed about these issues and, hence, are not able to give defendable answers about their willingness to pay to deal with them. In the case, for example, of touristic and recreational questions, this problem is not so evident. In this situation environmental complications are, of course, relevant, but they are not as significant as in the case, say, of the greenhouse effect. Therefore, there is a general opinion among environmental economists that this method can give relevant information when tourist issues are discussed.

Van der Linden and Oosterhuis (1988) did a study of the recreational values of Dutch forests and heaths using the contingent valuation method. They investigated the willingness to pay for the protection of Dutch forests and heath against the effects of acid rain. It turned out that the Dutch population is willing to pay Dfl. 1.5 billion every year for this purpose. This has nothing to do with the value of timber since this figure is only related to the recreational use of these nature areas.

The travel cost method has been used by Willis and Garrod (1992) in estimating the recreational value of the canals in the English Midlands. These canals were constructed during the Industrial Revolution as part of infrastructure for industries in the region. These canals are no longer used by industrial boats, as industries are using other modes of transport. The maintenance of these canals amounts to approximately 50,000 million UK. However, the travel costs realized by tourists visiting these canals is more than 60,000 UK. Therefore, the recreational value, estimated by using a travel cost method, is higher than the total level of the maintenance costs. So, we may conclude that the recreational value of these canals is much higher than is often assumed.

Of course, one cannot argue that in these cases 'the' economic value of nature has actually been determined as there are too many uncertainties in the methods. However, we cannot overlook the point that these investigations give an indication of the high economic value people are presumably willing to pay for the protection of these environmental assets. It can be concluded that most tourists are willing to pay for visiting high quality nature areas and landscapes. This is recognized by the national park system in Costa Rica where the authorities recently decided to charge visitors an entrance fee of $15. In the beginning, there were many objections, as people were afraid of a sharp decline in visitors to the national parks. It turned out, however, that tourists were willing to pay the fee. Since visitors have already spent a lot of money on the flight to Costa Rica and on lodging, they do not feel that another $15 makes any real difference. In many cases, tourists see the fee as an effective instrument for the protection of the rain forests. It is much more difficult to use the rain forest for timber purposes when a lot of money can be earned from the entrance fee.

These examples lead us to conclude that many modern tourists are fully aware of the high value of national parks. They see the label of a national park as a guarantee of high quality nature and landscapes. Therefore it makes no sense to concentrate on low prices for tourists in the marketing process.

6 Conclusion

The previous discussion allows us to draw the following conclusions:

- National parks should focus on the spatial and economic surroundings in which they function. Rural development is a concept which can embed national parks in the region. Sustainable tourism can be evaluated as a serious option for the development of national parks.
- In regions where sufficient economic and human resources for regional development are lacking, the establishment of big national parks without further rural economic development should be given serious attention, even in cases where ecological values are not extremely high.
- Due to the concept of negative external effect, environmental issues are easier to describe and analyze in economics than the value of nature and landscape.
- The concept of negative external effect is loaded with many assumptions which are not realistic when analyzing modern problems in the field of nature and the environment.

- Environmental disruption in general will negatively effect the ecological value of nature. Therefore we cannot take the economic value of nature as an independent category. The relationships between nature and the environment are varied and complicated.
- Concepts such as sustainable development and sustainable tourism, do not solve the conceptual and analytical problems in environmental economics and policies.
- Concepts such as the contingent valuation method do not give a scientifically correct answer to the economic value of nature. Nevertheless, they can give us relevant information about the value people in modern societies give to nature.
- The development of sustainable tourism in and around national parks should be promoted by national park authorities in close cooperation with the inhabitants and authorities of the region. The results of contingent valuation studies demonstrate that asking for money from tourists and visitors does not decrease their inclination to visit national parks.
- A good marketing strategy is a prerequisite for sustainable tourism.

7 References

Baumol, W.J. and W.E. Oates (1988). *The Theory of Environmental Policy*. Second edition, Cambridge University Press, Cambridge.

Coase, R. (1960). The Problem of Social Cost. *The Journal of Law and Economics,* October 3, 1966: 1-44.

Coccossis, H. and A. Parpairis (1992). Tourism and the Environment: Some Observations on the Concept of Carrying Capacity. In H. Briassoulis and J. van der Straaten (Eds.) *Tourism and the Environment*Kluwer Academic Publishers, . Dordrecht: 23-34.

Hanley, N. and C.L. Spash (1993). *Cost-Benefit Analysis and the Environment*. Edward Elgar, Aldershot.

Hoevenagel, R. (1994). *The Contingent Valuation Method: Scope and Validity*Institute for Environmental Studies, Free University, . Amsterdam.

International Union for the Conservation of Nature (1981). *World Conservation Strategy*. Geneva.

Linden, J.W. van der, and F.H. Oosterhuis (1988). *De Maatschappelijke Waardering van Bos en Heide* (The Social Valuation of Forests and Heath lands). Ministry of Housing, Physical Planning and the Environment. Leidschendam, The Netherlands.

Marshall, A. (1890). *Principles of Economics*. MacMillan London, UK.

Navrud, S. (1992). *Pricing the European Environment*, Scandinavian University Press, Oslo.

Pigou, A.C. (1920). *The Economics of Welfare*. MacMillan, London, UK.

Willis, K. and G.D. Garrod (1992). On-Site Recreation Surveys and Selection Effects: Valuing Open Access Recreation on Inland Waterways. In H. Briassoulis and J. van der Straaten (Eds.) *Tourism and the Environment*. Kluwer Academic Publishers, Dordrecht: 97-108.

World Commission on Environment and Development (1987). *Our Common Future,* Oxford University Press, Oxford.

Environmental Education in Protected Areas as a Contribution to Heritage Conservation, Tourism and Sustainable Development

Andrzej Biderman[1] and Wojciech Bosak[1]

[1] Ojców National Park Education Centre
32-047 Ojców, Poland

Abstract. The paper presents a new role for national park centred environmental education in relation to protected areas. The role involves both fostering an environmental consciousness among visitors and local people and using environmental education as a tool of protected area management. Research needs are discussed in the context of two karst landscapes in Poland and the U.K. Both are experiencing intense development stresses on account of their proximity to large urban centres and the large number of visitors.

Keywords. environmental education, interpretation, environmental consciousness, protected areas, protected area planning and management, communication, Poland, U.K.

1 Aims and Objectives

The main aim is to answer the following questions:
- What means are used to accomplish the present role of national park centred environmental education in protected areas?
- What range of actions are undertaken?
- What accomplishments can be seen?
- Who undertakes action?

2 Context

Environmental education has changed profoundly in protected areas in recent decades. We can distinguish a shift from interpretation of protected area features to the use of protected areas to foster environmental consciousness among visitors (Canadian Environmental Advisory Council 1991). In the broadest sense, environmental education refers to raising sensitivity, awareness, and understanding of the linkages and interdependencies among human beings and the natural world in which they live (Cornell 1989).

Moreover environmental education is now used as a tool of Protected Areas management in reducing environmental impacts on protected areas (IUCN 1994). Actions are focused on:
1. Fostering an general environmental consciousness among local people.
2. Changing local people's attitudes to and relationships with Protected Areas by:
   - fostering an appreciation of the environmental values of protected areas among local populations;
   - arguing for the commercial value of protected areas and training local stakeholders to exploit these;
   - fostering pride in protected areas.
   - changing visitor attitudes and behavior.

In the Czech Republic, Poland and other nations of Central Europe, the development of environmental education centres and park rangers with communication skills are a relatively new feature. Park staff are still predominantly assigned to specialized scientific research. Few protected areas have a communications or public relations officer. In recent years, however, the growing negative impacts of visitors, and the desire of many local communities and national parks to create infrastructure to accommodate still more of them for revenue purposes, have prompted a rethinking of the role of environmental education and a growing recognition that environmental education can be an effective management tool.

For many people in the countries of Central Europe, including the Czech Republic and Poland, national parks and protected areas with their uncommunicative scientific bias and their command-and-control approach to dealing with local communities and other stakeholders, continue to be seen as remnants of the old totalitarian regime which must be democratized and opened to the market. In this regard, the development and implementation of environmental education programmes confront the paradox that most individuals and institutions are more than ready to declare a commitment to safeguarding environment and nature conservation, but are not yet prepared to take practical action to improve environmental quality through changes in lifestyle, attitudes and behaviour. Symptomatic of this is the widespread and growing resistance to both existing and proposed protected areas by local communities, which perceive them as restrictive.

In the UK and other western countries, there is also growing concern that protected areas are constraining development and wealth creation. The challenge of environmental education in this context is to build a climate of public support for protected areas by drawing attention to the benefits that they bring to local communities, economy and society. This challenge has become all the more important at a time when conventional command-and-control approaches to park management based on policing and enforcement of restrictions - in effect separating people from nature - are being seen as too expensive and inadequate, especially in the face of growing development stresses associated with tourism, residential and other types of development.

2.1 Examples

2.1.1 Kraków Highland, Poland

Located on the edge of Kraków, this Highland is protected by the Ojców National Park, Jura Landscape Park System and an Area of Protected Landscape that together cover 57,610 ha in total. The national park includes the two biggest canyons of the Highland. The other canyons are protected by five Landscape Parks. There are also numerous side gorges, caves and other elements of karst landscape. The elevation of the Highland is about 470 m above sea level. The bottom of the canyons is at about 330 m above sea level. The features of the area are very diverse. The rich flora and fauna include species of extreme ecological and geographical character. The natural diversity of habitats is enhanced by traditional human activities like grazing and mowing. The natural and semi-natural plant communities are distributed in a mosaic-like pattern. This mosaic is further complicated by the cultural and historical monuments spread through the protected area.

2.1.2 The Peak District National Park, UK

Located between the Greater Manchester urban area and the city of Sheffield, the Peak District National Park covers an area of 140,400 ha. It lies at the southernmost extremity of the Pennines. This landscape consists of two parts. Elevated over 600 m above sea level the northern area is known as 'Dark Peaks', which is bedrock covered by a peat layer with broad moorlands. The southern part, the 'White Peaks', is a broad limestone highland elevated over 300 m above sea level with numerous gorges (dales), caves and other karst features. There are splendid calcareous flora and fauna among broad farmland. The natural values are neighboured by relics of human past.

2.2 Recreation and Tourism

2.2.1 Visitors

The Kraków Highland is one of the most popular recreational areas for the Kraków-Silesian urban area. The total number of visitors is unknown and their distribution is very uneven. The main tourist attraction is the Pr¹dnik and S¹spowska Valleys in Ojców National Park visited by 300,000 - 500,000 people a year. Visits are most frequent during May/June and September/October when Ojców is visited by 5,000 people a day. Lokietek Cave and Pieskowa Skala castle (120,000 visitors a year each) are the most busy sites. School groups following the national curriculum in geography, visit this area as a case study of the karst-highland landscape of Poland. Other visitors come at weekends and during vacations. Polish and foreign family groups or individuals visit the area as part of vacation tours. Such groups use local catering and accommodation services. Yet, nature is mentioned as a reason for visits only by 25% of park visitors. The majority declare that their motivation is just to get out of the city (about 40%).

Other busy sites at Kraków Highland are: Kobylanska, Bedkowska, and Mnikowska Valleys in the Landscape Parks. However, there is a significant difference between these sites and Ojców National Park. Ojców and Pieskowa Skala have well developed tourist facilities while the rest of the highland is very poor in infrastructure. This difference affects visitor behaviour. Landscape Parks mainly receive day visits of hikers and mountain-bikers. They are self-sufficient in terms of food and accommodation. Kobylanska Valley is also a main training area for free-climbing in Southern Poland. About 500,000 people visit the Landscape Parks each year.

The Peak District is visited by about 22,5 million people a year. The distribution of visits is also very uneven. Busy sites such as Castleton village, Dove Dale or Derwent Valley receive about 1.5 million people a year. Even the most fragile environments like the peat bog plateau of Kinder Scout, receive 250,000-350,000 visitors a year. As in Ojców the majority of the Peak District visitors come just to get out of the cities, i.e. for leisure. The favourite activity of visitors is hiking and rambling throughout the mountains. The new fashion is mountain biking. Some problems are caused by picnics in the most overcrowded areas like downhill Castleton or Kinder Scout high in the mountains.

2.2.2 Impacts

The main feature of Peak District and Ojców National Parks is overcrowding in the most busy places. This causes many impacts and diminishes the value of local landscape and heritage.

Erosion is one of major problems of the Peak District. The peat bogs and moors are specially sensitive habitats. Once destroyed the vegetation regenerates slowly or not at all. There are big surfaces of uncovered peat layers which are eroded quickly by frost, water and wind, down to bedrock. Destruction of vegetation is caused by trampling which is visible along the most busy footpaths like the Pennine Trail. This loss affects the entire habitat, spoils the landscape, and causes managerial problems for the park authority which is responsible for maintaining the footpaths. The karst part of Peak District as well as Kraków Highland, is much more erosion resistant. Nevertheless trampling causes significant destruction of vegetation along the most busy trails. Besides hikers, mountain-bikers, horse riders and Land Rovers are major sources of vegetation destruction and ground erosion in Peak District.

Fire is another very spectacular source of landscape damage. A single cigarette can cause vast devastation of the Peak District moors. Some sites have been burned and eroded to the mineral bedrock in this area. Hikers and picnickers often cause fires. There are some forest fires at Kraków Highland although they are rarely caused by tourists.

Traffic causes site management problems in places like Castleton, Edale, Bakewell, Goyt and Upper Derwent Valleys in the UK and Ojców and Pieskowa Skala in Poland. Cars fill all car parks, and stand along roads and streets. Schools still use coaches but the majority of visitors arrive in private cars now. The number of vehicles is magnified several times in comparison with 1980s. Air congestion causes significant air pollution, and vegetation damage.

2.3 Local Populations

The Kraków Highland is inhabited by a relatively dense population. The traditional occupation of local people is farming and farmland is the main element of the local countryside landscape. Other local occupations are small crafts and tourist services. The growth and industrialization of neighbouring cities have involved locals in taking new jobs over the past decades. New international airport and motorway networks attract big investments to sites on the edge of the protected area. Smaller investments are located farther into the countryside. Numerous new businesses have been developed recently, for example wholesale and retail establishments and modern small industry. These changes in turn, have caused development of new housing and related changes in traditional lifestyle. Many villages have been chosen for housing by people from Kraków.

The situation of the Peak District is very similar to that of Poland with some differences related to the fact that the moorlands on the north are wild. Farming is a traditional local occupation which has declined gradually in recent last years. Tourist services are stable and a significant source of local income. More locals are employed outside the region in industries in Greater Manchester and Sheffield. Many people from these cities have also settled in the numerous villages of Peak District.

The attitudes of local people to protected areas are usually negative. The legal status of the parks and their planning controls are regarded as a major limiting factor - stopping local developments. On the other hand local people appreciate the countryside they live in as well as the parks' efforts to maintain it. This ambiguity is also present in Poland and Ojców where the locals say the park should keep its hands on its forest and wilderness but not on 'our village'. Many local people in the Peak District National Park area claim that they accept the idea of nature conservation but they know a better way to its accomplishment than national park use. The declarations of a 'third way' between industrial development and 'orthodox' nature conservation is very typical for a majority of national parks in Poland as well.

2.3.1 Impacts

The heritage landscapes of the Kraków Highland and Peak District are based on traditional land use systems which were developed during a long history of human occupation. Surrounded by big industrial conurbation these regions are no longer the typical rural areas of yesteryear.

The most important source of change in the Kraków Highland is rapid urbanization of the southern part of the area. The traditional landscape is altered by expansion of villages and settlements. Inhabited areas become important centers of air and water pollution and of synanthropic species invasion and human penetration to the core zones of protected areas.

Similar rapid large scale urbanization is not seen in the Peak District but far reaching landscape changes occur there as well. The local villages undergo gradual transformation to dormitories of neighbouring cities or to recreational centres. There are fewer schools, shops and other local facilities. Local people become less aware of local issues and less interested in what is around them. Many farms are converted into countryside residences or leisure facilities. Numerous traditional ecosystems are eliminated through changes in land use practices, especially in farming. The best example are fresh or wet meadows being converted to grass or corn monocultures.

2.4 Public Relations

One of the major factors limiting management of protected areas in the Kraków Highland is the way they are perceived by the local population and visitors. People who live in the protected areas see their surroundings in different ways than do visitors. They are not specially interested in features of the region. They usually undervalue the local heritage. This situation causes numerous conflicts when local people try to enforce their own view of appropriate development.

3 Accomplishments

One of the main accomplishments to date with respect to environmental education and protected areas is that national park centred environmental education has been central to some change from seeing protected areas as 'islands of wilderness in a sea of development' to their being seen as opportunities for the realignment of economy and society towards sustainability. Environmental education set in the national park visitor experience has helped people to see beyond park boundaries and to strive to avert or at least to modify development stresses which are having an ever greater impact in terms of degradation of valued resources in the park area.

3.1 Target Groups

Three main groups are important in each of Protected Areas: visitors; local population; and the general public. Different Protected Areas address these groups in different ways.

3.1.1 Visitors

The visitors who come to look for leisure and tourism are still the main target of environmental education in most protected areas. Some efforts to address this group have been undertaken on Kraków Highland since the establishment of the national park in 1956 and the landscape parks in the early 1980's. One can now distinguish three major types of efforts: fulfilling the educational role of the protected areas; changing the pattern of visitor behaviour in course of the visit; and changing their environmental awareness. The last role is especially important given that many people visit from the large surrounding Kraków -Silesia urban area, which can be seen as part of the Greater Park Ecosystem. Similar approaches and views hold for the Peak District National Park area.

3.1.2 Local Population

The local population is declared as the major target group by the Peak and Ojców National Parks as well as the Jura Landscape Park System. The local population presents major challenges for both parks. One is that the local people often have development interests different than the parks. Another is the low environmental awareness of these people which makes access to them very difficult. To obtain any effect it is necessary to focus intense effort on some selected groups. These are mainly primary education and other school groups. Adults are targeted only to very limited extent.

3.1.3 Public Opinion

The wider public generally is not treated as a target audience as compared to visitors and local populations. Environmental education is generally seen as limited in effectively reaching or shaping a wider public opinion.

3.2 Forms of Education Used

3.2.1 Public Interpretation Events

The Peak National Park provides a broad range of such activities:

- guided 'Walks with a Ranger',
- 'Navigation and Hillcraft Skills Walks' facilitated by the Peak Park Ranger;
- 'Peak Hour Talks' in some park's Visitor Centres;
- 'Excavation Tours' guided by the archeologists;

- 'Leisure Breaks' and learning holidays at Losehill Hall; weekend and week long courses at Peak National Park Centre such as 'Landscape Painting for a Beginners', 'Bird Songs and Calls', 'Fantastic Ferns', and 'Autumn Rambling'.

The events are delivered by the Ranger and Environmental Education Service at Losehill Hall. It is estimated that only a small number of visitors get involved in these activities, mainly people from the middle-class who are already well committed to the national park ideas.

Ojców National Park and the Jura Landscape Parks provide a much smaller number of these activities. About 10 % of the visitors use these services. Ojców Park is still developing its facilities for residential courses.

3.2.2 Site Interpretation

Ojców National Park and the Jura Landscape Parks have developed a 20 km network of self-guided trails. The trails are interpreted by signs and wayside exhibits supported by several guidebooks and leaflets. There are also two interpretation posts along the trails in Ojców. One in Lokietek Cave is visited by about 30% of the total visitors to the Park, another in Ciemna Cave by about 10-20% of them. Big exhibits in the Park Museum and smaller ones in the castle in Ojców are both visited by about 10-20% of visitors. The Museum also has a small bookshop. All this facilities are managed by the Park authorities. There are also two interpretation posts in two caves in the Jura Landscape Parks. They are managed by private companies.

A broad network of interpretation facilities and opportunities is available in the Peak National Park. The Park Authority manages several Visitors Centres spread through the area. Small exhibits, bookshops and information posts are located in each of them. The exhibits provide interpretation of general park features as well as of specific local issues. A wide range of guidebooks is provided for individual walks through the Park's footpaths as well. Many of footpaths have wayside exhibits at some of the most interesting sites. The National Trust, English Heritage and English Nature own, manage, and interpret numerous features in the Peak District National Park . There are also some private information services focusing mainly on local caves. Unfortunately, as in Ojców, the majority of visitors miss this interpretation, focusing on leisure and recreation instead.

3.2.3 Publications

Besides the guidebooks, numerous printed materials are edited and distributed by the authorities of the protected areas of Kraków Highland and Peak District. A broad range of books, booklets and leaflets are distributed. They are usually addressed to visitors to: interpret the park; spread the general environmental message; and promote local heritage. The Peak National Park also distributes materials to advise local people on business and housing developments.

3.2.4 Cooperation with Schools

Cooperation with schools is a major activity for the Ojców National Park Education Centre. The general approach is to facilitate the teacher's outdoor environmental work within and around the park. To do this, three main objectives are set forth:

- to prepare field programmes, and the manuals and materials necessary to implement them;
- to prepare outdoor facilities;
- to train teachers in outdoor activities and related environmental education methods.

The Centre has worked on these issues since the beginning of 1990's. Numerous programmes have been developed and presented to the teachers. Hundreds have participated in training courses. The majority of participants have been from Kraków and the surrounding area. Perhaps 10-40% of local schools are involved in the work.

Numerous programs have been developed at Losehill Hall and the Peak District National Park to address the children in the schools of the surrounding area. These activities include school day visits and are delivered by park staff. The majority of the Rangers maintain contacts with primary schools in their areas. The range of activities delivered is very broad and depends on the number of schools in the area as well as on the commitment of the individuals concerned. The activities include:

- walks with a Ranger;
- miniprojects like 'wildlife garden', 'school reserve', 'a school pond';
- planting trees in the valley;
- monitoring of local resources;
- recording local wildlife by children;

About 30% of local schools are currently involved in this cooperation. The Peak District National Park offers similar actions and campaigns for locals.

3.2.5 Private Stewardship and Local Initiatives

A good example of this approach is the monitoring of local resources led by local landowners in Edale, Peak District. Facilitated by local area Rangers, monitoring enables the people to discover the values they possess. The role of the Ranger is to show the values and to facilitate the work, without any attempts to impose conclusions.

3.3 Effects Obtained

3.3.1 Successes

One major success is the broad accessibility of information and interpretation services. Everyone who is interested can find some help in both Peak District and Ojców parks. The challenge is in receiving and perceiving the message.

Contacts between the parks and teachers and schools is another big success. Some rangers of the Peak District National Park have worked with local primary schools and local communities for many years. However the involvement of the Rangers and the National Park in the collective and formal sense has occurred only in the last four years. This regular work is mentioned as a cause of real change in the attitudes of local people to the Ranger Service. The Rangers are no longer regarded as a representatives of an outside lobby but as somebody who is allied with locals. The gradual shift of public perception of the role of the whole National Park Service has been noticed by the park staff.

Vandalism of visitor facilities and deliberate destruction of heritage resources have decreased gradually in Peak National Park since the 1950s. This is regarded as a result of changing social attitudes to the park and educational efforts undertaken by the park authority. It is related to the changed visitor structure as well. The visitors who come to the park do so often, repeating visits.

3.3.2 Failures

Many important target groups are still not adequately addressed. The most important shortfall is the lack of commitment of locals to the idea of sustainable development. Very difficult to target groups include the mountain-bikers, off-road vehicle drivers and other 'adventure' groups in the Peak National Park. These strong and focused lobbies are very resistant to any targeting - there is no visible change in attitudes and environmental awareness in these groups as yet.

4 Challenges

The challenges ahead relate to the issue of whether national park centred environmental education should continue to be seen as a special kind of education programme or should develop a broader scope. How might environmental education be used in art, history or language schooling, for example? There are a growing number of practical examples of this kind of approach, but as yet protected areas and the learning opportunities they offer, remain outside the formal education curricula. Indeed, field-oriented education and learning is still the first to suffer in times of financial restraint in education. The challenge is to recognize environmental education as an agent of legal, administrative and institutional change. In this context it is essential to orient protected area programs to a more proactive role in sustaining social and economic innovation and change. It is important also to find ways of ensuring that the protected area mandate -- to conserve natural or landscape values - is not compromised. This is an ever present danger when protected area managers engage in consultative exercises. Environmental education must play a role in this regard, in raising awareness of the fact that it is protected area managers who have been entrusted to conserve natural and landscape resources on behalf of all, not just certain stakeholders.

4.1 Effectiveness

4.1.1 The Effectiveness of Communication

Environmental and heritage messages are addressed to everyone who visits the Parks but only some are keen to receive them. These are mainly people who already have high environmental awareness. The challenge is how to reach the rest.

Some major factors limit access to locals. One is their attitude to the park which inhibits messages addressed to them. A second is their conflicting development interests. A third is low environmental consciousness which makes the access to these folk very difficult. It seems to be a very complex matter to contact locals - and so it is in practice. But some rules of thumb can be offered based on experience to date. One is to never be too 'pushy'. Another is to be 'helpful' in solving local problems (which may not be your problems as a conservationist). Use language that facilitates communication - yet constant difficulties in communication prove that there is no simple answer to this question. Perhaps the answer simply is that much more work is needed in this regard.

4.1.2 The Effectiveness of the Environmental Message (real influence on peoples' attitudes)

The greatest challenge facing protected area professionals and all environmentalists today is to find new ways of demonstrating that the conservation of nature and the sustainable use of resources has a fundamental relevance to the daily lives of people, including those who may never visit a protected area. We will not open peoples' eyes to the vital importance of their natural environment as a life support system unless we also open our own doors as conservationists. Greater effort is required to provide the means, the opportunity and the motivation for more people to have access to parks and protected areas. Recreation should be recognized as re-creation - a way in which people can find refreshment in mind and spirit, escape the pressures of urban life, and re-discover themselves through direct contact with nature and the beauty of heritage landscapes. There is a need to promote greater public awareness of the broad range of opportunities offered by protected areas for enrichment of human life.

4.2 Threats to be Addressed

As the previous remarks indicate, these are numerous and include:

- Damage and conflict associated with user groups generally and especially groups such as mountain bikers, off-road vehicle drivers, down-hill skiers, and others employing newer technology.
- Identifying, zoning, maintaining and restoring sensitive or core areas, for example key wetland sites, also making greater use of strict reserves and international designations such as RAMSAR.
- Extensive soil disturbance and erosion by various users.
- The effects of convenient advertising which gives users messages that often conflict with national park messages.
- Low environmental awareness and stewardship, which in large part must be addressed by greater support for national park centred environmental education.
- The challenges offered by virtual reality - computer games and simulation - which can undermine consciousness of and concern for natural areas.
- Budget cuts, downsizing of staff, reduction and elimination of programmes.

4.3 Management

To respond to the foregoing threats and other challenges and opportunities, is a major task; the answers lie beyond the scope of this paper. However, some key elements include:

- Improving staff capability
- Increasing volunteer effort
- Customer Care - issue No 1 for protected areas.
- Coordination of local, national and international work for protected areas and especially environmental education in these areas including:
- national parks
- protected landscapes
- conservation authorities
- conservation trusts
- local, national and international funds and organizations
- NGOs
- schools and universities (national curriculum)
- outdoor education

5 Recommendations

In the light of the foregoing, major research challenges lie in developing effective communication approaches that enable protected area managers to work in partnership with other stakeholders. This interaction between protected areas and the mainstream of human society - and associated recognition of protected areas as an integral part of a healthy economy and society - are the major challenges for environmental education. It is evolution of sensitivity, attitudes and behaviour that is the measure of success and not just an appreciation of the functioning and importance of ecological processes, although this must remain an essential element. Research is needed to establish the content of environmental education programmes so that they can be linked to planning and management programmes and so that planning programs can be treated as education and learning opportunities.

6 References

Canadian Environmental Advisory Council 1991. *A Protected Areas Vision for Canada*. Ottawa.

Cornell J. 1989. *Sharing the Joy of Nature*. Dawn Publications, Nevada City.

IUCN. 1994. *Parks for Life: Action for Protected Areas in Europe*. Gland.

Kopp H. 1980. Education in National Parks and Equivalent Reserves. In *Experience of Nature, Awareness of Nature and Biological Education in National Parks and Equivalent Reserves*. Grafenau.

Martin P. 1990. *First Steps to Sustainability: the School Curriculum and the Environment*. World Wide Fund for Nature. Gland, Switzerland.

Watson M. D. 1980. The Educational and Interpretative Concept of Everglades National Park. In *Experience of Nature, Awareness of Nature and Biological Education in National Parks and Equivalent Reserves*. Grafenau.

Extending the Reach of National Parks and Protected Areas: Local Stewardship Initiatives

Jessica Brown[1] and Brent Mitchell[1]

[1] Atlantic Center for the Environment, Quebec-Labrador Foundation,
55 South Main Street, Ipswich, MA, 01938, USA

Abstract. Stewardship describes an array of approaches to enable responsibility of landowners and resource users to manage and protect land and natural resources. The concept offers a means of extending conservation practices beyond the boundaries of conventional protected areas, conserving heritage at the level of landscapes, and engaging local people in improving conservation practices. Techniques range from education, to management agreements, to full acquisition of properties. Private and public/private stewardship approaches have grown to play a significant role in heritage conservation in North America and are developing in Central Europe to address the unique challenges facing the region's landscape.

Keywords. Stewardship, conservation easements, public/private partnerships, natural areas, heritage conservation, landscapes, civil society

1 Introduction

In the face of declining public resources for land acquisition, and management of protected areas, there is a need to explore new options for land conservation and sustainable use. In areas facing rapid changes in land ownership and use, special challenges arise in protecting natural areas and maintaining the integrity of landscapes. To sustain working landscapes new approaches are needed that protect significant biological resources while maintaining local income from resource use.

The stewardship approach - predicated on encouraging individual and community responsibility for sound natural resource management - offers a means of extending conservation practices beyond the boundaries of conventional protected areas, to address needs in the 'land between'.

1.1 Aims

This paper will examine the application of stewardship to natural and heritage area management, with particular emphasis on its application beyond the boundaries of formal protected areas. It will provide an introduction to the concept of stewardship, its key elements, and practical techniques. It will examine how stewardship initiatives have developed to address needs and challenges in different settings, with special reference to northeastern North America and Central Europe. It will review the progress of selected local stewardship initiatives in both regions. It will discuss the challenges to advancing stewardship at the local level and internationally.

2 The Context: Why Stewardship?

2.1 Definition of Stewardship

When used with respect to natural resources, the term stewardship means - in its broadest sense - people taking care of the earth. The concept encompasses a range of private and public/private approaches to creating, nurturing and enabling responsibility in users and owners to manage and protect land and natural resources.

With education as its foundation, stewardship draws on an array of tools to conserve natural and cultural values of areas withheld from formal protection for economic or political reasons. In North America, methods generally focus on encouraging landowners - individuals as well as businesses - to manage areas to protect or enhance these values, or to allow others to so manage the resources.

Stewardship is an especially helpful concept in the many instances where sound management - rather than absolute protection or preservation - of natural resources is the objective. Though stewardship tools may be employed to preclude use of specific areas, they more often are used to restrict certain uses (strip development, for example) or to maintain others (extensive agriculture). A stewardship approach often fits the problem where a wilderness preservation approach does not. As techniques are introduced to a broader range of players, and adapted for use in new regions, stewardship can offer new ways of meeting conservation objectives in and outside of park boundaries.

The practice of stewardship draws on many disciplines, mostly from the natural sciences, social sciences, and law. Though the philosophy and many of the techniques of stewardship are not new, their application has become more frequent, better recognized and more formalized in the past two decades, as demonstrated by the explosive growth in private land trusts in the United States.

Land trusts often use the term in a limited sense, to describe their legal obligations to monitor and enforce deed restrictions. In this paper, stewardship is used in a broad way, though restricted to activities that have a direct impact on the land (*e.g.* not felling an important tree, or setting aside a woodlot) rather than actions with important but indirect impacts on the land (recycling).

While many stewardship approaches rely on market or financial leverage (*e.g.* tax incentives), stewardship should not be viewed as a solely private approach to conservation. Rather it rests on two elements coexisting in a civil society: private initiative (on the part of an individual landowner, a resource-user, a business and/or an NGO), and government control (which provides a framework in the form of tax and other incentives, land-use planning, and a supportive climate for private organizations). In addition, to be effective, stewardship approaches must take into account a third element: local-level and traditional resource management systems.

Stewardship can be adaptive, building on traditional means of management to meet changing needs. Recognizing that land management requires not only social will but *skills* within the society to do the job, the stewardship approach includes education for direct resource users and for decision makers in government and the private sector. Stewardship takes an overall landscape view, addressing conservation needs on land which cannot be separated from human existence and commerce. This approach can address the compatible objectives of biodiversity conservation, rural economic development and maintaining individual and community connections to the land.

2.2 Stewardship Tools

Specific stewardship tools vary according to social, legal, ecological and institutional constraints, but all operate to encourage, enable or formalize responsible management. Techniques include environmental education, technical information, demonstration projects, recognition of achievement, certification of standards, voluntary management agreements, subsidized management, deed restrictions, and full acquisition- through purchase, donation or many gradations of same - by private organizations. These tools (with many others and more variations), are listed in Figure 1 beginning with those that require little or no formal commitment or involvement and little per capita investment (*e.g.* education), to more permanent and specific protections (*e.g.* restrictions and acquisition). Though the latter tools are longer term and more certain forms of protection, they are often unattainable.

Figure 1. Stewardship tools

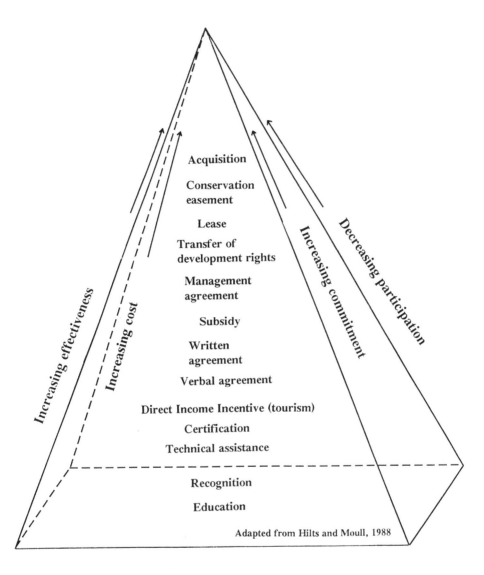

Adapted from Hilts and Moull, 1988

The fastest growing method for conservation is the conservation easement, or deed restriction prohibiting certain uses and permitting others. According to the Land Trust Alliance, the number of conservation easements written by its member organizations increased by two-thirds from 1990 to 1994. While securing conservation in perpetuity, an easement allows a landowner to retain ownership and rights to uses deemed appropriate to the land. A detailed description of these mechanisms as applied in North America is outside the scope of this paper. For more on the range of stewardship tools, see Diehl and Barrett 1988; Endicott 1993; Hilts and Moull 1984.

2.3 Stewardship and the Changing View of National Parks and Protected Areas

It is widely accepted that parks can no longer be treated as islands existing in isolation from their surroundings, but rather as part of a larger integrated system of land use which takes into account human activities on adjacent lands (McNeely et al 1990; Machlis and Tichnell 1985; Lucas 1984; Garratt 1984). In North America over the past century the focus has moved from preservation (*i.e.* setting aside blocks of land) to protection (*i.e.* establishment of enforcement capacity) to management (*i.e.* introduction of sustainable uses of some resources) to integrated management (involving greater management of resources in lands within, as well as outside of, protected areas). The past decade has seen a growing interest in ecosystem management, protection of greenways and biological corridors and landscape preservation.

The long-term security of a protected area depends on its relationship with neighboring communities, and yet many factors contribute to non-acceptance of protected areas. A common reason is the role of the central government in the establishment of a protected area, transferring control of land and resources away from a community. As a result, local people may relinquish responsibility for management of their natural resources, and may view the park as taking care of nature, thus leaving surrounding lands free for any use (McNeely et al 1990). Stewardship initiatives - to the extent that they respond to local interests and concerns, cultivate responsibility for conservation, and actively engage people in management - can serve as a powerful antidote to these problems.

Stewardship offers a means of extending the reach of conventional protected areas because i) it addresses conservation needs on lands outside park boundaries, and ii) it cultivates local responsibility for sound resource management. It offers the potential to conserve heritage at the level of ecosystems and landscapes. By engaging resource-users, landowners, civic organizations and municipalities, a local stewardship initiative can help to build a strong constituency for conservation, thus strengthening the position of protected areas.

2.4 The Changing Status of Protected Areas in Central Europe and Other Trends Shaping the Rural Landscape

The political changes of 1989 have brought new challenges and opportunities for Central Europe's national parks and protected areas. The region's protected areas systems are generally well developed, and comparable to those in Western Europe in terms of extent, criteria for selection and management (IUCN 1993). However, Central European protected areas now face changes in land ownership, increasing public scrutiny of management measures, development pressures, and declining budgets for management and operations. At the same time, there are new opportunities for: expansion of the protected areas estate, due to the opening of borders; the designation of new protected areas as part of the land redistribution process; the upgrading of current designations in some cases; and improved management through bilateral cooperation and NGO-government partnerships (Brown and Mitchell 1994).

In its 1994 Action Plan for Protected Areas in Europe, IUCN - or the World Conservation Union - makes several recommendations which support the need for stewardship approaches in Central Europe. These include: i) broadening the conservation focus from national parks and protected areas to the rural landscape as a whole; ii) mitigating the effects of land redistribution on protected areas in the region; iii) ensuring better representation of managed areas under traditional and other low-intensity farming systems in Category V Protected Landscapes; and iv) promoting joint management of protected areas (IUCN CNPPA 1994).

In addition to the changing status of protected areas, a variety of trends shaping the rural landscape in Central Europe present both challenges and opportunities for heritage conservation in the region. These are presented in Table 1.

3 Accomplishments

3.1 Meeting Heritage Conservation Needs in Northeastern North America

Private approaches to conservation are particularly developed in northeastern North America for a variety of historical reasons. Not least among them is that most of the land in the region is privately owned. Only four percent of land in the New England states is publicly owned (compared to 20 percent federal ownership on the national scale). Private landowners control over half of the forested land in the United States (353 million acres). Similarly, there is a high degree of private land ownership in the Atlantic Provinces of eastern Canada.

A key force in land protection in this region is land trusts: private conservation organizations which acquire land, through purchase or donation, and/or negotiate and monitor conservation easements. Their proliferation over the last two decades has been explosive. There are now over 1,200 land trusts in the US, one-third of them in New England. This is double the number ten years ago and ten times the number from 1965 (Land Trust Alliance 1995). Excepting Newfoundland, all the Atlantic Provinces of Canada have provincial-level and smaller land trusts, whereas a decade ago there was only one. These organizations range from large national organizations, to those working on a state or regional level, to local land trusts, many of which focus on a single county or community. Half of all land trusts have no paid staff and budgets of less than $10,000 CAN.

Land trusts are able to work effectively because: i) many landowners are more willing to donate or sell their properties to a private organization than to the government; ii) donation of land or conservation easements can confer significant tax advantages to landowners; and iii) private organizations can generally respond quickly and flexibly to threats and opportunities. While precise figures are difficult to compile, it is estimated that private nonprofit conservation organizations now directly protect more than 12 million acres in the United States. This is an area the size of the states of Vermont and New Hampshire. By contrast, the National Park Service owns approximately 77 million acres, Fish and Wildlife Service 92 million, and the Forest Service controls almost 200 million. The total land area of the United States is more than 2.3 billion acres.

Estimates of protected acreage do not alone tell the full story of the contribution private organizations are making to heritage conservation in northeastern North America. Private organizations i) conserve lands not available to public agencies; ii) often convey other lands to public agencies; iii) often provide buffer zones around or linkages between public lands; iv) respond to social interests not included in the mandates of public agencies; and v) conserve land effectively with methods not recorded as land transfers. Perhaps most significant, land protection by private organizations is increasing exponentially.

Of course, government agencies also work as stewardship agents. For example, the US National Park Service works directly with landowners to develop alternatives to land acquisition: since the 1930s the National Park Service has acquired easements on more than 80,000 acres in at least 86 park units (Brown 1993). Through its Partners for Wildlife Program, the US Fish and Wildlife Service pays for wetland restoration on private lands if the landowner enters a management agreement. The US Department of Agriculture pays farmers to set aside land for conservation purposes under the recently authorized Conservation Reserve Program.

Table 1. Trends in Shaping the Central European Landscape

A variety of trends shaping the rural landscape in Central Europe present both challenges and opportunities for heritage conservation in the region (Brown and Mitchell 1994). These include:

1. *Redistribution and Reprivatization of Land* - Changes in land tenure and the reprivatization process initiated since 1989 continue to play a major role in shaping the Central European landscape. The complex process of land redistribution is linked to other trends shaping the rural landscape. For example, changes in land tenure will affect protected areas and agricultural patterns; at the same time economic and demographic trends in rural areas are influencing the pace of land redistribution. Privatization's impact on the planning process places new responsibilities on local and regional government authorities. The scale of land reprivatization varies according to country, and is proceeding unevenly throughout the region (Mundy 1992). However, its inexorable progress, and the opportunities and threats this progress presents, contribute to a sense of urgency for those concerned with heritage conservation.
2. *Changing Attitudes Toward Planning and the Emergence of a New Culture of Land Ownership* - The transition from communism to a free-market economy has brought with it a shift in public attitudes toward land use planning, evident in a growing resistance to government controls and wide-ranging interpretations of private property rights. These changing attitudes reflect Central Europeans' reaction against the centralized planning characteristic of the former regimes, as well as their efforts now to come to terms with the rights and responsibilities of private land ownership. In a growing aversion to land-use planning controls are the seeds of a new culture of land ownership and the emergence of a private property rights movement not unlike the 'Wise Use' movement which has gathered momentum in the United States in the last decade.
3. *Changes in Rural Life* - Today, unemployment in rural areas is increasing rapidly, due to the scaling back of industrial activity no longer viable in a free-market economy, and the reorganization of large-scale agricultural enterprises. Rural residents have been hit hard by inflation and the cutting back of government subsidies for transportation and housing, resulting in increasing migration out of rural areas, as people relocate in search of employment.
4. *Changes in Agricultural Patterns* - Agriculture has played a major role in shaping the scenic qualities, cultural features and biological diversity of much of the Central European landscape. Now agricultural patterns which had changed - dramatically in many countries - under socialism, are set to change again, due to land redistribution, the introduction of free-market policies, and integration into the European Union. While governments are beginning to develop programs to sustain and revive traditional farming methods in rural areas, there are many practical obstacles. These include: a shortage of government funds for incentives and subsidies, as well as demographic, cultural and psychological obstacles (e.g., an aging rural population, the severed connection to the land among many rural people, a shortage of skills and experience in running a private farm, and reluctance to invest in farming in the face of uncertain conditions).
5. *Increasing Development Pressures* - Many rural areas face a surge in the construction of new housing, recreational and commercial developments, often financed by outside investors. The opening of borders has brought new pressures in the form of increased tourism, and the construction of hotels and second homes. With the devolution of planning responsibility, regional and local authorities face new challenges in balancing economic development with environmental protection. For local authorities facing massive unemployment and other economic problems, the promise of jobs and tax revenue which a development proposal might bring can be a far more compelling argument than its environmental impact.
6. *Conservation and Changing Political and Economic Priorities* - For a brief historical moment (1989 to about 1992), environmental protection was at the top of national agendas in Central Europe, but it is now a much lower priority, relative to pressing concerns of economic development, employment, public health and safety.
7. *Integration into the European Union* - Poland, the Czech Republic, the Slovak Republic and Hungary are working concertedly to join the European Union. Integration into the EU will have many impacts on land use, including: redirection and improvement of highways and other transportation routes; globalization of agricultural markets; and increased outside investment in development projects. On another level, EU membership will bring access to pooled assistance funds to meet environmental standards, and to Europe-wide incentive schemes (e.g. for management of fallow land). Member countries are subject to EU directives related to landscape management (e.g. the Environmentally Sensitive Areas program).

Among the heritage conservation challenges being addressed by local- and national-level stewardship initiatives are those described below, with selected examples.

3.1.1 Protecting Open Space and Fragile Natural Areas in the Face of Development Pressures, Especially in Areas where Planning Controls are Weak

While many conservation organizations are active in this area - in education, lobbying and activism - land trusts play an important role in direct protection. In a 1994 survey that excluded the large, national trusts, the Land Trust Alliance found that regional, state and local trusts had conserved slightly over 4,000,000 acres, an increase of nearly 50 percent from 1990 to 1994. In addition, national organizations have directly protected at least 8,000,000 acres over the past 50 years.

3.1.2 Conserving Biodiversity Through Protection of Habitats

Though it has one of the most specific missions - biodiversity and habitat conservation - The Nature Conservancy owns the largest system of private nature reserves in the world. The organization currently has operations in each of the fifty states, Canada and many Latin American and Caribbean countries. In the United States, it has protected over 9,500,000 acres, including acquiring 3,549,000 acres and obtaining easements on a further 626,000 acres.

Another major stewardship organization, Ducks Unlimited, has conserved over 7 million acres of habitat, primarily waterfowl breeding and wintering areas, in North America (nearly 5 million acres in Canada, 1.5 million in Mexico and 1 million in the United States).

3.1.3 Maintaining Landscape Integrity in Times of Changing Land Uses and Declining Rural Economies

As the land trust movement matures in the United States, it is moving away from the protection of isolated land parcels and toward a recognition of the importance of preservation at the landscape level. Examples include The Nature Conservancy's emphasis on protecting whole ecosystems, the Trust for Public Land's watershed protection activities, and the greenways movement to protect scenic and recreational corridors (Endicott 1993).

The Trustees of Reservations, the oldest land trust in the US, is an example of an organization concerned with natural and cultural heritage protection within a defined geographic area (the state of Massachusetts). It owns and manages 75 reservations totaling 19,000 acres, with an additional 8,000 acres protected through conservation restrictions (Abbot 1993).

3.1.4 Sustaining Traditional Land Uses, Such as Farming and Small-scale Forestry, Important to Ecological, Economic and Scenic Values

The state of Vermont has seen large-scale changes in land use. Largely forested before European colonization, it had only 20 percent forest cover by the beginning of this century. Today, the ratio to open land has reversed, with 80 percent forest cover regrown as agricultural land has been abandoned. With family farms important to the scenic values, biological diversity and tourism appeal of the landscape, and to the local economy, the Vermont Land Trust sets the preservation of farmland as its priority. The organization has conserved 103,000 acres through acquisition and restriction across the state.

3.1.5 Creating Biological Corridors, Greenways and Trails Across Privately Owned or Managed Land

Of increasing concern to conservationists is the fragmentation of wildlife habitat. While scientific evidence of the need for and technical options for corridors mounts, NGOs and public agencies are working to secure connections for biodiversity conservation and public recreation. Outside of Los Angeles, California, the National Park Service and Santa Monica Mountains Conservancy are linking public and private lands through urban areas to connect habitat in the Santa Monica Mountains, Simi Hills, and Santa Susana Mountains.

3.1.6 Through Partnerships, Enhancing the Ability of Government Agencies to Acquire and Manage Publicly Owned Parks and Protected Areas

Private organizations can often react more quickly, more flexibly and often more cost-effectively than government agencies to land conservation opportunities. For example, three of the largest private conservation organizations in the United States - The Nature Conservancy, the Trust for Public Land and Ducks Unlimited - are responsible for assisting government in preserving more than 4 million acres (Endicott 1993).

As the US National Park Service contends with limited resources for land acquisition, it is turning increasingly to partnerships with land trusts and other private organizations because they allow the agency to address protection needs in and around parks, avoid adverse impacts of adjacent land uses, respond quickly to opportunities, meet landowners' objectives, and devise creative arrangements (Brown 1993).

The Trust for Public Land is an example of a land trust that works entirely through partnerships with federal, state and municipal governments, local land trusts, civic associations, and does not hold land, but always conveys it to a client organization. Since its establishment 24 years ago, the Trust for Public Land has protected over 820,000 acres in the US and Canada, much of it as parks, community gardens and historic landmarks in urban areas.

3.2 Meeting Heritage Conservation Needs in Central and Eastern Europe

A variety of stewardship approaches are developing to address the unique heritage conservation challenges facing the Central European landscape. Some of these initiatives draw on the long tradition among conservation NGOs of practical conservation work in assessing and managing natural areas (*e.g,* mowing upland meadows) and maintaining cultural features (*e.g.* restoring folk architecture). Others represent new approaches, such as those that seek to sustain rural landscapes by revitalizing rural communities.

In 1993 the authors conducted a survey of stewardship needs and initiatives in the Central European countries of the Czech Republic, Hungary, Poland and the Slovak Republic (Brown and Mitchell 1994). This study has helped to guide the development of a multi-year program of stewardship training, technical assistance, professional exchange and community-based planning projects being conducted by QLF/Atlantic Center for the Environment in partnership with several Central European NGOs.

Drawn from this experience, selected examples of stewardship approaches in the region follow.

3.2.1 Acquisition by NGOs of Natural Areas

A number of NGOs are trying to take advantage of the conservation opportunities presented by privatization in rural areas where land values are low. An early example of a private nature reserve is the Somogy Environmental Association's Borunka Reserve in Hungary. Organizations such as the Hungarian Ornithological Society are experimenting with purchase of certain land use rights, as an alternative to ownership of land. Recent participants in Atlantic Center fellowships are experimenting with land trust approaches; for example the Adonis Center in the Czech Republic has launched a project to establish and support land trusts in southern Moravia.

3.2.2 NGOs Cooperating with Government to Designate and Manage Protected Areas

Unlike many conservation organizations in the US, NGOs in Central Europe rarely work on land they own. Most in the Czech Republic, for example, are somehow connected to protected areas (especially Protected Landscape Areas) and are gradually taking over some tasks from the government administrations. Often they are contracted by the government for specific tasks, saving money with volunteer and in-kind support.

It is also common for conservation NGOs to help designate areas and secure property for government protection. The Mazurian Landscape Park, in the heart of the so-called 'Green Lungs' area of Poland, has been significantly enlarged through the efforts of an associated NGO. In ways not dissimilar to many examples in the US, the private organization was able to secure lands the government could not.

3.2.3 Working with Private Landowners to Improve Management

In the White Carpathians Protected Landscape Area, chapters of the Czech Union of Nature Protectors (Kosenka and White Carpathians on the Czech side; Koza on the Slovak side) are working with landowners to sustain meadow management. Traditional mowing is important to the biodiversity of the areas, especially orchid species, but the economic justification for mowing has disappeared. The groups are looking for new ways to encourage traditional uses, including payments and reinforcing the cultural significance of the practice. With the help of the NGO Veronica, these groups and others are also trying to provide markets and processing for fruits in order to reestablish local cultivars.

3.2.4 Pursuing New Opportunities to Create Protected Areas

While maintaining and managing existing protected areas is currently a concern, new areas are being added to the protected area estate in the region. The initiative, Ecological Bricks, seized on the opportunity to retain the *de facto* conservation areas of former depopulated borderlands and other military reserves. Similar efforts continue, for example the work of the Polish NGO ProNatura to create the privately managed reserve at Przemków.

New lines of political and economic cooperation also may allow advances in bi- and multi-national conservation areas. NGOs in all four countries affected continue to push to better stewardship of the Danube-Morava floodplain.

3.2.5 Promoting Conservation Through Eco-Tourism and Enhanced Public Access to Natural Areas

Inspired by the Hudson River Greenway, an NGO in southern Moravia hopes to develop a footpath from Vienna to Prague. *Greenways* (Zelené Stezky) is exploring: questions of public access to privately owned natural areas; economic development incentives for conserving biocorridors; and restoration and promotion of cultural features. It plans to promote 'soft' tourism along the greenway, for example, restoring old houses into pensions which would provide traditional meals and entertainment. As a first step, the Greenways project is focusing on the Mikulov/Valtice area adjacent to the Palava Protected Landscape Area.

In the same region, an arm of the environmental NGO Adonis operates a tourism information center. The group hopes to promote agro-tourism in the wine-producing country, a concept that would develop a local industry of small accommodation, recreation and interpretive services based on the viniculture of the region. The group aims to keep tourism facility developments at an appropriate scale and promote better viniculture practices.

In the Tatra Region of Slovakia, the Beneficial to the Public Fund has developed from an campaign opposing a winter Olympics proposal to a comprehensive community development organization. The group is focusing on developing local support for village-scale tourism facilities. The group is working to convert the large number of empty houses in the area (abandoned during post-war urbanization programs) into bed and breakfast units. Other activities include a regional tourism promotion campaign, and support for local environmentally friendly businesses.

Other approaches being undertaken by NGOs include: working with government to influence policy; promoting sustainable agriculture; creating biological corridors; community-based planning; and education.

Atlantic Center fellowships and exchanges have heightened awareness of opportunities in stewardship, while technical assistance and workshops across the region have begun to grapple with the hard realities of application.

4 Challenges

4.1 General Challenges

Stewardship techniques offer great potential to strengthen and extend the impact of conventional protected areas in conserving natural and cultural heritage. Challenges to developing stewardship initiatives in any context include:

4.1.1 Creating a Legal Framework Conducive to Private Initiatives

Incentives (*e.g.*, tax advantages) for conservation and best management practices on private lands must be incorporated into national legislation. As key actors in stewardship, NGOs require a stable legal basis for establishment and legitimacy as an important sector in any civil society.

4.1.2 Creating the Climate for Productive, Enduring Partnerships Among Sectors

Government agencies charged with protected areas management must have the flexibility to develop appropriate partnerships with NGOs and other private interests. To create an atmosphere of trust and cooperation, government must view these NGOs as true partners, rather than subcontractors; NGOs must be willing to engage in non-adversarial relationships with government; all parties must be committed to ongoing communication and coordination of efforts.

4.1.3 Integration into Land Use Planning and Protected Areas Management

Private stewardship efforts, however extensive, are no substitute for a strong government role in land-use planning and protection of natural areas. These efforts should reinforce land-use planning and policy at all levels. At the same time, private initiatives should be viewed not as an afterthought, but as central to meeting protection and management objectives. To this end, coordination among private and public actors is essential.

4.1.4 Ensuring Participation by All Interested Parties

Stewardship relies on public support and participation. Whether through landowner contact or public fora, opportunities must be created for those most affected by land use decisions to voice their concerns. Value must be placed on local knowledge and traditional resource management systems.

4.1.5 Marshalling the Necessary Resources

Funding is necessary for land acquisition and compensation for certain development rights or uses. Often NGOs are in a strong position to raise private funds for these purposes. Fiscal incentives, such as reduced property taxes, carry a cost in terms of lost revenues to municipalities.

4.2 Challenges Specific to Central Europe

Many conservation groups in Central Europe are recognizing the need to work with private landowners and resource users to achieve conservation objectives. They face special challenges in developing stewardship approaches, which include:

4.2.1 Reprivatization of Land and New Development Pressures

Land reprivatization and development pressures are having a far-reaching impact on the landscape, at the same time that governments are formulating land use policies. In the Czech Republic, for example, the government initiated land redistribution at the same time that it began a comprehensive planning process to set conditions for ecological stability. This has raised concerns that it will be difficult to restrict land uses on properties that have already been restituted.

4.2.2 A Prevailing Sense of Urgency

Ongoing changes in land uses and tenure contribute to an atmosphere of urgency among conservation agencies and NGOs, creating the risk that public and private actors will focus on acquisition without developing long-term management plans, fail to build public support for new protection measures, or operate without a coordinated strategy.

4.2.3 The Changing Position of NGOs

At the same time that the region is experiencing a rapid growth and diversification of new NGOs, the non-profit sector faces challenges including: declining membership and volunteerism; limited funding; changes in enabling legislation; and a need to develop legitimacy in the eyes of government and the general public.

4.2.4 A New Culture of Land Ownership

As noted earlier, collectivization of land, large-scale agriculture and rural industrialization have severed traditional ties to the land in many areas. With the transition to a free-market economy, land is often viewed solely as a commodity, and there is a growing aversion to planning controls.

5 Recommendations and Conclusions

5.1 Recommendations for Research and Practice

To advance stewardship in Central Europe, and in other regions where these approaches are just starting to be tapped, there is a need for research and adaptation of practice in the following areas:

5.1.1 Legal and Institutional Tools to Encourage Stewardship on Private Lands

Available legal and institutional tools in each country must be assessed. What public and private institutions are currently performing stewardship functions? What precedent exists for management agreements with landowners? How might legal tools being used effectively in other regions (*e.g.*, conservation easements and management agreements in North America; ESA designation in Western Europe) be adapted for use in Central European countries?

5.1.2 Incentives for Conservation and Best Management Practices

A priority in IUCN's Action Plan for Protected Areas is to identify policies that could be used to provide effective incentives for conservation other than purchase (IUCN CNPPA 1994). What provisions exist in national legislation to provide landowners with incentives for conservation of natural or cultural heritage? Non-financial incentives for protection, such as landowner recognition, must be explored.

5.1.3 Public Attitudes Toward Land Ownership and Conservation

Research is needed on public attitudes toward private land ownership and conservation, especially in rural areas. Studies should endeavor to gather local knowledge of traditional land use practices and recent changes. Understanding the basis of individual land use decisions will guide in developing incentives for best conservation practices on private lands, and may yield clues on how to restore people's traditional ties to the land.

Public attitudes toward parks and protected areas in Central Europe are another area where further study is needed. This information could guide management policies within protected areas and outreach to neighboring communities and landowners.

5.1.4 Alternative Institutional Options for Protected Areas Management

There is a growing body of literature on examples of partnerships between protected areas agencies, other government agencies, NGOs and communities (IUCN 1995; Endicott 1993). What lessons do these case-studies offer Central European conservationists on creating institutional mechanisms for effective public-private partnerships? What precedents exist in the region for this kind of cooperation? Research on co-management approaches will provide new perspectives on how to enlist the cooperation of those who use and manage the land.

A related question is how stewardship approaches can enhance management of protected areas, particularly IUCN categories V and VI. The experience of protected areas authorities in cooperating with landowners, resource users, and other government bodies needs to be documented.

5.1.5 New Approaches to Encourage Public Participation in Resource Use Decisions

Given the lack of a recent tradition of public participation in the region, new approaches must be developed to involve local people in resource use decisions. Drafting of new legislation should include provisions for public right-to-know and public review of certain development proposals. Methodologies for bringing diverse stakeholders together, such as participatory rural appraisal, should be further adapted for possible use in rural communities in Central Europe.

5.1.6 The Impact of Europe-Wide Policies on Land Use in Central Europe

A few conservation organizations in the region are researching the potential impacts of EU membership on land use policy and practices in Central European countries, and this information should be widely disseminated. What will be the impact of new transportation routes or the globalization of markets for agricultural goods and forest products? How can EU directives, environmental standards, and incentive schemes be used to improve conservation practices in member countries?

5.1.7 Promoting an Exchange of Ideas and Practice

Just as stewardship practice in North America is evolving to meet new challenges, stewardship approaches are developing that are appropriate to the special needs and conditions of the Central European context. Conservationists working in both regions face many similar problems and challenges, but often work in isolation. The experience of recent fellowships, workshops and exchanges between the US and Central Europe has demonstrated that these programs can contribute to a transfer of innovations and the improvement of practice on both sides. More opportunities to exchange ideas and learn from the successes and failures of national and international counterparts are needed to strengthen this growing movement.

5.2 Conclusions

The concept of stewardship can offer valuable tools for addressing the heritage conservation challenges facing landscapes in diverse regions of the world. In addition to the direct benefit of improving heritage conservation, stewardship approaches can serve to strengthen local leadership and institutions, encourage citizen participation and bring together diverse stakeholders to address problems at a local level. These contributions, in turn, reinforce the characteristics of a civil society.

6 References

Abbot, G. 1993. *Saving Special Places: A Centennial History of the Trustees of Reservations, Pioneer of the Land Trust Movement.* Ipswich Press. Ipswich, MA, USA.

Brown, J. and B. Mitchell. 1994. *Landscape Conservation and Stewardship in Central Europe.* Nexus Occasional Paper. No. 10. QLF/Atlantic Center for the Environment. Ipswich, MA, USA.

Brown, J. and B. Mitchell. 1994. *Stewardship in Central Europe.* Unpublished report. QLF/Atlantic Center for the Environment. Ipswich, MA, USA.

Brown, W. 993. Public/Private Land Conservation Partnerships in and Around National Parks. *Land Conservation through Public/Private Partnerships.* Lincoln Institute of Land Policy. Island Press, Washington, DC., USA.

Diehl, J. and T. S. Barrett. 1988. *The Conservation Easement Handbook.* Trust for Public Land, San Francisco, CA and Land Trust Exchange, Alexandria, VA, USA.

Endicott, E. (ed). 1993. *Land Conservation through Public/Private Partnerships.* Lincoln Institute of Land Policy. Island Press, Washington, DC., USA.

Garrat, K. 1984. The Relationship Between Adjacent Lands and Protected Areas: Issues of Concern for the Protected Area Manager. In *National Parks, Conservation and Development: The Role of Protected Areas in Sustaining Society.* McNeely and Miller (eds.), IUCN-The World Conservation Union. Smithsonian Institution Press. Washington, DC., USA.

Hilts, S. and T.C. Moull. 1988. *Protecting Ontario's Natural Heritage Through Private Stewardship.* University of Guelph. Guelph, Ontario.

IUCN Protected Areas Programme. 1995. *Parks.* Institutions for Parks. Vol. 5 No. 3. IUCN-The World Conservation Union. Gland, Switzerland and Cambridge, UK.

IUCN Commission on National Parks and Protected Areas (CNPPA). 1994. *Parks for Life: Action for Protected Areas in Europe.* IUCN-The World Conservation Union. Gland, Switzerland and Cambridge, UK.

IUCN. 1990. *Protected Areas in Eastern and Central Europe and the USSR (An Interim Review).* IUCN-The World Conservation Union. Gland, Switzerland.

Land Trust Alliance. 1995. *1994 National Land Trust Survey, Summary.* Land Trust Alliance, Washington, DC.

Lucas, P.H.C. 1984. How Protected Areas Can Help Meet Society's Evolving Needs. In *National Parks, Conservation and Development: The Role of Protected Areas in Sustaining Society.* Jeffrey McNeely and Kenton Miller (eds.). Smithsonian Institution Press. Washington, DC., USA.

Machlis, C., Tichnell, G. E. and D. L. Tichnell. 1985. *The State of the World's Parks.* Westview Press, Boulder, CO, USA and London, UK.

McNeely, J. A., K.R. Miller, W.V. Reid, R.A. Mittermeier and T.B. Werner. 1990. *Conserving the World's Biological Diversity.* IUCN, Gland, Switzerland; WRI, CI, WWF-US and the World Bank, Washington, DC., USA.

Mundy, J. 1992. *Central and Eastern Europe: Land Redistribution and Nature Conservation.* IUCN European Programme Discussion Papers 1. IUCN-The World Conservation Union. Gland, Switzerland.

The Role of Banff National Park as a Protected Area in the Yellowstone to Yukon Mountain Corridor of Western North America

Harvey Locke[1]

[1] Past President, Canadian Parks and Wilderness Society
700, 401 - 9th Avenue S.W. Calgary, Alberta, Canada T2P 3C5

Abstract. Banff National Park is a mountainous area of the Canadian Rockies which forms part of the larger ecosystem running from Yellowstone to the Yukon. Recent research has demonstrated that to maintain wide-ranging species like grizzly bears and wolves, a landscape approach must be taken to ensure connections between populations. Like all Western North American national parks, Banff is subject to external stresses in its ecosystem. It is also subjected to internal stresses caused by intensive tourism infrastructure and the national transportation corridors which traverse it. The park's ecology is now acknowledged by Parks Canada to be impaired. Public controversy over development in Banff National Park has existed for over 25 years. The controversy escalated to an international issue in 1993 when the Canadian Parks and Wilderness Society, a non-government organization, launched a campaign to halt further development in the park due to the proliferation of commercial development and decline in ecological integrity in Banff Park. The Minister of Canadian Heritage intervened and established a $2 million CAN Banff Bow Valley Study headed by a Task Force independent of Parks Canada. The Task Force was charged with studying the health of the park and finding an honourable solution to the issues facing it. The Task Force used a round table process in an effort to bring together the competing interests. The Banff Bow Valley Study reaffirmed Banff's role and responsibilities as a national park. The diverse group of interests in the round table process agreed on a common vision for Banff's Bow Valley. Goals for ecological integrity were agreed to which recognize Banff's role as a protected area providing source populations of wildlife to the broader Yellowstone to Yukon corridor.

Keywords. National parks, protected areas, conservation biology, Yellowstone to Yukon mountain corridor, Banff National Park, Banff-Bow Valley Study, tourism, wildlife, ecological integrity, biodiversity, bioregions, ecosystems, research and planning

1 Introduction

The Yellowstone to Yukon corridor in the Northern Rockies of the United States and the Canadian Rockies increasingly is being viewed as a common ecosystem (Figure 1). It is a mountainous area which generally runs north-south. It is the last refuge for grizzly bears in settled North America. It contains many protected areas ranging from designated wilderness areas in the United States to provincial parks in Canada. Two of the world's oldest and most famous national parks, Yellowstone in the United States and Banff in Canada, are located there. Banff National Park is an appropriate focus for the discussion of the role of protected areas in the Yellowstone to Yukon corridor as it faces all of the issues which affect the other protected areas in the corridor as well as unique challenges of its own. This paper explores the ecological character and reasons for creation of Banff National Park; its role in the broader Yellowstone to Yukon region of which it forms part; the range of issues which affect it; how these issues developed and have been addressed by planners, managers and decision-makers; what accomplishments have made and what challenges remain; and, finally, the research and planning needs of the park into the future.

Figure 1. Yellowstone to Yukon Mountain Corridor

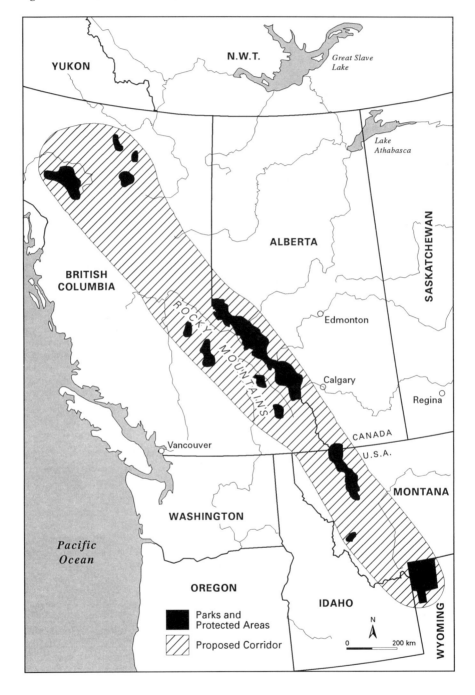

2 Ecological Character and Reasons for the Establishment of Banff National Park and its Role in the Yellowstone to Yukon Mountain Corridor.

Banff National Park began in 1885 as a 10 square mile reserve focused on a hot spring discovered while Canada's first transcontinental railway was being pushed through the mountains of western Alberta. Current conventional thinking subscribes to the view that the park was created as a commercial opportunity to bring tourists to the hot springs rather than for any higher purpose of conservation. However, this view is only partly accurate. By 1887 it had become the 260 square mile Rocky Mountains National Park pursuant to legislation very similar to that which created Yellowstone National Park in the United States and efforts were underway to preserve the park's wildlife. Thus conservation as well as commercial thinking was present when the park was first established.

Rocky Mountains National Park grew quickly. By 1902 it was 4,400 square miles. These expansions were motivated at least in part by the desire to protect the area from commercial exploitation. In the same period, other national parks like Waterton Lakes were being established for conservation reasons. It was not until 1930, however, that Canada's commissioner of national parks, J.B. Harkin, succeeded in having a National Parks Act passed by Canada's parliament. Modeled after the 1917 U.S. National Park Service Organic Act, it provided that the national parks of Canada be dedicated to the benefit, education and enjoyment of Canadians and, subject to the Act, the parks were to be maintained and used in a manner that would leave them unimpaired for future generations. Around the time the Act was passed, Rocky Mountains Park was renamed Banff National Park and was reduced to 2585 square miles to facilitate industrial development on the excluded lands.

Banff National Park includes good representation of the alpine, and upper and lower subalpine natural regions. It also represents a small but critical portion of the montane natural region. The montane is less than 5% of the park but it is the most biologically productive area as it is found in a low elevation valley bottom which is often free of snow even in winter. The Park contains several broad, deep valleys which were excavated by large glaciers and which run generally north-south. The most prominent of these is the Bow Valley where is found montane habitat as well as the Town of Banff and the Lake Louise Visitor Centre. The Park's spectacular mountain scenery is world famous and the park is a UNESCO World Heritage Site.

Recent research on wolves and grizzly bears has demonstrated that they move freely in and out of Banff National Park. Wolves from Banff range as far south as Montana where they have only recently re-established themselves after being extirpated. Grizzly bears, which once roamed widely throughout western North America, now have a very limited distribution. In the settled regions grizzlies are now confined to an isolated population in the Greater Yellowstone Ecosystem, some areas immediately adjacent to the Canadian border, and a thin peninsula running down the Canadian Rockies from Banff to Montana's Bob Marshall Wilderness. By contrast, in 1922 there were numerous isolated grizzly populations in the United States but they all went locally extinct because once subjected to population reducing stresses, they could not replenish themselves. Development in Banff cuts into the remaining peninsula running down the Canadian Rockies, and if continued, will lead toward isolation of the southern population of grizzlies from the larger population to the north.

Large carnivores, like wolves and grizzly bears, are considered 'umbrella species'. If the habitat needs of these carnivore are met, then so are the needs of most other species because they require less range. To be effective in maintaining viable populations of umbrella species, current conservation biology analysis suggests that large networks of core protected areas, connected by movement corridors, are required. This in turn, leads to a view of Yellowstone, Waterton, Glacier, Banff, Jasper, Kootenay and Yoho National Parks as a series of core reserve areas which are part of an integral whole (see Figure 1). Management of this vast area as a whole has been proposed by many Canadian and American conservationists and biologists through the Yellowstone to Yukon Biodiversity Strategy.

3 Issues Facing Banff National Park

Banff is Canada's most popular park and, in terms of tourism infrastructure, it is without doubt the most heavily developed nature oriented national park in North America. Like most western North American national parks, it is also subjected to external pressures from land use activities on adjacent lands which affect the park's ability to maintain ecological processes. However, Banff is unique in that its major ecological stress is human activity inside the park. This has led Parks Canada to admit it has contravened its own mandate to maintain ecological integrity and biodiversity in a portion of Banff[1]. A group of prominent North American ecologists brought together in 1995 concluded that the Bow Valley in Banff National Park has already been impaired by human activity and development and that there was an urgent need for remediation and mitigation if the park's ecological integrity was to be sustained[2]. Yet, pressures for increased development in the park exist at the three ski hills in the park, at the golf course adjacent to Banff townsite, at Lake Louise Visitor Centre and on the highway which traverses the park. Facilities like shopping malls have been built in recent years. They cater to a shopping market for luxury items which are inappropriate under Canada's national park policy.

4 How the Issues Affecting Banff Developed and are being Addressed

It has taken 30 years for these issues to develop. In the late 1960s, Banff had a busy summer season of park visitors with relatively little visitation outside of the months of June, July and August even though it had a permanent community and three small ski hills. In the 1960s, Parks Canada encouraged ski hill development to promote year round use and Imperial Oil Corporation proposed to build a year round resort called Village Lake Louise at the base of the Lake Louise ski hill. Some Canadians felt this kind of development was inappropriate in a national park. The issue lead to a very public, Canada-wide, debate over the Village Lake Louise proposal and the purpose of national parks. The pro-conservation side was led by the National and Provincial Parks Association of Canada (NPPAC), a non-government organization. Ultimately, the Federal government revoked the approval. That issue politicized national park management. But it was not a definitive end to ski hill expansions in Banff National Park as Canada's federal government was prepared to respond to demands of the growing ski market[3].

In the 1970s, the Sunshine ski hill, one of three in the Park, sought to expand and was given approval to do so. Similarly, the Lake Louise ski hill was allowed to expand in 1981. Offsetting these expansions was a new national park policy issued in 1978 for all Canadian parks which specified there would be no new ski hills. It also contained a strong emphasis on protection while accommodating visitor use appropriate to a national park setting. However, notwithstanding the clear thrust of the policy, parks administrators continued to interpret the dedication clause in the National Parks Act as creating a dual mandate of competing goals - preservation and use. Such an interpretation is highly questionable because the Act is clear that the parks must be made use of in a manner that leaves them unimpaired.

The Canadian Parks and Wilderness Society (formerly NPPAC) sought to eliminate any doubt about the existence of a dual mandate for all Canadian national parks. Its efforts were successful in 1988 when amendments were made to the National Parks Act so that when considering visitor use in a national park, ecological integrity is the first consideration. Similarly, a Banff based non-government organization, the Bow Valley Naturalists, succeeded in their long-term effort to prevent incremental expansion of the Banff townsite by obtaining an amendment which required a fixed town boundary to be set by law. Ski hill boundaries were also to be fixed after accommodating the approvals for expansion previously given.

---

[1] Parks Canada. 1995. *Initial Assessment of Proposed Improvements to the TransCanada Highway in Banff National Park*, Phase IIIA, p. 3-28.

[2] Banff Bow Valley Study. 1995. *Ecological Outlook, Cumulative Effects Assessment and Futures Modeling Workshop 1, Summary Report*, p. 1.

[3] Bella, L. 1987. *Parks for Profit*. Harvest House, Montreal, p. 126.

The foregoing would suggest that by 1988 the legal and policy framework was in place to ensure Banff met national park conservation goals. However, other forces were at work in the 1980s. Banff is the cornerstone of Alberta's tourism economy and one of only a few Canadian travel destinations of world renown. The Province of Alberta and the Canadian Federal Government wanted to increase tourism and in the mid-1980s initiated the Canada-Alberta tourism agreement to promote the development of tourism infrastructure. During the years of that program, over $11 million CAN of public money was given to Banff's businesses as an inducement to expand. The 1978 national park policy was also ignored on more than one occasion to facilitate such development. Since 1927 Banff has contained an 18 golf course. The 1978 park policy provided that no new golf courses would be allowed in national parks.

Yet in 1987 a new nine hole course was allowed to be built in Banff. The 1978 policy also provided that only unobtrusive structures would be permitted inside national parks. Yet Parks Canada approved the 11 storey Rimrock Inn expansion in 1988. The 1978 policy stipulated that only essential services would be provided inside national parks and yet several shopping malls were approved in the 1980s, one of them which called itself 'Banff's Great Indoors'. The proliferation of luxury stores prompted a travel writer for the London, England Sunday Times to write '... Banff's City fathers should hang their heads in contrition ... Mammon has set up stall all the way from Bow River almost to the foot of Mount Rundle, a hundred gift shops dispensing life's identical duty free necessities obtainable at any international airport: cashmere, crystal, Cartier, Caleche.'[4] From 1984 to 1994, just under $500 million CAN in building permits were issued for developments in Banff National Park. Electric power consumption more than tripled.

Another problem for the park was created when a decision was made in the early 1980s to 'twin' the two lane TransCanada Highway which runs through Banff, by creating a divided highway. Wildlife mortality, particularly for elk (*Cervus elaphus*) was very high along the old two lane road so Parks Canada decided to put up high wire fences to keep animals off the enlarged road. Small overpasses and culverts were scattered under the road to provide movement corridors for wildlife. The result was disastrous for some species while it favoured others. Elk, which habituate easily to humans when they are not hunted, used the underpasses very successfully. But wolves did not accept them well. Some were observed going around to the end of the divided road where the fence ended, crossing there and returning back to the point opposite where they started. Coyotes learned they could trap bighorn sheep against the fence and wiped-out the Mt. Norquay bighorn sheep herd, which herd previously was the subject of an interpretive display for park visitors on the highway. Elk became superabundant and browsed the poplar forest so heavily that it did not regenerate.

The legislated boundary which followed the 1988 amendments was so generous that the Town of Banff grew to the point where it almost totally straddled the low elevation Bow Valley bottom. As a result of the Town's size and adjacent developments, movement corridors for wolves and grizzly bears were effectively blocked, leaving high quality habitat in the Park east of the town devoid of those large carnivores. Elk usage of the town area increased, and conflicts between these animals and park visitors became a major management issue.

Parks Canada also further contributed to the problems when it drew up the 1988 Banff Park Management Plan. This plan excluded many parts of the park like the ski hills and the Town of Banff from the plan. Thus projects were considered in isolation rather than in the context of their cumulative effect on the park's ecology. The plan did include a promise to close the Banff airstrip, which is in the middle of one of the wildlife movement corridors referred to previously, but later Parks Canada broke that promise in response to pressure from small plane pilots.

In the early 1990s some research and conservation biology analysis was done on wildlife in the park. Concerns emerged about the decline of the black bear and moose population as well as the barrier created by the fenced highway to animal movements, particularly grizzly bears and wolves. Some courageous park wardens began to speak out that some species of wildlife might not continue to persist in Banff National Park due to all the development.

---

[4] The Sunday Times. April 30, 1995. *Into the West.* London

Concerned about the continued proliferation of commercial development and the health of wildlife, in 1993 the Canadian Parks and Wilderness Society (formerly NPPAC), began a major campaign to halt commercial development in Banff National Park. It argued that the activities which were creating the problems in Banff National Park could spread to other parks unless they were stopped there. The campaign gained international attention. National television and radio programs and international publications did feature stories about Banff and the problems that development had caused. Parks Canada commissioned a public opinion survey which showed Canadians overwhelmingly wanted their parks to protect wilderness and wildlife for future generations and that Canadians were concerned about the level of development in Banff townsite. A comparative study of North American parks published in the U.S. described Banff as 'unique in its excess'[5]. Meanwhile, development interests responded to the conservation campaign by establishing their own Association for Mountain Parks Protection and Enjoyment and launched their own public campaign to protect their 'traditional' uses. The issue of development in Banff had become a major political problem.

5 How Decision Makers Responded to the Issues Facing Banff

In response to growing public pressure, in the spring of 1994 the Minister of Canadian Heritage, who is responsible for national parks, announced a partial moratorium on development in Banff National Park. He also appointed a Task Force and commissioned it to do a $2 million CAN Banff Bow Valley Study. However, pro-development forces succeeded in having a number of proposals exempted from the moratorium. The Bow Valley Study Task Force consisted of five people with tourism, business, environmental management and biology experience as well as a secretariat of seconded Parks Canada employees. Their job was to report in 19 months with recommendations for an 'honourable solution' to the problems facing the Banff Bow Valley. The Bow Valley Study was independent of Parks Canada and charged to report directly to the Minister. This represented a loss of control for Parks Canada and tacitly recognized Parks Canada had failed to live up the challenge of successfully managing Banff National Park.

In an effort to bring opposing forces together, the Task Force established a Round Table of various sectoral interests including commercial operators, the transportation industry, environmental groups, and Banff residents concerned about social issues, government, outdoor recreationists and infrastructure-intensive tourists. The Round Table was asked to help craft a shared vision for the Banff Bow Valley, to set goals for ecological integrity and to grapple with what constitutes an appropriate use of a national park. The Round Table had a strong local bias with most participants living in the Park. The Task Force compensated for this somewhat by holding meetings in Vancouver, Toronto and Ottawa. The Task Force also commissioned a variety of commercial, historical, and environmental consultants to help in their work. It organized workshops for internationally recognized ecologists to consult on the problems facing Banff. The latter was necessary because an initial tactic of the pro-development lobby was to deny there were any environmental problems. Perhaps most important of all was that the Task Force and Round Table were bound by the National Parks Act and the new 1994 park policy. This ensured all discussions began from the constraints inherent in a national park setting.

6 Accomplishments

The multi-stakeholder Round Table process used by the Bow Valley Study served to bring competing interests closer together and provided valuable education to some sectors which had not previously paid much attention to ecology or parks policy. Backed by the partial moratorium, the study had enough importance to be relevant to reticent participants. It used consensus decision-making rather than win-lose decision-making. In the end, a common vision for the Park was drafted by the Round Table. The Round Table recognized national park policy, the need to maintain ecological integrity and that change may be necessary to achieve that goal. The Round Table set strategic goals. These

---

[5] Lowry, W.R. 1995. *The Capacity for Wonder*. Brookings Institute, Washington D.C. , p. 179

included maintenance of wolf and grizzly bear populations within Banff National Park as a source population which is a part of a viable and connected population of large carnivores within the Yellowstone to Yukon corridor. The Round Table also made recommendations for dealing with appropriate visitor uses in the park. It agreed that any proposed activity expressly prohibited by national parks policy should not merit further consideration by Parks Canada. This was no small achievement in a park where the policy had not always been followed.

The final report of the Task Force is due for presentation to the Minister of Canadian Heritage as this paper is being prepared so its contents are not known. However, to be true to the ecological integrity goals set, the Task Force's report must come to grips with restoring the degraded ecosystem in the Banff Bow Valley. Given Parks Canada's uneven record of enforcing its own policies in Banff, it will take political commitment to ensure meaningful remediation measures are carried out. The challenge of managing Banff National Park will have been successfully met only when the national park policy is fully enforced and the Park constitutes a source rather than a sink for wildlife populations in the Canadian Rockies. This will involve removal of some facilities and an end to continual expansions of existing facilities.

7 Future Research and Planning Needs

Much work remains to be done to undo the damage to wildlife movements created by fencing the highway through Banff National Park. No substantial work is planned to address that concern although in the new phase of road development presently underway, two 30 m wide animal overpasses built on top of artificially created tunnels are being installed to see if they facilitate large carnivore movements. The rail line through Banff is also a major source of wildlife mortality. No work or research has been done on how to mitigate its effect. The highway and railway issues are not only relevant to Banff. There are several other places in the Yellowstone to Yukon corridor where the Rockies are bisected by roads and railways. Banff National Park is an ideal place to develop effective mitigation that can be exported.

Ecological restoration work and monitoring are required to restore the health of all native species in the park. Better data on park wildlife populations and trends are required. Park visitors to Banff often do not have access to a national park message due to the number of commercial concerns competing for their attention. Parks Canada needs to plan and implement a program to build awareness of the purpose of national parks.

In some places, Banff is overused by people enjoying activities which are appropriate. An active plan is required to control use in order to maintain the quality of experience and environmental values. Finally, Banff needs a park management plan which deals with the park as a whole. The plan should be consistent with national park policy and consistently implemented. If these recommendations are followed, Banff, as Canada's flagship national park, will be an example to the world of how to restore a damaged national park and will perform a key role in the maintenance of biodiversity in western North America. If not, Banff National Park will play the ironic role of a major ecological problem in the Yellowstone to Yukon corridor and tarnish Canada's otherwise fine reputation in the national parks field.

8 References

Angus Reid Group Inc. 1993. *A Study of Canadian Attitudes Towards Canada's National Parks.* Calgary, Alberta.

Banff Bow Valley Study. 1995. *Ecological Outlook: Cumulative Effects and Futures Modelling, Workshop 1, Summary Report.* Banff, Alberta.

Banff Bow Valley Study. 1995. *State of the Banff Bow Valley.* Compiled by C. Pacas, Banff Bow Valley Study Secretariat, Banff, Alberta, David Bernard, ESSA Technologies Ltd., Vancouver, B.C., Nancy Marshall, Praxis Inc., Calgary, Alberta, and Jeffrey Green, Banff Bow Valley Study Task Force, Banff, Alberta.

Banff Bow Valley Study. 1996. *Banff Bow Valley Round Table Summary Report.* Banff, Alberta.

Bella, L. 1987. *Parks for Profit*, Harvest House Press, Montreal, Quebec.

Christian Science Monitor, International Daily Edition, September 7, 1994. *A Natural Wonder Under Seige*, Boston, Mass.

Dearden, P. and Rollins, R. eds. 1993. *Parks and Protected Areas in Canada: Planning and Management*. Oxford University Press, Toronto, Ontario.

Globe and Mail, December 24, 1995. *Banff's Outlook Not a Pretty Picture*, Toronto, Ontario.

Heuer, K., 1994. *Large Carnivore Movements Around the Town of Banff* Progress Report, unpublished. Prepared for Parks Canada for submission to the Interjurisdictional Wildlife Corridor Task Force, Ecology Base Research, Banff, Alberta.

Hildebrandt, W., 1995. *An Historical Analysis of Parks Canada and Banff National Park*, Banff Bow Valley Study, Banff, Alberta.

Locke, H, 1994. *Preserving the Wild Heart of North America: The Wildlands Project and the Yellowstone to Yukon Biodiversity Strategy*, Borealis Magazine. Spring 1994 issue, Canadian Parks and Wilderness Society, Toronto, Ontario.

Lothian, F. 1976-1981. *A History of Canada's National Parks*, Volumes 1-4, Supply and Services, Ottawa, Ontario.

Lowry, W.R., 1995. *The Capacity for Wonder*, Brookings Institute, Washington, D.C.

Marty, S., 1984. *A Grand and Fabulous Notion; The First Century of Canada's National Parks*, Supply and Services, Ottawa, Ontario.

Noss, R.F. 1992. *The Wildlands Project Land Conservation Strategy*, Wild Earth Special Issue, Cenozoic Society, Vermont.

Noss, R.F. 1995. *Maintaining Ecological Integrity in Representative Reserve Networks*, World Wildlife Fund Canada/World Wildlife Fund, Toronto, Ontario.

Paquet, P. and Hackman, H. 1995. *Carnivore Conservation Strategy for Rocky Mountains*, World Wildlife Fund Canada, Toronto, Ontario.

Parks Canada, 1979. *Parks Canada Policy*, Ottawa, Ontario.

Parks Canada, 1994. *Guiding Principles and Operational Policies*, Ottawa, Ontario.

Parks Canada, 1995. *Initial Assessment of Proposed Improvements to the TransCanada Highway in Banff National Park Phase IIIa*, 1995, Calgary, Alberta.

The Sunday Times, April 30, 1995. *Into the West*, London, UK.

White, C. and Otton J., 1993. *Research Links: Montane Ecoregion: Banff's Key Ecosystem Management Challenge*, Research Links, Canadian Parks Service Western Region, Spring 1993, Calgary, Alberta.

White, C.A., Hurd, T., Gibeau, M.L., Pengelly, I.R., Pacas, C., 1993. *Showdown with the Evil Seven: Banff National Park Ecosystem Group Focus (1994-1998)*, Research Links, Parks Canada Western Region, Winter 1993 issue, Calgary, Alberta.

White, C., Gilbride, D., Scott-Brown, M.S., and Stewart C., 1995. *Atlas of the Central Rockies Ecosystem*, Komex International, Calgary, Alberta.

The Danube Challenge: Protecting and Restoring a Living River

Philip Weller[1]

[1] WWF International, Green Danube Programme Coordinator
Ottakringer Strasse, 114-116
A-1162 Wien, Postfach 1, Austria

Abstract. The 2,840 km Danube River is one of the major unifying ecological features of Central and Eastern Europe. The river begins in the Black forest of Germany and eventually winds its way to its discharge in the Black Sea through the Danube Delta. Historically human activity (channelization, dams, land-use changes, and pollution) has damaged the Danube Basin ecosystem. Of particular concern is the dramatic loss of floodplain forests and wetlands that has occurred within the basin. Since 1992, the World Wide Fund for Nature (WWF) has been carrying out a major programme to address the problems associated with this loss of habitat. The WWF Green Danube Programme aims to protect the remaining natural areas, to preserve biodiversity, restore damaged habitat and thereby reduce everyday pollution. Important success has been achieved in this effort and is described in this paper. Demonstration projects have begun to reshape thinking towards coordinated wetland and floodplain protection in the Danube Region.

Key Words. Danube River and Delta, Danube basin ecosystem, floodplain and wetland restoration, biodiversity, parks and protected areas, World Wide Fund for Nature, Green Danube programme

1 Introduction

From its beginnings in the Black Forest to the impressive Danube Delta in Romania and Ukraine, the Danube river is 2,840 km long and traverses nine European countries. It is Europe's greatest river - a river that binds together a multitude of different cultures and peoples. It is also a lifeline that connects manifold ecosystems and natural areas. It combines diverse areas such as the Alps, the Puszta, and the Delta. An amazing diversity of plants and animals live in the region including such rare and seriously threatened species as the White-tailed Eagle, the Black Stork, the Sturgeon and the Dalmation Pelican.

Through channelization, building of dams and other landscape changes - including draining of wetlands - the Danube as a living river has been dramatically damaged. In response to these problems, the World Wide Fund for Nature (WWF) has since 1992 operated a programme to ensure protection, conservation and restoration of natural habitat along the Danube and its tributaries. The WWF Green Danube programme is a contribution to maintaining the ecological health of this important river through demonstration projects that increase the conservation of wetland habitat or bring about rehabilitation of wetland functions and values. The paper that follows presents information on this programme, on the ecological condition of the Danube, and in conclusion poses various questions related to research that needs to be done in relation to this topic.

2 The Ecological Character of the Danube River

The Danube river itself flows 2,840 km from the Black Forest of Germany to its outflow in the Black Sea. Nine countries have territory on the Danube and an additional eight are within the basin. The entire basin covers 817,000 km^2 (Figure 1). Thousands of tributaries feed the Danube including a number of large and significant rivers, such as the Inn, Morava, Sava, Tisza and the Yantra.

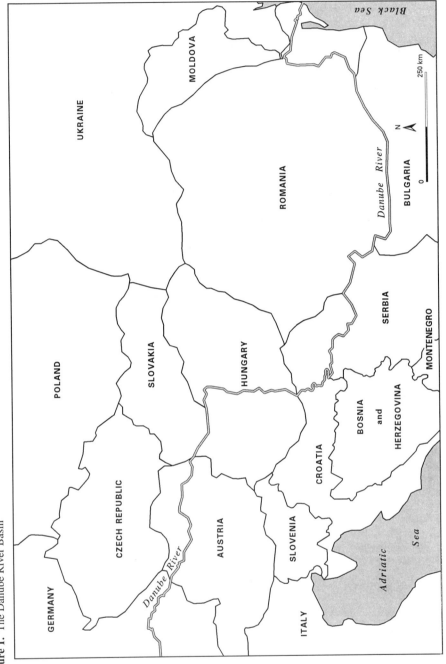

Figure 1. The Danube River Basin

The source of the Danube and its tributaries is Central Europe's major mountains - the Alps, Tatras and Carpathians. The average discharge upstream from the Delta is 6,500 m/s^2. Through the first 1,000 km the Danube has the character of a mountain river but then begins to flow more slowly across the Hungarian plain before winding its way through the Danube Delta into the Black Sea. See Bayerische Akademie Für Naturschutz und Landschaftsplege and WWF (1991) for details on character of the Danube.

The human population of the basin is large - approximately 80 million. The river and the basin ecosystem have greatly influenced and affected human history and in turn, the Danube region has been intensively altered by human settlement and activity. Although the river has been an important factor in supporting human development, the river as has been severely damaged. The river as a living system has suffered the consequences of:

1) Channelization and straightening,
2) Building of dams,
3) Land-use alterations in the watershed (draining of wetlands, reduction of forest etc), and
4) Discharge of pollutants from industry and human settlement.

3 Changes to the Danube Ecosystem

All of the activities mentioned in the previous section have significantly reduced the biodiversity and natural heritage values of the river and the basin. In this section we will look at these factors in more detail.

3.1 Channelization

Of all the factors which have reduced the natural heritage in the Danube probably none is more significant than channelization. In a short one hundred to one hundred and fifty year period the Danube and tributaries have lost the major portion of their tributaries and backwaters. Prior to this period the Danube and its tributaries were composed of an extensive network of channels which were continually being altered in response to changing water levels. Beginning in the early 1800's, however, flood protection works and channelization for shipping began to destroy the once diverse network of vein-like side channels and tributaries, particularly in the middle and lower portions of the river. Instead a straight and deeper main channel was created.

Along the Tisza river in Hungary, for example, over 2,590,000 ha of land were previously flooded by the continually changing water levels. However, during the last century, regulation of the river and its confinement in a straight channel has dramatically reduced this former floodplain habitat. Today only 100,000 ha of active floodplain remain (IUCN 1995). A similar situation exists in other areas. In the Regelsbrunner Au east of Vienna, for example, the side channel system in 1914 contained over 190 ha of water bodies. Today that has been reduced by over half to 80 ha (WWF Austria 1996). Confinement of the river in a narrow deep channel has caused continual deepening of the channel through scouring of faster moving water. This is in part the result of the incomplete replenishment of stones blocked from their natural movement by dams. The deepening of the channel has further reduced the water which reaches the side channels of the river.

3.2 Dams

Between 1950 and 1980 a total of 69 dams, each with a volume of at least 1 million m^3 were constructed in the Danube (Haskoning 1994). Their total volume is 7.3 billion m^3 which is equivalent to the total flow of the Danube into the Black Sea during about 2 weeks. In the first 1000 km of the Danube there are 58 dams. As a result only 4 % of the total stretch of this portion of the river has somewhat natural floodplain forests. The major portion of this area, located east of Vienna, was also planned for damming but a campaign of opposition forced a stop to construction in 1984. Another natural floodplain habitat along the border between Slovakia and Hungary, was however, recently destroyed by the Gabcikovo dam scheme. The Gacikovo power plant and its 30 km channel downstream from Bratislava went on line in 1992 and as a consequence diverted over 80% of the water from the Danube into the power canal. One of the largest remaining floodplain forests in the Danube along this stretch of river is therefore slowly dying as a result of a lowering of the water table and reduction of the frequency of flooding (Zinke 1996) .

An important consequence of construction of dams has been a reduction in the load of suspended solids that are reaching the Black Sea. Over the last century the yearly load of suspended solids has been significantly reduced. In the period 1931-1964 the total suspended solids at Izmail in the Ukraine were 53 million tons per year. In the period 1965-1970 the amount was 42 million tons and in 1971-1984 38 million tons. From 1988 to 1991 the suspended solids were only 12 million tons yearly (Haskoning 1994). This reduction has slowed the rate of land formation in the Danube Delta.

Another of the major consequences of the building of dams has been the blockage of the spawning runs of migratory fish. The sturgeon in particular has suffered population declines as a result of dam building. Of the five species of Sturgeon which live in the Danube all of the populations are now seriously threatened with extirpation.

3.3 Land Use Changes

In addition to the factors mentioned above, conversion of wet meadows, grasslands and forest land to agriculture, cutting of forests and changes in forest species composition, and building of towns and villages, have all affected the natural heritage of the Danube ecosystem. Some of these changes are clearly negative for natural heritage values, while others, such as the grazing of Puszta, have been positive for nature and wildlife. While there has been no overall assessment of European landscape changes, or comprehensive assessments of the consequences of these changes, it is clear that some of the land use trends are worrisome in the context of the Danube ecosystem health. In particular the loss of floodplain habitat (forest and meadows etc) is of significant concern. In Bulgaria, for example, only 10% of the original Danube wetland areas remain. In other areas of the Danube the picture is similar.

Along the Morava River in Austria the major portions of flooded meadows have been converted to agriculture and as a result a unique habitat is being lost. In the Danube Delta the grand schemes of the former dictator brought about the damming and conversion to agriculture of about eight per cent of the total area of the highly unique Danube Delta. Fortunately plans to convert up to forty per cent of the Delta to agriculture were stopped by the dramatic political changes. And although throughout Europe, there is currently an increase in forest cover, the quality of the forests is vastly different than it was in previous centuries. One of the major changes that has taken place in the Danube Basin is the loss of natural floodplain forests. As a report on the Danube by the Cousteau Society (1993) notes:

'The damage done by modern silviculture, which tends to simplify the forest structure by eliminating undergrowth, and whereby foreign varieties of trees are introduced and clear felling is practiced over vast surface areas, has greatly reduced the surface area of natural forests. ... The naturally structured forest's decreasing surface area (99% since 1800) considerably alters the overall functioning of the floodplain ecosystem. Forests older than 250 years are extremely rare in Europe'.

As the foregoing report indicates, planting of fast growing foreign trees (hybrid poplar) along many floodplain areas has been an important change in the Danube ecosystem. Although productive from the perspective of foresters, these areas are biologically impoverished. The loss of the quantity and quality of floodplain ecosystems is one of the most dramatic and significant alterations that has taken place in the European context in the last 150 years. As the Dobris Assessment of Europe's Environment concluded (EEA 1995), alluvial forests are 'one of the most endangered habitat types of Europe.'

3.4 Pollution

In addition to all of these changes the river itself has been polluted. Nutrient inputs in particular have altered the chemical and biological composition of the water of the river and the north west corner of the Black Sea. According to the major summary report of water quality in the Danube River prepared by the firm Haskoning (1994), 'The effect of the eutrophication on the northwestern shelf of the Black Sea is generally recognized as disastrous. Concentrations of nutrients are presently 4-5 times higher compared to 30 years ago'. The Danube is the major source of nutrients to the Black Sea and if the problems of the Black Sea are to be addressed, they must be addressed in the Danube.

In some areas of the Danube chemical pollution has also been significant although perhaps not as severe as in other areas of Europe. Data on micro pollutants in sediment samples show that, with the exception of lindane, the levels of chlorinated hydrocarbons and PCBs are lower than those in the more industrialized western rivers. According to Haskoning (1994), 'pollution of sediments (especially with metals), deposited in reservoirs before dams, will become a major problem in coming decades'.

4 Response to the Problems

The Danube River has been dramatically altered by human activity. Although the Danube has enabled many remarkable human developments, there are significant signs of ecological ill health and threats to the sustainable nature of human activity in the region. Fortunately, the dramatic political changes that affected the countries of Central and Eastern Europe at the end of the decade of the 1980s and early 1990s presented an expanded opportunity to begin cooperative actions to address these problems.

In the early 1980s the governments of the region had begun to be concerned about environmental problems. In 1985 they signed the Bucharest Declaration on Water Management of the Danube. In the Declaration they agreed to protect the Danube and its tributaries from pollution (Danube Environmental Program (DEP) 1994; 1995). In 1991, further initiatives were begun to collectively address Danube problems. The countries began the preparation of the Convention on Cooperation for the Protection and Sustainable Use of the Danube River. The Convention was eventually signed in Sofia on 29 June 1994. In addition to the development of the Danube Convention the countries of the region, together with international financial agencies and non-governmental organizations, initiated further specific response to the environmental problems of the Danube River. They established the Danube Environment Programme (DEP 1994).

The Danube Environment Programme has been developed to strengthen the operational basis for environmental management in the Danube River Basin. A work plan for the Programme was developed and has focused on identifying pollution hotspots and early warning systems for pollution events. These actions and activities are important in the longterm improvement of environmental conditions in the Danube Basin. However, it is the belief of World Wildlife Fund (WWF) that many of the problems and challenges for sustainable development are not limited to water quality alone.

WWF has recognized that the water quality concerns cannot be separated from land use considerations. In particular the protection and restoration of water quality requires a healthy and properly functioning floodplain ecosystem. In response to the threat to the Danube ecosystem and wetlands in particular, WWF International has developed and is carrying out a major programme to address the problems associated with this loss of habitat. The WWF Green Danube Programme aims to stop the increasing destruction of the Danube, protect the remaining natural areas, to preserve biodiversity, restore damaged habitat and thereby reduce everyday pollution.

5 The WWF Green Danube Programme

WWF projects in Germany, Austria, the Slovak and Czech Republics, Hungary, Bulgaria and in the Danube Delta (Ukraine and Romania) demonstrate model floodplain forest management, ensure the survival of endangered and threatened species of animals and plants, restore drained areas and demonstrate and promote sustainable economic development. The projects are carried out in cooperation with governments, WWF national and regional offices, and local partner organizations. The programme is composed of the following five major projects which are intended to serve as examples of restoration of wetland habitat.

5.1 The Mouth of the Isar

Threatened by water management schemes, WWF purchased land at the mouth of the Isar River to protect this natural floodplain forest and demonstrate model conservation management practices that can restore and sustain this area of national importance. The region, which is regularly flooded and rich with oxbow lakes, is a critical habitat for a number of threatened animals and birds including nesting Bluethroats, and birds of prey, such as Black Kite and Osprey. It is the goal of WWF to ensure long-term protection and sustainable management of this upper Danube floodplain forest.

5.2 Central Danube Multilateral Park

WWF is promoting transfrontier conservation for this large and relatively undisturbed area of floodplain forest and meadows along the Danube, Morava and Dyje rivers with partner organizations in the Czech Republic (Veronica), Slovakia (Daphne) and Hungary (Reflex). Hundreds of White and Black Storks, Beaver, and rare and unusual fishes and plants benefit from WWF reserves and projects. Major steps to long-term conservation have been achieved through floodplain restoration along the Morava and Dyje rivers and a successful 10 year WWF campaign to convince Austrian officials to open, in 1996, an 11,500 ha floodplain national park along the Danube east of Vienna.

5.3 Gemenc-Beda Protected Areas

This 24,000 ha area in southern Hungary is a focal point for a WWF project involving restoration, rehabilitation and conservation for the largest remaining floodplain forest along the Danube. White-tailed Eagles, Black Storks and numerous fish depend on this important forest wetland for survival. A WWF campaign has been undertaken to gain designation of the area as a national park, to carry-out the necessary rehabilitation of original river dynamics, and ensure adoption of WWF proposals for long-term protection. In addition, WWF is attempting to ensure the cooperation of Austrian and Hungarian authorities in management of the twin Danube Floodplain national parks.

5.4 Bulgarian Islands

In cooperation with the Bulgarian Committee of Forests and local NGOs, WWF is carrying out a floodplain forest restoration project for this complex of Danube River islands. The project involves eliminating non-native mono-culture plantation trees and re-establishing natural forest conditions. Despite past human disturbance, the islands, which range in size from a few hectares to up to 2,000 ha, contain important reserves of threatened biodiversity. In addition to small stands of old oaks on drier ground and white willow in the wet areas, the islands host large colonies of nesting herons and are a breeding site for endangered White-tailed Eagle. Cooperation between WWF, government officials, and local communities and organizations will ensure long-term preservation of this important floodplain habitat.

5.5 Danube Delta

The Danube Delta is the second largest wetland complex in Europe and a critical habitat and refuge for a number of rare and endangered plants and animals. Over 60 species of fish and 300 bird species depend on the Delta, including a major portion of the world's population of the endangered Dalmatian Pelican. Despite the significance of this unusual habitat, drainage and channelization threaten the long-term health of this natural jewel. Together with the Biosphere Reserve Authority, the Danube Delta Institute (Romania), the Dunaiskie Plawni Authority (Ukraine) and local NGOs, WWF aims to protect the natural values of the Delta and provide sustainable economic opportunities for local peoples. A central component of this effort has been the leadership of the WWF Floodplains Institute in carrying out restoration of wetlands unsuccessfully drained for agriculture. Over 2,000 ha on the Island Babina have already been restored.

6 Danube Government Protected Areas Strategies

Through the work of WWF and other NGOs (i.e. IUCN) there has been a recognition by Danube countries of the importance of floodplain habitat and the need to retain these areas. In particular the roles that these areas perform in water purifying, reducing flooding hazards, and preserving dwindling biodiversity, have been acknowledged. The governments of the region formally recognized the important role of wetlands in the Danube Strategic Action Plan (SAP), endorsed in Bucharest on 6 December 1994 by the Danube Environment Ministers and the EU Commissioner responsible for the Environment (DEP 1995). The SAP states that conservation, restoration and management of riverine habitat and biodiversity is important for maintaining the natural capital of the basin (its biodiversity) and to establish its natural purification and assimilative capacity.

The SAP identified 65 wetland areas along the Danube and its tributaries that are in need of protection and restoration. Although the SAP is only now beginning to be implemented, the recognition of the value of wetland habitat has had a major role in governmental decisions to protect three major areas of wetland habitat - the 500,000 ha Danube Delta Biosphere Reserve, the 11,500 ha Danube National Park in Austria, and the 50,000 ha Danube Drava National Park in Hungary. In addition, Bulgaria has declared some of the small islands of the Danube as nature reserves. These are all important steps in increasing the extent of natural floodplain habitat that is retained along the Danube.

Each of these decisions has involved long hard fights to convince decision makers of the value of these habitats. Economic considerations have been important in each case in swaying decisions. A WWF study released in 1995 (WWF 1995) concluded that the annual economic value of riverine wetlands ranged from 223 ECU to 1,300 ECU per ha. Although studies have demonstrated the economic value of wetland habitat, it is important that the strategies for management enhance local economic understanding and benefit from the protection of the areas.

As has been previously mentioned, the Danube Environment Programme has promoted restoration of areas of wetland previously damaged. A project to restore wetland forests and meadows along the Thaya and Morava Rivers is in preparation and will likely be funded by the European Union's PHARE programme. In addition, WWF has initiated a proposal for wetland restoration along the Tisza River, NGOs in Romania are promoting restoration of the Brailia Islands, and the Mur River between Slovenia and Austria is the topic of restoration and protection proposals. Encouraging in this context has been the remarkable restoration of Babina Island in the Danube Delta through a combined project of the Danube Delta Institute and WWF.

Clearly, important changes have taken place with respect to floodplain habitat in the Danube Basin and the efforts underway to restore and protect this dwindling aspect of the European environment have borne some positive fruit. Important questions remain, however, about how to maximize success in these efforts and what additional efforts are needed for protection of natural heritage values in the basin and in particular the implementation of sustainable development strategies.

7 Research Questions and Planning Needs

The efforts to protect and restore the remaining wetland habitat in the Danube Basin have been described in previous sections of this paper. Visible success has been achieved in a number of high profile projects. It is the hope of WWF that these projects will serve as models for restoration and protection in other less well known areas. Questions remain, however, about the extent to which restoration of wetlands in the Danube basin can be achieved and how much restoration and protection of this habitat is necessary to maintain ecological health in the system?

Restoration of wetland habitat logically has important benefits for the area where the project occurs. One of the key issues associated with the floodplain restoration projects in the Danube, however, is that the major impacts of this effort will often be most significant far from the site of the project. The Black Sea environmental problems are clearly related to Danube river concerns. The question arises as to whether there is a way to measure the impact of restoration on the entire system? Funding decisions for environmental improvement in the Danube and Black Sea will in part be based on having better answers to this question. The Global Environment Facility recently decided to support an additional one year information collection phase before committing upwards of $12 million to address priority Danube environmental problems.

The efforts of WWF have been concentrated on protecting and restoring riverine wetland habitat along the Danube and its tributaries, but what is the role of protected areas outside of the river floodplain on the ecological health of the river ecosystem and how do these areas interact? Importantly as well, how can small protected areas be built into a protected areas system that crosses so many political jurisdictions?

At present there exists no comprehensive watershed planning in the Danube region. Through the Danube Environment Programme and the signing of the Danube River Convention there have been improvements in cooperation among Danube states in addressing water quality related issues. There has, however, been little recognition of the need to coordinate and develop watershed planning that involves systems of protected areas. This challenge remains and is worthy of greater attention.

8 References

Bayerische Akademie Für Naturschutz und Landschaftsplege and WWF. 1991. *Conservation and Development of European Floodplains*. Bayerische Akademie Für Naturschutz und Landschaftsplege, Germany.

Equipe Cousteau. 1993. *The Danube...For Whom and For What?*. European Bank for Reconstruction and Development, London, UK.

Environmental Programme for the Danube River Basin (DEP). 1994. *Strategic Action Plan for the Danube River Basin 1995-2205*, Vienna, Austria.

Environmental Programme for the Danube River Basin (DEP). 1995. *Action for a Blue Danube*. Vienna Austria.

European Environmental Agency (EEA). 1995. *Europe's Environment: The Dobris Assessment*. (D. Stanners and P. Bourdeau, eds.). European Environmental Agency. Copenhagen, Denmark.

Haskoning (Royal Dutch Consulting Engineers and Architects). 1994. *Danube Integrated Environmental Study - Report Phase 1*. Copenhagen, Denmark.

IUCN (The World Conservation Union). 1995. *River Corridors in Hungary*. IUCN. Budapest,Hungary.

World Wide Fund for Nature Austria (WWF Austria). 1996. *Restoration of Network of Waterbodies, Oxbow Lake System Between Maria Ellend and Regelsbrunn*. Press Briefing Notes. Vienna, Austria.

World Wide Fund for Nature (WWF). 1995. *Economic Evaluation of Danube Floodplains*. WWF. Gland, Switzerland.

Zinke, A. 1996. Gabcikovo: Argumente der Kritiker. In *Der Donauverkehr: Möglichkeiten einer grenzüberschreitenden Zusammenarbeit*. Südosteuropa Aktuell 21. Südostereupa Gesellschaft. Munich, Germany.

Exploratory Planning for a Proposed National Marine Conservation Area in Northeast Newfoundland

Paul A. Macnab[1]

[1] Terra Nova National Park,
Glovertown, Newfoundland, Canada

Abstract. A representative marine area along the fjorded coast of Northeast Newfoundland is being considered for designation as a National Marine Conservation Area. Early investigations have highlighted the lack of available information on spawning areas and fishing activities. To complement scientific studies, Parks Canada is working with area fishers to document local knowledge. Regional dependence on the traditional fishery dictates a planning approach that builds on community management practices. Research is described that will help to further the establishment process.

Keywords. Newfoundland, marine protected areas, marine resource management, sea use planning, sustainable fisheries, local knowledge, public participation, community planning and management, collaborative planning

1 Introduction

Parks Canada is exploring the possibility of establishing a National Marine Conservation Area in the waters of Northeast Newfoundland. As part of this investigation, documented marine resource information has been assessed against anticipated planning needs. Informal meetings have also been initiated with area fishers to discuss local priorities for conservation and approaches to collaborative planning. This case study begins with a description of the National Marine Conservation Area Program and a sketch of the Newfoundland Shelf Marine Region. Next, information requirements and data gaps are considered. The importance of community involvement and the potential contributions of local knowledge are then reviewed. Finally, the paper examines research priorities.

2 National Marine Conservation Areas

Despite a long and proud history of National Park establishment, Canadian efforts to protect representative and sensitive marine areas lag far behind achievements in terrestrial conservation (Duffus and Dearden 1993; Graham et al. 1992). The National Marine Conservation Area (NMCA) designation has been developed to help redress this situation. NMCA's are intended to 'protect and conserve for all time national marine areas of Canadian significance that are representative of the country's ocean environments and the Great Lakes, and to encourage public understanding, appreciation and enjoyment of this marine heritage so as to leave it unimpaired for future generations' (Parks Canada, 1994 p.49).

From its antecedent, the National Marine Park concept (Mondor 1992), Canada's NMCA designation has evolved along with the international marine protected area (MPA) experience to incorporate sustainable resource use:

'National Marine Conservation Areas are marine areas managed for sustainable use and containing smaller zones of high protection. They include the seabed, its subsoil and overlying water column and may encompass wetlands, river estuaries, islands and other coastal lands. They are owned and managed by the Government of Canada. While activities such as undersea mining, oil and gas exploration and extraction would not be permitted within the boundaries of NMCA's, most traditional fishing activities, managed on a sustainable basis, would continue'. (Parks Canada 1995, p.8)

National Marine Conservation Areas are meant to:

- Represent the diversity of Canada's marine ecosystems
- Maintain marine ecological processes and life support systems
- Preserve biodiversity
- Serve as models of sustainable utilization of species and ecosystems
- Facilitate and encourage marine research and ecological monitoring
- Protect depleted, vulnerable, threatened, or endangered species and populations and preserve habitats considered critical to the survival of these species
- Protect and maintain areas critical to the lifecycles of economically important species
- Provide interpretation for the purposes of conservation, education and tourism

NMCA's are managed in partnership with other government departments and local stakeholders. While existing agencies continue their roles in marine management (e.g. Fisheries and Oceans in capture fisheries and Transportation in marine shipping), Canadians are encouraged to become stewards of their marine heritage so as to ensure long term sustainability and the preservation of ecological integrity.

Zoning instruments are the principal mechanisms used in NMCA's for implementing conservation measures. The main purpose of zoning is 'to define and map the different levels of protection and use that will occur in the marine conservation area and to separate potentially conflicting human activities' (Parks Canada 1994, p.53). In order to harmonize different activities in an NMCA, Parks Canada utilizes the flexibility provided by vertical and temporal zoning[1] and three levels of protection: Zone I, for strict preservation; Zone II, for non-extractive uses; and Zone III, for sustainable harvesting of renewable resources, navigation and aquaculture. These categories are offered as guidelines only; alternative zoning schemes may be recommended during the assessment of candidate NMCA's.

Although the NMCA establishment process has a fair amount of flexibility, there is a likely sequence of five key steps: identifying representative marine areas; selecting a potential national marine conservation area; assessing marine conservation area feasibility; negotiating a marine conservation area agreement; and establishing a new national marine conservation area in legislation. Parks Canada has divided the country into 29 marine regions, each of which is to be represented by an NMCA. Three sites are currently designated: Fathom Five National Marine Park in Ontario; Saguenay-St. Lawrence Marine Park in Quebec; and Gwaii Haanas National Marine Conservation Area in British Columbia.

2.1 Newfoundland Shelf Marine Region

LeDrew, Fudge and Associates (1990) completed the first step in the establishment process for the Newfoundland Shelf Marine Region (Figure 1) by identifying three representative marine regions. Parks Canada then convened an experts workshop to provide input for the selection of a potential NMCA (Mercier 1995). Consensus at the experts session helped Parks Canada select the landward

---

[1] Vertical zoning might regulate use at different levels in the water column (e.g., bottom fisheries and surface recreation). Temporal zoning could restrict activities during specific times (e.g., spawning seasons).

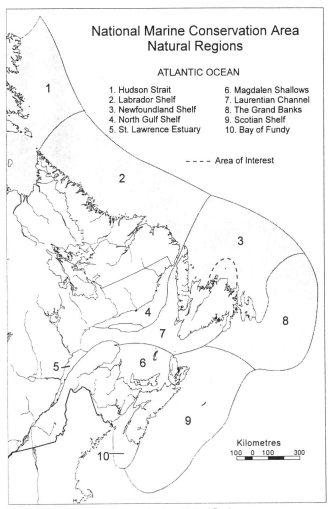

Figure 1. National Marine Conservation Area Natural Regions

extent of an NMCA study area: Bonavista Bay out to Funk Island, and the eastern portions of Notre Dame Bay (Figure 2). In terms of offshore extent, no outer boundary has been defined, but there is optimism that an NMCA, in combination with other federal programs, might extend protection, and thus representation, to the edge of the continental shelf. Parks Canada staff working from Terra Nova National Park since 1993 have initiated discussions with other government agencies and local user groups to promote and discuss the NMCA concept. These preparatory outings have explored the interest in proceeding to the formal launch of an NMCA feasibility assessment, the third step in establishing an NMCA.

Within the NMCA study area, there already exist a number of statutory protected areas with coastal and marine components. Terra Nova National Park, established in 1959, protects 165 kilometres of shoreline along Inner Bonavista Bay. Six small Provincial Parks in the area contain coastal segments while several Provincial and National Historic Sites preserve elements of the region's cultural heritage. The Terra Nova Bird Sanctuary, established by the Canadian Wildlife Service, protects important marine staging grounds for migratory waterfowl in two embayments adjacent to the National Park. Funk Island, once a major colony for the extinct Great Auk, has been designated a Seabird Sanctuary by the Province of Newfoundland to protect nesting gannets and a large breeding population of common murres. Finally, the federal Department of Fisheries and Oceans has implemented inner bay closures to net fisheries throughout the study area in an effort to avoid entrapment of returning salmon. Such protective measures, it should be mentioned, are entirely compatible with the proposed NMCA designation.

2.2 Bonavista - Notre Dame - Funk Island Study Area

The candidate NMCA encompasses a wide range of coastal and marine environments including banks and deep troughs, exposed shorelines, numerous archipelagos and sheltered fjords. The Labrador Current generally supports Arctic benthos in the region, however, the inner reaches and areas of sheltered water are noted for temperate species and a higher diversity. The area is home to a wide variety of pelagic, demersal, crustacean, and anadromous fish populations as well as strong communities of North Atlantic seabirds. Many of the same species that sustain birds and larger fish, particularly the diminutive capelin, also attract several varieties of whales and seals to the region.

The area's rich living resources have supported human occupation for over seven thousand years as evidenced by prehistoric Indian and Eskimo sites, and more recently, by Beothuk records. Although the region was only settled permanently by the English in the eighteenth and nineteenth centuries, five hundred years of European presence will be commemorated during 1997 in the Cape Bonavista area, the reputed 1497 landfall of John Cabot. The early communities were coastal and most, with the exception of a few towns built during periods of logging activity, were populated by fishing families who carried out their trade in small boats operating close to shore. Shipbuilders produced a range of vessels for the fishery including the sturdy schooners used by local crews to pursue the distant water fishery along the coast of Labrador.

Coastal communities' modern day dependence on the fishery has become painfully evident since 1992 when the Atlantic Groundfish Moratorium was announced.[2] With thousands out of work as a result of the closures, the strengthening of other sectors such as tourism and aquaculture has been promoted, but many assert that community survival is inextricably linked to a renewed fishery (National Round Table 1995). Fishing continues to be an important component of the regional economy with herring, capelin, lobster and crab remaining viable, but redirection of effort has placed heavy demands on these and other species prompting great concern amongst fishers and conservationists.

---

[2] For a review of the fisheries collapse, see Hutchings and Myers (1995).

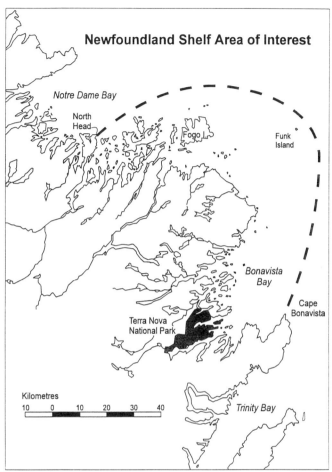

Figure 2. Parks Canada Area of Interest on the Newfoundland Shelf

3 Information Requirements

Unlike most of their Canadian National Park counterparts, NMCA's permit the sustainable use of renewable marine resources. Other uses of the sea, such as navigation and recreation, also continue within NMCA's. These realities, coupled with the complexities of marine ecosystems, dictate the central management principle for NMCA's, or indeed for marine resource management in general; that is, to manage activities rather than attempting to manage resources. In order to facilitate planning, management and decision making, it is therefore essential to supplement conventional forms of biophysical information with information on existing and potential sea uses (Kelleher and Kenchington 1992; Laffoley 1995).

Given the spatial nature of most coastal and marine resource information, mapping and electronic storage in geographic information systems (GIS) has been suggested as the best means to organize and compile different information sources (Ricketts 1992). Table 1 presents a list of potential data parameters for a preliminary GIS inventory of the Bonavista/Funk NMCA study area. The themes are presented as a set of minimum information requirements for a feasibility assessment; upon NMCA establishment, a greater range of information and baseline studies would be required to complete the preparation of a management plan.

A large number of the suggested data types have been uncovered at different locations and in a wide range of analog and digital formats. Owing to this diversity of information sources, and the early stage of the Parks Canada exploration, efforts have been directed towards communication with other agencies to initiate co-operative research and data partnerships. Current marine research activities in the NMCA study area include:

Agency	Activity
Canadian Hydrographic Service	Digital chart production
Natural Resources Canada	Shoreline classification
Department of Fisheries and Oceans	Coastal resource inventories
Environment Canada	Oil spill sensitivity mapping
Province of Newfoundland	Digital topographic mapping
Memorial University of Newfoundland	Fisher's ecological knowledge, juvenile cod, zooplankton

The entire range of information described above and in Table 1 will be important for further investigations of the proposed NMCA, but some categories will be more essential than others in helping to identify and define conservation priorities with local communities.

In the continuing review of available information, two crucial categories have been identified as lacking in documentation: spawning areas and inshore fishing grounds. Despite a long history of sampling and scientific investigation in offshore areas, spawning sites and inshore waters are only beginning to be studied and understood (e.g., Rose 1993). As for fishing activities in the candidate NMCA, little information has been documented at the scale of inshore fishing grounds, an appropriate level for collaborative planning with local communities. Head (1976) describes the paucity of information in this regard:

'While inshore, the various cod populations become further concentrated and it is these concentrations that become the fishing grounds of Newfoundlanders. Although these grounds are well known to the fishermen, even today they are little known to the scientific observer. In very few places in Newfoundland have they been identified and mapped'.
(Head 1976, p.21)

These data gaps have been targeted as a research priority for Parks Canada staff. Information of this nature will be required if the agency is to move from NMCA concepts towards conservation measures on the water.

Table 1. Potential Data Parameters

Base layers	Fishing grounds	Pollution sources
coastline	lobster	sewage outfalls
soundings	cod trap berths	ocean dumpsites
bathymetry	jigging grounds	fish plants
contours	herring seine/fixed gear	industry
drainage	offshore otter trawl	agriculture
Administration	hook and line	saw mills
baseline	sea urchins	pulp and paper
territorial seas	salmon berths	Minerals and hydrocarbons
inland waters	crab fisheries	active wells
200-mile limit	lumpfish	capped wells
fisheries divisions	capelin traps	placer mining
shellfish closures	winter flounder	Recreation
net fisheries closures	turbot	dive sites
protected areas	Oceanography	sailing
seabird hunting zones	currents	anchorage
military zones	tides	sea kayaking
navigation channels	salinity	fishing
mineral leases	chemistry	hunting
aquaculture sites	temperature	onshore facilities
Infrastructure	ice cover	tour boats
wharves	Birds	birding
navigation aids	tern rookeries	Shoretype
underwater cables	eagle nests	geology
pipelines	eider duck concentrations	intertidal communities
roads	reserves and sanctuaries	sensitivity ratings
buildings	Spawning Areas	sea caves
shoreline development	cod	arches
seawalls	herring	Bottom type
breakwaters	capelin beaches	geology
Cultural features	lumpfish	geomorphology
shipwrecks	turbot	benthos
archaeological sites		

4 Local Involvement

A growing body of literature attests to the importance of having local communities play an active role in protected areas planning and management. In the context of marine protected areas, the imperative for meaningful participation can not be overstated; only with user input and community support is success likely (Brunckhorst 1994; Kelleher et al 1995; Wells and White 1995).

Two examples contrast the success of exclusive and inclusive approaches to MPA establishment in Canada. In the terminated West Isles Marine Park proposal for New Brunswick, the public was invited to comment on the project only after 7 years of planning (Butler 1994). Confusion over the continuance of commercial fisheries in the proposed park had meanwhile built strong opposition to the concept (Ricketts 1988). In a marked contrast, the Saguenay-St Lawrence Marine Park was much quicker in collecting the views and opinions of area residents (Dionne 1995). The circumstances were also quite different; owing to strong local voices and pressing environmental concerns, namely, the threatened beluga whale population of the St. Lawrence River, there was already significant momentum in conservation. Clearly, the early involvement of the community hastened local acceptance of the Saguenay proposal.

Although mechanisms have yet to be put in place to formalize local involvement in the NMCA proposed for Newfoundland, there has been an informal exchange of ideas between planning staff and area fishers. Local conservation priorities, as voiced in meetings with Inshore Fishermen's Committees, turn out to be compatible with many of the NMCA objectives; fishers have long called for sustainable resource utilization and the protection of spawning fish, juveniles and supporting habitats. Considerable interest has also been expressed in the potential for spillover effects and larval disbursement from closed areas for lobster.[3] The challenge for Parks Canada will be to implement marine protected area concepts through community-directed measures.

4.1 Local Knowledge

The rich knowledge base of traditional resource users has been recognized as an important complement to scientific modes of inquiry for environmental management and protected areas planning (e.g. Harmon 1994; Inglis 1993; Johnson 1992; Sadler and Boothroyd 1994). Mailhot's (1994, p.11) brief definition encompasses the basic idea of local knowledge[4]: 'the sum of the data and ideas acquired by a human group on its environment as a result of the group's use and occupation of a region over many generations'. The complexity of marine ecosystems and gaps in scientific understanding often necessitate the use of local knowledge in approaches to marine conservation (Agardy 1995; Kenchington 1990; Norse 1993). Initially recognized for its potential contribution to MPA planning and management in Pacific tropical realms (e.g. Johannes 1984), local marine knowledge is beginning to attract strong research proponents in Canada (e.g., Graham and Payne 1990; Graham et al 1992; Neis 1995; Younger et al. 1996)

Parks Canada staff and fishers are working together to document local knowledge in the NMCA study area. Using semi-structured interviews and nautical charts, several types of information are being captured by and with inshore fishers: local toponyms and taxonomies; spatial/temporal harvesting patterns; species distribution; spawning locations (Potter 1996); community-based management practices; suggestions for conservation; and additional features deemed important by the participants. Visualization of this knowledge engenders local pride and is an important step towards resource stewardship; it also serves as a catalyst for discussions of site-specific conservation priorities and co-management with outside agencies. Newfoundlanders have long-regulated fishing space within their communities (Anderson and Stiles 1973; Martin 1979; Matthews 1993). Incorporating such local management practices into MPA's through user-related zoning schemes is broadly supported by international experience (Lafolley 1995; Wells and White 1995).

---

[3] For a discussion of the stock enhancing benefits of marine harvest refugia, see Dugan and Davis (1993), FRCC (1995) and Rowley (1994).

[4] The term local knowledge has been adopted here after Ruddle (1994). Other commonly used terms include traditional ecological knowledge and indigenous knowledge.

5 Research Priorities

In order to work towards the establishment of a National Marine Conservation Area that balances conservation and sustainable resource use in Northeast Newfoundland, certain key areas require enhanced attention from researchers and planners.

Spawning areas and critical habitats. Protection for these components of coastal and marine ecosystems will require a greater knowledge of fish behaviour and associated biophysical preferences. Continuing identification of possible locations by fishers should be paralleled by scientific investigation.

Marine harvest refugia. The application of closed areas for sedentary species (e.g. Atlantic lobster) remains largely untested in Northern regions. There is a clear need for scientific studies to help resolve questions of scale and optimal placement. Voluntary closures could provide an opportunity for study and experimentation.

Data integration. There is a need to facilitate the compilation and synthesis of existing and emerging marine resource information so as to make it useful for planning, management, and decision making. Geographic information systems hold the most promise for the storage, analysis and display of spatial marine data.

Participation. Appropriate mechanisms for meaningful public participation need to be developed for an NMCA. Parks Canada should uncover and promote existing marine stewardship activities and examine successful models of fisheries co-management.

Fishing patterns. The way fishers perceive, use, delimit and regulate marine environments must be understood and incorporated into planning and management. This will involve documenting and mapping the lived-in experience of coastal ocean space.

Local Knowledge and Science. Systematic capture and geo-referencing will make local knowledge more compatible with technical approaches to marine resource management. What is most required, however, is the two-way exchange of information among fishers, scientists and managers. As a starting point for constructive dialogue, this will help to build positive working relationships and common planning objectives. Ultimately, a National Marine Conservation Area in Northeast Newfoundland will need to achieve science-based conservation through measures managed in partnership by communities and government agencies.

6 Acknowledgements

The views expressed are those of the author; full responsibility is accepted for any errors or omissions. Helpful comments from Ted Potter, Marianne Ward, Francine Mercier and Suzan Dionne are greatly appreciated. Ian Joyce and Larry Nolan supplied base materials for the figures. Dr. Gordon Nelson, University of Waterloo, must be thanked for his enthusiasm and keen insight. Parks Canada and the Department of Fisheries and Oceans provided financial assistance and other forms of support for these investigations.

7 References

Agardy, T. (ed) 1995. *The Science of Conservation in the Coastal Zone: New insights on how to design, implement and monitor marine protected areas.* IUCN, Gland.

Anderson, R. and G. Stiles. 1973. Resource Management and Spatial Competition in Newfoundland Fishing: An Exploratory Essay. In *Seafarer and Community: Towards a Social Understanding of Seafaring,* P. Fricke (ed), pp. 44-66. Croom Helm, London.

Brunckhorst, D.J. (ed) 1994. Marine Protected Areas and Biosphere Reserves: Towards a New Paradigm. In *Proceedings of the 1st International Workshop on Marine and Coastal Protected Areas.* Canberra, Australia. August 1994.

Butler, M. 1994. When You Can't Build Fences: The West Isles Marine-Park Proposal. In *The Sea Has Many Voices: Oceans Policy for a Complex World,* C. Lamson (ed), pp. 125-141. McGill-Queen's University Press, Montreal.

Dionne, S. 1995. Creating the Saguenay Marine Park: A Case Study. In *Marine Protected Areas and Sustainable Fisheries.* N.L Shackell and J.H.M. Willison (eds), pp. 189-196. Science and the Management of Protected Areas Association, Wolfville, Nova Scotia.

Duffus, D.A. and P. Dearden. 1993. Marine Parks: The Canadian Experience. In *Parks and Protected Areas in Canada: Planning and Management.* P. Dearden and R. Rollins (eds), pp. 256-272. Oxford University Press, Toronto, Ontario.

Dugan, J. E. and G. E. Davis. 1993. Applications of Marine Refugia to Coastal Fisheries Management. *Canadian Journal of Fisheries and Aquatic Sciences* 50, pp. 2029-2042.

Fisheries Resource Conservation Council. 1995. *A Conservation Framework for Atlantic Lobster.* Report to the Minister of Fisheries and Oceans. Supply and Services, Ottawa, Ontario.

Graham, R. and R. J. Payne. 1990. Customary and Traditional Knowledge in Canadian National Park Planning and Management: A Process View. In *Social Science and Natural Resource Recreation Management,* J. Vining (ed), pp. 125-150. Westview Press, Boulder.

Graham, R., N. Stalport, D. Vanderzwagg, C. Lamson, M. Butler and D. Boyle. 1992. The Protection of Special Marine and Coastal Areas. In *Canadian Ocean Law and Policy,* D. Vanderswagg (ed), pp. 341-390. Butterworths, Toronto, Ontario.

Harmon, D. (ed) 1994. *Coordinating Research and Management to Enhance Protected Areas.* IUCN, Cambridge.

Head, C.G. 1976. *Eighteenth Century Newfoundland: A Geographer's Perspective.* McClelland and Stewart, Toronto, Ontario.

Hutchings, J. A. and R. A. Myers. 1995. The Biological Collapse of Atlantic Cod off Newfoundland and Labrador: An exploration of historical changes in exploitation, harvesting technology and management. In *The North Atlantic Fisheries: Successes, Failures and Challenges,* R. Arnason and L. Felt (eds), pp. 37-93. Institute of Island Studies, Charlottetown, PEI.

Inglis, J.T. (ed). 1993. *Traditional Ecological Knowledge: Concepts and Cases.* International Development Research Council, Ottawa, Ontario.

Johannes, R.E. 1984. Traditional Conservation Methods and Protected Marine Areas in Oceania. In *Proceedings of the World Congress on National Parks,* Bali, Indonesia, October 11-22, 1982, pp. 344-347.

Johnson, M. (ed) 1992. *Lore: Capturing Traditional Environmental Knowledge.* International Development Research Council, Ottawa, Ontario.

Kelleher, G., C. Bleakley and S. Wells. 1995. *A Global Representative System of Marine Protected Areas.* Volume 1. The Great Barrier Reef Marine Park Authority, The World Bank and The World Conservation Union. Gland, Switzerland.

Kelleher, G. and R. Kenchington. 1982. Australia's Great Barrier Reef Marine Park: Making Development Compatible with Conservation. *Ambio,* 11(5), pp. 262-267.

Kelleher, G. 1992. *Guidelines for Establishing Marine Protected Areas.* IUCN, Gland.

Kenchington, R. 1990. *Managing Marine Environments.* Taylor and Francis, New York.

Laffoley, D. 1995. Techniques for managing marine protected areas: zoning. In *Marine Protected Areas: principles and techniques for management,* S. Gubbay (ed), pp. 103-118. Chapman and Hall, London.

Ledrew, Fudge and Associates Ltd. 1990. *Identification of marine natural areas of Canadian significance in the South Labrador Shelf Marine Region. 2 Volumes.* Prepared for the Canadian Parks Service, Ottawa, Ontario.

Mailhot, J. 1994. *Traditional Ecological Knowledge: The Diversity of Knowledge Systems and Their Study.* Background Paper No. 4. Great Whale Public Review Support Office. Montreal, Quebec.

Martin, K. 1979. Play by the Rules or Don't Play at all: Space Division and Resource Allocation in a Rural Newfoundland Fishing Community. In *North Atlantic Maritime Cultures: Anthropological Essays on Changing Adaptations,* R. Anderson (ed), pp. 277-298. Mouton Publishers, The Hague.

Matthews, D.R. 1993. *Controlling Common Property: Regulating Canada's East Coast Fisheries.* University of Toronto Press. Toronto, Ontario.

Mercier, F. 1995. Report of a workshop to identify a potential national marine conservation area on the NE coast of Newfoundland. In *Marine Protected Areas and Sustainable Fisheries,* N.L Shackell and J.H.M. Willison (eds), pp. 240-248. Science and the Management of Protected Areas Association. Wolfville, Nova Scotia.

Mondor, C. 1992. Canada's National Marine Park Policy, Evolution and Implementation. In *Marine, Lake and Coastal Heritage,* Occasional Paper 15, R. Graham (ed), pp. 57-70. Heritage Resources Centre, University of Waterloo, Waterloo, Ontario.

National Round Table on the Environment and the Economy. 1995. *The Report of the Partnership for Sustainable Coastal Communities and Marine Ecosystems in Newfoundland and Labrador.* Ottawa, Ontario.

Neis, B. 1995. Fishers' Ecological Knowledge and Marine Protected Areas. In *Marine Protected Areas and Sustainable Fisheries,* N.L Shackell and J.H.M. Willison (eds), pp. 265-272. Science and the Management of Protected Areas Association. Wolfville, Nova Scotia.

Norse, E.A. (ed) 1993. *Global Marine Biological Diversity: a strategy for building conservation into decision making.* Island Press. Washington D.C.

Parks Canada. 1994. *Guiding Principles and Operational Policies.* Ministry of Supply and Services Canada. Ottawa, Ontario

Parks Canada. 1995. *Sea to Sea to Sea: Canada's National Marine Conservation Areas System Plan.* Ministry of Supply and Services Canada. Ottawa, Ontario.

Potter, A. 1996. *Identification of Inshore Spawning Areas: Potential Marine Protected Areas?* Unpublished Master's Project. Marine Affairs Program, Dalhousie University. Halifax, Nova Scotia.

Ricketts, P.J. 1988. Use conflicts in Canada's National Marine Park Policy. *Ocean and Shoreline Management,* 11, pp. 285-302.

Ricketts, P.J. 1992. Current Approaches in Geographic Information Systems for Coastal Management. *Marine Pollution Bulletin,* 25(1-4), pp. 82-87.

Rose, A.G. 1993. Cod spawning on a migration highway in the north-west Atlantic. *Nature,* 366 (6454), pp. 458-461.

Rowley, R.J. 1994. Marine reserves in fisheries management. *Aquatic Conservation,* 4 (3), pp. 233-254.

Ruddle, K. 1994. Local Knowledge in the Folk Management of Fisheries and Coastal Marine Environments. In *Folk Management in the World's Fisheries, Lessons for Modern Fisheries Management,* C.L. Dyer and J.R. McGoodwin (eds), pp.161-206. University of Colorado Press. Boulder.

Sadler, B. and P. Boothroyd. (eds) 1994. *Traditional Ecological Knowledge and Environmental Impact Assessment.* University of British Columbia. Vancouver, B.C.

Wells, S. and A.T. White. 1995. Involving the community. In *Marine Protected Areas: principles and techniques for management,* S. Gubbay (ed), pp. 61-84. Chapman and Hall. London, UK.

Younger, A., K. Healey, M. Sinclair, and E. Trippel. 1996. Fishers Knowledge of Spawning Locations on the Scotian Shelf. In *Report of the Second Workshop on Scotia-Fundy Groundfish Management,* D.L. Burke, R.N. O'Boyle, P. Partington and M. Sinclair (eds), pp. 147-156. Can. Tech. Rep. Fish. Aquat. Sci. 2100, Halifax, Nova Scotia.

Experience in Cross-Border Cooperation for National Park and Protected Areas in Central Europe

Zygmunt Denisiuk[1], Stepan Stoyko[2] and Jan Terray[3]

[1] Institute for Nature Conservation, Polish Academy of Sciences, Kraków, Poland
[2] Institute of Ecology, Ukrainian Academy of Sciences, Lviv, Ukraine
[3] Ustredie Statnej Ochrony Prirody, Slovakia

Abstract. The European continent has made itself famous in the world for its rapid economic development, particularly industrial development which has involved natural environment devastation. Yet Europe is also the place where the greatest number of national parks have been created (about 240 of a worldwide total of approximately 850). The division of Europe into two political and economic blocs lasted until 1990. The post-World War Two period has been characterized by unprecedented economic development in the West European countries. But also in these countries, achievements in nature conservation were greater, finding expression in the number of established national parks - nearly twice as many as in the countries of the Eastern Bloc. Remarkable achievements were noted in Sweden where 22 national parks were created up to 1990 and in Norway with 19 national parks. Central and Eastern European countries belonging to former communist bloc, were much less active in this respect. New and more effective methods of nature conservation have therefore recently been sought. One of the more important initiatives is transfrontier protected areas. Traditions in this field are poorly established, but the method is promising, notably in regard to large transfrontier national and landscape parks and associated biosphere reserves.

Keywords. National parks, protected areas, cross-border co-operation, transfrontier protected areas, Central and Eastern Europe

1 Introduction

Europe has made itself famous in the world for its rapid economic development, particularly with respect to industrial development which has led to devastation of the natural environment. However, it is also in Europe that the greatest number of national parks have been created, about 240 of 850 in the world.

The division of Europe into two political and economic blocs, until 1990, was marked by the rapid economic development of western countries. Achievements in nature conservation were greater in these countries and found expression in the number of established national parks - nearly twice as many as in the countries of the Eastern Bloc. Remarkable achievements were noted in Sweden where 22 national parks were created by 1990, as well as in Norway with 19 national parks. The position of Central and Eastern European countries belonging to the former communist bloc, was much less favourable in this respect. Particularly serious failures in nature conservation took place in the former German Democratic Republic (GDR) where not a single national park was created up to 1990. In the former Soviet Union, the first national parks were established 15 years ago, and in Romania the only national park (Reterat) was created more than 60 years ago. On the other hand, some countries did not lag behind Western Europe and are known for their achievements in nature conservation. Particularly worthy of notice are: Yugoslavia with 22 national parks by 1990 and Poland with 17 national parks.

2. Parks and Protected Areas in Central and Eastern Europe

Political, economic and social changes in Central and Eastern Europe following the communist period have contributed to the considerable activity in the nature conservation field. Over 25 new national parks and other protected areas have been created and many existing ones have been extended in this post-communist period. Especially favourable changes have taken place in Romania where shortly after the overthrow of Ceausescu's regime, 11 national park projects were at last legally approved, having waited for completion for many years. Similarly, in the former GDR, the new authorities approved five new national parks. In the 20 countries of Central and Eastern Europe there are now 95 national parks covering about 2,236,100 km^2 in total, about 0.34% of the eastern part of the continent. More than one third of all national parks were created in two countries: Poland (22 parks) and Rumania (12 parks).

The percentage of area protected in national parks varies among Central European countries from a fraction of 1 per cent to over 5% of the total area. The lowest index is found in Russia (only 0.03%) and the highest one, in Slovakia (over 5%). With Russia excluded, an index of protected area for the remaining countries increases to 0.98% of the total area. Poland with the index of 0.92% occupies quite a good position among Central European countries.

The greatest number of national parks in Central and Eastern Europe was created after the Second World War. Only 7 national parks date from the pre-war period, among them 4 Polish national parks (Białowieża NP and Pieniny NP from 1932, Babia Góra NP and Wielkopolski NP from 1933), 1 park in Rumania (Reterat, 1935) and 1 park in Ukraine (Carpathian NP created in Czarnohora in 1936). The youngest national parks include those established in the former GDR in 1990, almost all the parks approved in Rumania in the same year, 8 Polish national parks created in the period 1989-1996, 2 Slovakian parks, and 1 park in Ukraine.

Different legal regulations in the various countries of Central and Eastern Europe greatly influence their systems of nature conservation and their achievements in this field. This applies particularly to large protected areas. In the countries of Central Europe they are represented by national parks and landscape parks, while in Eastern Europe (Russia, Ukraine, Bielorussia) the main form of protection is large strictly managed reserves, so-called zapovedniki. They exist along with landscape parks and national parks which are considered to be categories of lesser importance.

In some countries, landscape parks belonging to category V in the classification of IUCN are considered as the most practical and effective method of protection of both landscape and wildlife. That is why landscape parks constitute a large fraction of the total area of many countries. New and more effective methods of nature conservation, coordinated with economic programmes are now being sought. One of the more important initiatives in this respect is protected areas in transfrontier regions belonging to two, or even three countries. Traditions in this field are still poorly established, but the approach is showing promise. Recent joint bilateral programmes of natural heritage conservation in transboundary regions have resulted in the creation of large protected areas. These are most often transfrontier national parks and landscape parks, and associated biosphere reserves.

In the cross-border areas of the countries of Central and Eastern Europe (Poland, Germany, Denmark, Czech Republic, Slovakia, Ukraine, Bielorussia, Austria, Hungary and Rumania), 14 national parks have been created through bilateral agreements. This includes 8 Polish parks situated at the borders of our eastern, southern and western neighbour countries. Twenty-eight borderland landscape parks, also were created. As many as 17 are Polish parks. The foregoing national and landscape parks include eight International Biosphere Reserves, of which five are Polish reserves situated in the borderland areas. Particularly worthy of notice is the tripartite Eastern Carpathians International Biosphere Reserve which includes protected areas at the junction of the Polish, Slovakian and Ukrainian borders (Figure 1 and Tables 1 and 2). Polish experience in borderland nature conservation is associated with old national parks such as the Białowieża National Park, Pieniny National Park, Babia Góra National Park and the Tatra National Park.

Figure 1. Cross-border National Parks and Protected Areas in Central and Eastern Europe

Table 1. National Parks in Poland

NATIONAL PARKS IN POLAND
(1-st August 1996)

Ord. No.	Name of the National P.	Being under protect.	Established in:	Aktual area (ha): general	Aktual area (ha): strict protect	Forest area ha	Forest area %
1.	Babia Gora N.P.*	1933	1955	1 734	1 061	1 585	91.4
2.	Białowieza N.P.*	1932	1948	5 317	4 747	4 904	92.2
3.	Bieszczady N.P.*	-	1973	27 064	18 536	22 746	84.0
4.	Drawa N.P.	-	1990	11 019	225	9 119	83.0
5.	Gorce N.P.	-	1981	6 750	2 968	6 406	94.9
6.	Kampinos N.P.	-	1959	35 486	4 303	26 979	76.0
7.	Karkonosze N.P.*	-	1959	5 579	1 717	3 766	67.7
8.	Ojcow N.P.	-	1956	1 592	338	1 307	82.1
9.	Pieniny N.P.	1932	1955	2 346	804	1 564	67.2
10.	Polesie N.P.	-	1990	9 648	428	4 131	42.8
11.	Roztocze N.P.	-	1974	7 811	808	7 333	93.9
12.	Slowinski N.P.*	-	1967	18 247	5 935	4 631	25.4
13.	Swietokrzyski N.P.	-	1950	7 626	1 741	5 590	73.3
14.	Tatra N.P.*	-	1955	21 164	11 514	15 046	71.1
15.	Wielkopolski N.P.	1933	1957	5 198	221	4 387	84.4
16.	Wolin N.P.	-	1960	10 937	162	4 422	40.4
17.	Wigry N.P.	-	1989	14 840	386	9 155	61.7
18.	Biebrza Valley N.P.	-	1993	59 223	2 569	15 444	26.1
19.	Stolowe Mts. N.P.	-	1993	6 280	48	5 606	89.2
20.	Magura N.P.	-	1994	19 962	-	17 731	88.8
21	Tuchola N.P.	-	1996	4 789	-	4 248	88.7
22	Narew Valley N.P.	-	1996	7 350	-	269	0.04
Joint area (ha)				287 602		176 369	
Average (ha)				13 073		8 017	63.0

* biosphere reserve

Table 2. Experience with cross-border cooperation for national parks and protected areas in Central Europe

I. Cross-border national parks and biosphere reserves

    A.   With Poland
- 1. Białowieża N.P. (Poland/Belorus)
- 2. Bieszczady N.P. (Poland/Ukraine/Slovakia)
- 3. Magura N.P. (Poland/Slovakia)
- 4. Pieniny N.P. (Poland/Slovakia)
- 5. Tatra N.P. (Poland/Slovakia)
- 6. Babia Góra N.P. (Poland/Slovakia)
- 7. Stołowe Mts. N.P. (Poland/Czech Republic)
- 8. Karkonosze N.P. (Poland/Czech Republic)

    B.   With other countries
- 9. Sachsiche Schweiz N.P. (Germany/Czech Republic)
- 10. Aggtelek N.P. (Hungary/Slovakia)
- 11. Schleswig-Hollsteinisches Wattenmeer N.P. (Germany/Denmark)
- 12. Slovensky Kras N.P. (Slovakia/Hungary)
- 13. Podyji N.P. (Czech Republic/Austria)
- 14. Sumava N.P. (Czech Republic/Germany)

II. Cross-border landscape parks

    A.   With Poland
- 1. Low Odra Valley L.P. (Poland/Germany)
- 2. Snieżnicki L.P. (Poland/Czech Republic)
- 3. Orawskie Mts. L.P. (Poland/Czech Republic)
- 4. Żywiec (Poland/Slovakia)
- 5. Poprad L.P. (Poland/Slovakia)
- 6. Jaśliska L.P. (Poland/Slovakia)
- 7. Cisna-Wetlina L.P. (Poland/Slovakia)
- 8. San River Valley L.P. (Poland/Ukraine)
- 9. Słonne Mts. (Poland/Ukraine)
- 10. Przemyśl Upland L.P. (Poland/Ukraine)
- 11. Southern Roztocze L.P. (Poland/Ukraine)
- 12. Latorica L.P. (Poland/Ukraine)
- 13. Jizerske Hory L.P. (Czech Republic)
- 14. Orlicke Hory L.P. (Czech Republic)
- 15. Broumovsko L.P. (Czech Republic)
- 16. Kysuce L.P. (Slovakia/Czech Republic, Slovakia/Poland)
- 17. Horna Orava L.P. (Slovakia/Poland)

    B.   With other Countries
- 18. Labske Piskovce L.P. (Czech Republic/Germany)
- 19. Beskidy L.P. (Czech Republic/Slovakia)
- 20. Bile Karpaty L.P. (Czech Republic/Slovakia)
- 21. Sumava L.P. (Czech Republic/Austria)
- 22. Maramoresh L.P. (Ukraine/Rumania)
- 23. Gutin L.P. (Ukraine/Rumania)
- 24. Vihorlat L.P. (Ukraine/Rumania)
- 25. Tissian L.P. (Ukraine/Hungary)
- 26. Cerova Vrchovina L.P. (Slovakia/Austria)
- 27. Zachorie L.P. (Slovakia/Austria)
- 28. Trebon Basin L.P. (Czech Republic/Austria)

3. Opportunities and Prospects

Over the past ten to twenty years, the prospect emerged of creating Biosphere Reserves under the auspices of UNESCO MAB, which provide a new forumula for international co-operation. The system of Biosphere Reserves comprised national and landscape parks, including many transfrontier parks. Four Polish national parks, Bialowieża, Tatra, Babia Góra, and Karkonosze have now been included in the Biosphere Reserve network. The Bialowieża, Bieszczady, Tatra, and Karkonosze National Parks have the greatest experience in the field of international co-operation. In the near future new protected areas may be created in the borderlands. Opportunities arise through possible co-operation between Poland and Ukraine in the area protected as Szacki NP (Ukraine) and Poleski NP (Poland). A prospect has also emerged in the region of the lower Odra River where Germany and Poland are both interested in the protection of this borderland area.

Research is urgently needed on opportunities for borderland protected areas of various kinds. In this respect research is also required on an ecosystem framework for the creation, planning and management of these protected areas, including biosphere reserves. Evaluative research and assessment is also needed for the planning, management and institutional arrangements for these protected areas. Comparative study and assessment of these arrangements for existing borderland protected areas would be very useful.

Parks for Life: An Action Plan for the Protected Areas of Europe

Marija Zupancic-Vicar[1]

[1] Regional Vice-Chair, IUCN/CNPPA, Rodine 51,
4274 Zirovnica, Slovenia

Abstract. There is great diversity in Europe's natural heritage. However, the protected area coverage in Europe is very uneven and the protected area systems are rather weak by international standards. Most protected areas in Europe are under strong pressure and the management of many protected areas is not as strong as it needs to be to combat the powerful threats that these areas now face. Therefore Parks for Life - the first action plan for protected areas in Europe - was prepared, and its implementation is underway. It is a document intended to help governments and protected area managers to ensure an adequate, effective and well managed network of protected areas in Europe.

Key words. Action plan, protected areas, management, Europe

1 Introduction

The issue of conservation of nature and the environment has been discussed at numerous world and regional meetings and conferences in recent years (Rio 1992; Caracas 1992; Luzern 1993; Sofia 1995). The messages and appeals arising from these meetings and conferences are dramatic enough and they require us to choose ways of life and courses of development that respect the limits of nature and enable us to function within these limits.

We are becoming more and more conscious of the fact that the conservation of nature and the environment is not only an inalienable right and responsibility of nations and their states, but also a common responsibility, first of all in our relation to neighbouring countries, and then also in individual regions, as well as a common concern of all of humanity. In this respect, new links are being forged in Europe, e.g. in the area of Central and Eastern Europe, the Baltic, and the Mediterranean. In this sense one should also understand the endeavours of the IUCN (The World Conservation Union), WWF (The World Wide Fund for Nature), FNNPE (The Federation of Nature and National Parks of Europe) and BirdLife International to adopt Parks for Life, as the first action plan for protected areas in Europe. The plan is the result of a remarkable partnership between countries, agencies and individuals throughout Europe. It represents a consensus across Europe about what needs to be done.

2 Why are Protected Areas so Important?

Europe may be one of the smaller continents but it has remarkable diversity, not just of nature, but also of civilizations, cultures and languages, reflecting its long and complicated history. Europe is also a young continent, in that most of its present ecosystems, at least those north of the Alps, developed only after the retreat of the ice sheets some 10,000 years ago. Nevertheless, there is great diversity in Europe's natural heritage, varying from the Arctic tundra to the evergreen oak forests of the Mediterranean. Europe has spectacular mountains, outstanding wetlands and valuable forests. Its coastline is complex, with numerous inland seas and islands.

Europe's naturally diverse vegetation has however, been profoundly influenced by human activity. The changes over most of Europe began thousands of years ago. These long-established changes led to a varied and biologically diverse continent until only a generation or so ago. In the past 50 years, however, there has been a steady degradation in much of this rich landscape, for example in Northern

Europe. Mechanized agriculture has reduced the rich patchwork quilt of woodlands, hedges and small fields to an agro-industrial prairie largely devoid of wildlife. In Southern Europe, the massive expansion of tourism is causing great damage to the fragile Mediterranean coast (McNeely, Harrison and Dingwall 1994).

Europe has had protected areas for centuries, for example as royal hunting reserves or as forest reserves. Since the first part of the 19th Century, small areas or even single features such as trees and rocks have been declared as protected. Yet the national park concept emerged later in Europe than in other parts of the world. Even so, Europe has a complex and extensive system estimated at up to 20,000 protected areas, of which about 2200 are on the UN List (McNeely, Harrison and Dingwall 1994). There are dramatic differences from one country to another. Of the IUCN Categories of protected area (IUCN, 1994) those mostly applied in Europe are: Category V: protected landscape/seascape; Category IV: habitat/species management area, and, Category II: national park.

In the protected areas field, Europe is perhaps best known internationally for its rich heritage of cultural landscapes - those areas where people and nature have lived together in harmony for centuries. Indeed, the idea of protected landscapes - IUCN management category V - started in Europe and according to the data of the World Conservation Monitoring Centre, nearly 60% - by number - and 24% - by area covered - of all protected landscapes are in Europe. But many of these protected landscapes are not well managed. Many lack management plans and staff. Many have been set up without legal protection or other means by government authority. And there are still many cases where local people are not yet seen as vital allies in conservation. There is a need for much improvement in our protected landscapes.

Yet it is a mistake to think that all of Europe is a managed landscape and that the only opportunities open to us are creating protected landscapes and small nature reserves. Despite its small size and large population density, Europe does still have some areas of wilderness. They may be smaller than the wilderness areas in, say, North America and Africa, but they are still substantial. And with the declining pressures on land for food production and military land, there is an opportunity to create national parks in IUCN Category II - where plant and animal species, habitats and geomorphological sites are of special spiritual, scientific, educational, recreational and tourist significance. Lapland in northern Scandinavia, some areas of forest in Poland, much of the Balkan Mountains, the Danube Delta and parts of the Pyrenees all come to mind as wilderness areas in Europe. There is also an opportunity to rebuild whole ecosystems.

Protected areas are certainly not the only way of conserving nature and landscapes. However, they are the pinnacle of conservation efforts, acting as models for others to follow in the wider countryside. They are particularly important in maintaining biodiversity and the best way - in most cases the only way - of conserving the jewels of Europe's natural heritage. Nature reserves and national parks are the best way to conserve the rare and endangered species, and ensure that representative samples of all the different types of habitats remain in good condition. In addition, protected landscapes are an effective way of maintaining ways of life that are in harmony with nature and the environment, and also in finding forms of sustainable rural development. Protected areas also enrich the quality of human life, in particular as places for recreation. They offer opportunities for inspiration, scope for peaceful enjoyment, and a place for understanding and learning. Above all, they are a source of mental, physical and spiritual renewal.

3 What are the Major Threats to Protected Areas?

Too often the protected areas in Europe are regarded as something separate or apart from the sustainable development of a nation. Indeed, in the past it was too often assumed that the aims of local communities and those of protected areas were in conflict. As a result, the management of protected areas often emphasized controls and regulations. In Europe, there is a lack of policies to encourage and include the widest possible array of partners in establishing and managing protected areas. And, this is crucial for mobilizing public and political support to achieve an effective and well-managed network of protected areas in Europe.

Following the recent political changes, the Central and East European countries are in the process of democratization and the establishment of market economies. On the one hand, the political, economic and social changes pose difficulties for the conservation of nature. On the other hand, the greater openness of governments to public opinion and the growing contacts between East and West could provide great opportunities for conservation. Possibly the greatest threat to protected areas in most parts of this region is land distribution. The greatest needs in Central and East European countries are, besides financial assistance, in mitigating the effects of land redistribution on protected areas, as well as the development of environmental legislation, and the support of partnerships and exchanges between East and West and the Central and East European countries.

The major threats to European protected areas have been identified as:

- agricultural activities including changes in farming practices such as hay to silage-making, animal stocking densities, arable rotation, mowing and harvesting dates, and changes in land-use such as irrigation, abandonment of agriculture, and afforestation
- air pollution (acid rain)
- dam construction
- forest exploitation
- hunting
- invasive species (introduced species)
- military activities
- mining
- over-grazing
- tourist pressures
- transport links
- water extraction
- water pollution
- war and conflicts

4 Why an Action Plan for Protected Areas?

When we prepared the European Regional Review of the Protected Areas (McNeely, Harrison and Dingwall 1994) for the IVth World Congress on National Parks and Protected Areas in Caracas, Venezuela in 1992 we were struck by three main conclusions:

- We found out that the protected area coverage in Europe was very uneven; some countries had very well developed systems with all the main types of protected areas, but others had protected area systems that were rather weak by international standards.
- Most protected areas in Europe were under strong pressure and the qualities of many were being destroyed; also we found that the management of many protected areas was not as strong as necessary to combat the powerful threats that these areas were facing.
- We found out that a coherent policy for protected areas in Europe was lacking, despite the many good initiatives like FNNPE, UNESCO through MAB (Man and Biosphere), the Council of Europe and the European Commission.

The main problems facing the protected areas of Europe are:

- integration of protected areas into larger planning frameworks;
- inadequate coverage of protected areas;
- management of protected areas;
- unreliable political and public support.

These were the reasons why we prepared Parks for Life, the Action Plan for Protected Areas in Europe.

5 What is the Aim of the Action Plan?

The aim is to ensure an adequate, effective and well-managed network of protected areas in Europe (IUCN Commission on National Parks and Protected Areas 1994). The vision for the Action Plan was to develop an adequate, effective and well managed network of protected areas by 2002 (the Vth World Congress on National Parks and Protected Areas). This Plan was built on the following policies:

- integrate the protected areas network into all parts of national life; this means that the protected areas would be embedded in regional planning and that policies for related sectors such as agriculture, forestry, and tourism would be environmentally benign;
- increase the extent of Category I and II protected areas, with an emphasis on countries in which coverage is currently inadequate;
- strengthen management capacity and conservation status of Category V protected areas;
- extend protected areas in each country to encompass at least one population of every species or habitat threatened on a European scale;
- protect in each country a representative of each ecosystem found within it;
- adopt management plans for all protected areas;
- establish effective laws in each country to enable the establishment and management of protected areas;
- assure staff of sufficient quality, quantity and training for appropriate protected areas;
- provide adequate funding where funding currently limits the extent and value of protected areas;
- improve co-operative arrangements between protected area managers and local communities;
- establish a single, easily-accessible, Europe-wide system of information on the protected areas.

6 What Activities are Covered by the Action Plan?

Covering an area from Portugal to Bulgaria, Romania and Baltic States, the Parks for Life Action Plan sets out the policies and actions each country should take to improve its protected areas, as well as outlining the action needed at the international level.

The Plan contains three types of action. Firstly, it endorses many existing initiatives, recognizing that much has already been done in Europe. Secondly, it makes recommendations, mainly to governments, about protected areas and their relationships with other policies, such as those for agriculture, forestry, and tourism. Thirdly, it outlines 30 Priority Projects to be undertaken by conservation bodies as a way of mobilizing support and providing a multiplier effect.

In the Plan all relevant problems and threats to protected areas in Europe are addressed and actions are suggested. The audience is organizations with influence over protected areas and individuals with direct responsibility for planning and managing protected areas, that is: all those who are working with or in protected areas, including governments, planning authorities and protected area managers.

The Parks for Life Action for Protected Areas in Europe consists of four parts:

6.1 Part I: Placing Europe's Protected Areas in their Wider Context

- Protected Areas in their Global and Regional Context (the lessons of Rio, of the IVth World Parks Congress, European strategic initiatives)
- Protected Areas and Environmental Planning and Management (protected areas and national planning, and local planning, land-use planning for sustainable development, protected areas and pollution control)
- Protected Areas and Key Sectors of Public Policy (agriculture, forestry, tourism, transport, energy and other industries)

6.2 Part II: Addressing Priorities at the European, Sub-regional and National Levels

- A Europe Wide Approach (EECONET initiative, coverage of protected areas to meet the needs of habitats, landscapes, species)
- Priority Terrestrial and Marine Sub-regions (Central and Eastern Europe, Southern Europe and the Mediterranean Sea, The Baltic, The North-East Atlantic)
- The Needs of Countries (countries in great economic and social difficulties, countries suffering from armed conflict, countries where political support needs reinforcement, countries where decentralization creates problems, countries needing an improved coverage of protected areas)

6.3 Part III: Strengthening the Planning and Management of Europe's Protected Areas

- The Legal Framework (improving national law, developing Natura 2000 through implementation of the Birds Directive and the Habitats Directive of European Union (EU) Member States, adhering to global conventions, strengthening regional conventions, developing a Convention on Conservation of Rural Landscapes of Europe)
- Protected Area Management (management planning, broadening partnerships in protected area management, funding, Category II National Parks, Category IV habitat/species management areas, Category V protected landscapes and seascapes, transfrontier protected areas, protected areas as models for the future)
- Training (action at the national level, action at international level)
- Monitoring and Information

6.4 Part IV: Creating the Climate for Success

- Public Support for Protected Areas (building greater public awareness, formal education, providing good information and facilities for visitors, enlarging the base of supporters for protected areas, building the support of local communities)
- Working Together to Implement the Plan (national actions to implement the plan, international actions to implement the plan, encouraging and maintaining the partnerships developed in preparing the plan)

    The actions undertaken under the Plan are:
    - region-wide activities
    - sub-regional activities
    - national activities

At the international level, the most important actions have been identified in terms of 30 Priority Projects. These projects should be/are undertaken by conservation bodies and organizations. Co-operation between all conservation bodies and international funding agencies will be vital to implementing this Action Plan. Only through a synergy among the many initiatives in Europe can its aims be achieved. The Priority Projects are high profile, international projects, designed to fill the gaps and enhance the prospects for protected areas in Europe. They are catalytic in nature, designed to encourage shifts in policy and to lever the substantially greater sums needed to implement the Plan.

The Action Plan determines the priorities of terrestrial and marine sub-regions. On land, the Plan covers two sub-regions: Central and Eastern Europe and Southern Europe. They both have special needs and requirements, though these should in no way detract from the need for action in Western Europe also. At sea and along the coastline, three sub-regions are included, which between them cover all the marine and coastal areas of Europe - the Mediterranean Sea, the Baltic and the North-East Atlantic. This full coverage reflects the paucity of marine protected areas in the region.

As the most important decisions about protected areas are taken at national level, the most important actions to implement Parks for Life should also be undertaken at that level. National governments should review their actions and policies against the principles and recommendations in the Plan and make any necessary changes. Different countries will find different ways of doing this, but it is recommended that every country set up a forum for discussing and carrying forward the Plan. Some countries may wish to go further and prepare an action plan for protected areas in their country.

National non-governmental organizations active in nature conservation should carry out activities in the plan and where appropriate, should lobby government bodies for effective implementation of the plan, especially at the level of policy and financial commitment. These organizations should also report on cases where implementation is not succeeding, and raise public awareness.

7 Who Will Oversee and Co-ordinate Implementation?

Parks for Life is an attempt to develop a truly European approach to the establishment and management of protected areas - whether they be national parks, protected landscapes or nature reserves. The action plan was developed by Europe's nature conservation experts - governmental and non-governmental - using IUCN as their international union. Over 50 members of the Commission on National Parks and Protected Areas (CNPPA) expressed their wish to cooperate and assist in the implementation of the plan, Through CNPPA and through the European Programme, IUCN will oversee and coordinate the implementation of the Plan.

The aim of CNPPA in Europe - about 180 members from nearly all European countries - is simple: to promote and co-ordinate the implementation of Parks for Life. A Steering Group is assisting in this task. A professional Programme Co-ordinator has been appointed and CNPPA is in a good position to carry forward the Plan. CNPPA is producing a CNPPA newsletter, Lifeline Europe, which is focusing on the Parks for Life programme and which should maintain contact with all members of CNPPA and other networks in Europe.

An annual forum will be organized which will be open to all international partners and others to monitor, plan and co-ordinate the implementation of the Plan, especially in carrying out the Priority Projects. At the 1996 Parks for Life Partners Forum, the partner organizations expressed their wish to co-operate and contribute to the implementation of the Plan.

So far Project Outlines for 30 Priority Projects have been prepared. Six Priority Projects are underway:

* Priority Project 3 - Sustainable Tourism Charter and Service. The lead organization is the French Federation of Regional and Nature Parks.
* Priority Project 8 - Study visits from East to West. This project fits into the strategy and work of the FNNPE, which is currently working on a PHARE sponsored project of exchange and training.
* Priority Project 10 - Training for protected areas staff from Mediterranean countries. Based on the results of the previous courses, Tour de Valat, France, has the intention to prepare a more detailed training programme and to start it in 1997.
* Priority Project 22 - Support to transfrontier protected areas. The lead organization is FNNPE and the Austrian Ministry for the Environment has supported this project.
* Priority Project 25 - Guidance, standards and promotion of training for protected area staff. FNNPE is the lead organization of this PHARE
* Priority Project 30 - Partnership and Exchange programmes. Parks for Life has to be integrated within the IUCN European Programme and Pan-European Biological and Landscape Diversity Strategy activities.

Project proposals for four projects has been prepared and applications for funding have been submitted to funding organizations:

* Priority Project 14 - Identification of potential Natural World Heritage Sites
* Priority Project 18 - Promotion of joint (collaborative) management in protected areas
* Priority Project 19 - Service to support upgrading the management of Category II sites
* Priority Project 21 - Conference on the protected landscape heritage of Europe

Clearly, not every project will start at the same time, but there is a wish to start with as many projects as possible. Organizations and individuals are offering their assistance in carrying out some of the projects. So far leading organizations have been identified and the following projects proposals are in preparation:

* Priority Project 2	-	Guidelines on conservation zones in managed forests
* Priority Project 6	-	Identification of important plant areas
* Priority Project 9	-	Conference on the regeneration of rural economies through national parks
* Priority Project 13	-	Implementation of the Ramsar Convention in Europe
* Priority Project 16	-	Feasibility of a Rural Landscapes Convention
* Priority Project 17	-	Guidance on how to apply the IUCN management categories in Europe
* Priority project 20	-	Study of the use of micro-reserves
* Priority Project 27	-	Improving management of information on protected areas and streamlining international protected area databases
* Priority Project 28	-	Book on working with the Arts Community

A Parks for Life package for fund-raising will be designed. IUCN will provide it to potential donors to secure funds for starting the Priority Projects, for their implementation, and for oversight and co-ordination of the Plan.

8 References

IUCN Commission on National Parks and Protected Areas. 1994. *Parks for Life: Action for Protected Areas in Europe.* IUCN, Gland and Cambridge.

IUCN. 1994. *Guidelines for Protected Area Management Categories.* CNPPA with the assistance of WCMC. IUCN, Gland and Cambridge.

McNeely, J.A., Harrison, J. and Dingwall, P.(eds.). 1994. *Protecting Nature, Regional reviews of Protected Areas.* IUCN, Gland and Cambridge

PART 2

Information and Communication

The papers in Part 1 support the active flow of many kinds of information among scientists, scholars, professionals and civic decision-makers. The papers in Part 2 address the issue of information and communication more directly, for example:

- Useful knowledge can be provided by the social or human sciences and specifically by disciplines such as anthropology, economics, geography, sociology and history individually or in combination; one possible means of promoting the use of such information is through networks of protected area centres or institutes in universities and similar institutions in North America, Europe and other countries.

- Detailed computer-based information systems such as CORINE in Europe, can store analyze, summarize and interpret an array of useful knowledge relatively quickly and effectively.

- Improvements in communication are needed among scientists, scholars, professionals and concerned citizens. Computer systems and collaborative institutions, such as MAB (Man and Biosphere) can assist here but attitudinal changes are needed as well, for example among scientists, about use of complex technical terminology or language.

- Regional, national and local networking in North America, Western, Central and Eastern Europe is playing and can play a stronger role in communicating and applying useful information, as for example in Hungary.

- More careful monitoring and assessment and regular reporting of activities, programs, policies and plans is needed to determine whether what is happening was or is desired, and what changes or adaptations should be made in future, as for example in the Yorkshire Dales in the UK and the Niepołomice Forest, Poland.

Usable Knowledge for National Park and Protected Area Management: A Social Science Perspective

Gary Machlis [1] and Michael Soukup [2]

[1] Visiting Chief Social Scientist
US National Park Service, Washington, DC, USA

[2] Associate Director for Natural Resource Stewardship and Science
US National Park Service, Washington, DC, USA

Abstract. Scientific understanding of the relationship between people and parks is critically important to protected area management. The social sciences can and must provide usable knowledge to managers, at the appropriate scale and dealing with issues relevant to protected area management. In this paper, the social sciences are defined and described, as is the concept of usable knowledge. The ecological context of protected area management is discussed, and a research agenda for the social sciences is presented. To accomplish this agenda, an international network of Cooperative Protected Area Studies Units (CPASUs) is proposed.

Keywords: Usable knowledge, scale, social science, Cooperative Protected Area Studies Unit

1 Introduction

The management of national parks and protected areas[1] is necessarily the management of people. Visitors, employees, tourism operators, nearby communities, interest groups, industries, local governments--all affect parks and protected areas. An accurate understanding of the relationship between people and parks is critical to preserving resources unimpaired, providing for public enjoyment, and sustaining human populations dependent upon park resources. Such understanding requires a sound scientific basis. Hence, social science is a necessary partner in protected area management.

This paper has three main points: 1) the social sciences can best aid park managers by focusing on usable knowledge -- a special kind of scientific information, 2) usable knowledge must be sensitive to scale, and appropriate to the various ecological and organizational levels that characterize protected area management, and 3) to conduct the necessary scientific research requires an innovative effort at building the institutional capacity of park agencies and universities in Latin America, Africa, Asia and Eastern Europe. In developing countries especially, the effectiveness and economy of close partnerships between park management and universities can overcome the critical shortage of funding and scientific resources available to parks. Therefore, we propose the development of an international network of Cooperative Protected Area Studies Units (CPASUs) as an innovative and cost effective mechanism for delivering needed research to park and protected area managers.

1.1 The Social Sciences Defined and Described

Social science is defined as the application of the scientific method to the study of human behavior. Several disciplines can be included under this definition: anthropology; economics; geography; psychology; political science and sociology. Each focuses upon certain units of study and driving forces important in understanding human behavior. Table 1 describes each discipline's key unit(s) of analysis, and examples of potential contributions the discipline can make to protected area management. In addition, the social sciences are important partners in interdisciplinary fields such as environmental economics, conservation biology, and landscape ecology.

---

[1] Throughout this paper, the terms parks and protected areas are used interchangeably, except where specific mention is made to national park status.

Table 1. Social science disciplines, their focus and potential contributions to park management

Discipline	Key unit(s) of analysis	Examples of potential contributions
anthropology	human cultures	understanding traditional land and resource use patterns, mechanisms of dispute settlement, cultural variation in attitudes toward protected areas
economics	markets, industries, economies	study of economic impacts of parks, cost-benefit analyses of park policies, role of parks and tourism within larger economy
geography	landscapes, spatial units (political, ecological, etc.)	understanding tourist travel patterns, regional development, spatial perspective on human impacts on park resources
political science	the state, government institutions	studies of public participation in planning, role(s) played by local communities and interest groups in park management, improving organizational effectiveness
psychology	individuals	studies of visitor experiences, interpretive media, other forms of park communication
sociology	social groups, organizations, communities	studies of demographic trends affecting park management, visitor behavior, public opinion on park policies; organizational analyses of park management structures
interdisciplinary fields	varied	studies of economic uses of park flora by local populations; economic impact assessments of ecosystem management policies; analyses of visitor impacts on wildlife

2 The Ecological Context, with South Florida and Poland as Examples

Social forces are driving forces that can threaten the sustainability of park systems (McNeely 1992, 1995; Machlis 1985). Table 2 illustrates some of the social problems that threaten park resources around the globe. The specific problems may vary from one region to another, but the driving forces are similarly socioeconomic: population growth and movement, human settlement patterns, policies or laws (governing development, private property rights, land use and tenure), capital accumulation and industrial activity, and so forth.

An example in the United States is South Florida. The US National Park Service (NPS) manages four protected areas in South Florida: Big Cypress National Preserve; Biscayne National Park; Dry Tortugas National Park; and Everglades National Park. The South Florida region and its people face extraordinary challenges. These challenges include a human population expected to triple within 50 years, intense competition for water supplies among residential, agricultural and preservation uses, cumulative and severe ecological stresses resulting from urban growth and agricultural activity, and the need for sustainable economic development simultaneous with restoration of the South Florida ecosystem. As a result, protected areas in South Florida confront serious problems: an 81% increase in park visitors since 1980 (including dramatic increases at Biscayne and Dry Tortugas national parks); significant resource impacts (from exotic species to acute and chronic water concerns); complex natural resource management issues (from recreational vehicle use to commercial fishing); and limited financial and human resources (Machlis et al. 1996).

South Florida can be considered a human ecosystem - a complex, bounded, human-dominated ecosystem that includes socioeconomic as well as biophysical systems. The South Florida NPS units are critical elements in this human ecosystem. Perhaps nowhere else in the US is the long-term future of protected areas and a geographical region so intertwined. Hydrological flows, weather patterns, ecological cycles, urban growth, local and regional economies, government jurisdictions, an agricultural heritage, cultural values (including tribal interests)--all link the South Florida parks to the larger region, and make social science a necessary partner to ecosystem management in the region.

Table 2. Social problems threatening protected areas in different world regions

Region	Selected social problem(s) threatening park resources	Source(s)
North America	increasing volume of visitors; regional development leading to enormous population increases near parks	National Park Service 1992; Light et al. 1995
SubSaharan Africa	human encroachment, overgrazing, expansion of extensive agriculture	Olindo & Mbaelele 1994
Southeast Asia	marine ecosystems suffering from oil exploration, and fishing activities employing explosives and cyanide	Mishra 1994
Eastern Europe	agricultural encroachment and diversion of water for irrigation; air pollution from electric power generation and other industry; pressures to privatize state lands	Bibelriether & Synge 1994; Nikol'skii et al. 1994
Region	Selected social problem(s) threatening park resources	Source(s)
Central America	land tenure conflicts; concessions for tourist development; deforestation; land colonization policies, practices; habitat fragmentation caused by land ownership patterns	Ugalde & Godoy 1994; Neumann and Machlis 1989
Western Europe/Mediterranean	dam construction; overgrazing; road construction; water pollution	Bibelriether & Synge 1994

Poland's national park system is another example. The country is undergoing rapid political, economic and social change, including rapid turnover in ministerial positions, a 30% inflation rate with 16% unemployment, and adoption of western lifestyles and consumption patterns (REC 1994). The overriding concern of many citizens and decision-makers is now economic growth at all costs (REC 1994). Such a development strategy will likely impact the nation's already stressed environmental resources. The Polish environment has suffered decades of abuse caused by extensive industrialization and use of natural resources (Nowicki 1992). One-third of the Polish population lives in areas where government pollution standards are exceeded; one-third of Poland's forests are threatened, and only 4% of the watercourses still carry clean waters (Nowicki 1992).

In the face of both ecological stress and socioeconomic change, the Polish national parks face significant problems. The 1991 annual report on the National Parks in Poland (Lubczyński 1992) lists numerous risk factors as threats to Polish national parks (Table 3). Many of these threats (such as 'excessive tourism' and 'long-range pollution' due to industrialization) are driven by socioeconomic forces. For example, the increased availability of automobiles has overwhelmed Ojcow National Park's parking facilities and road network (Transport for Leisure LTD 1995). The growing communities adjacent to Tatra National Park are demanding that the park expand its facilities for tourism (Gill pers. com.). As in South Florida, the long-term future of these protected areas and the socioeconomic conditions of people are intertwined, making social science a necessary partner in the nation's protected area management.

3. Usable Knowledge Defined and the Importance of Scale

Park managers often face a complex set of decisions, most of which must be made relatively quickly, simultaneously, without complete information and with consequences that require additional decision-making. Many decisions involve or affect people--visitors, employees, local communities, interest groups and others. Hence, there is continual opportunity for social science to assist protected area management, if it can provide usable knowledge.[2]

In the social sciences, there are several categories of usable knowledge. The first is information. Monitoring data collected on visitors and resource impacts are an example. The second are insights, such as understanding how visitor use impacts park resources. A third form of usable knowledge is predictions, such as forecasts of visitation and which visitor impacts are likely to increase. Finally, there are solutions, such as suggested ways that visitor impacts can be reduced. Table 4 outlines these categories and illustrates the kind of research products needed by managers.

---

[2] For a full discussion of the concept of usable knowledge, see Lindblom and Cohen (1979).

Table 3. Risk factors to Polish national parks

Risk factor	Number of parks affected
Long range air pollution	11
Local air pollution	5
Water pollution	7
Building of water reservoirs	2
Endangered water balance	3
Communal dangers	6
Unauthorized construction work	6
Communication threats	6
Excessive tourism and recreation	8
Poor state of forest health	3
High density of game animals	17
Poaching, stealing, removal of park resources	6
Private property inside parks which disrupt proper protective activities	6
Farming and fishing activities	6
Lack of sewage treatment plants in or near parks	11
Eroding seashore	2
Poor relations between park and communities	2

(from Lubczyński 1992)

Table 4. Categories of usable knowledge

Category	Typical products	Examples
Information	data sets, maps, monitoring tools	monitoring data on level of visitor impact on resources
Insights	analyses, theories, interpretation of data	description of how and why visitors impact resources
Predictions	models, scenarios, hypotheses, estimates	forecasts of visitation increases, estimates of future impacts
Solutions	impact assessments, innovations, new techniques, management alternatives	levels of acceptable change assessments, management alternatives

To be usable knowledge, research must meet certain requirements. The development of research objectives should be jointly undertaken with managers, so that realistic goals and expectations are set. The research must be provided at the proper point in the decision-making process. It must directly address the park manager's needs and at a level of detail appropriate to the decision. The manager must understand the limitations of the research, the degree to which it can be applied, the probability of successful application and the reliability of the research results. The results should provide managers with relevant information, insights, predictions, or solutions. Such research may sometimes be controversial, for it must provide scientific input to significant issues and difficult problems. The delivery of scientific information in the form of usable knowledge is a requirement of scientists; so is the expectation that managers consider this knowledge in their decision-making. And this decision-making takes place at several scales.

3.1 The Critical Importance of Scale

Protected area management takes place at significantly different scales, and the issue of scale is central to the partnership of science and park management (for a full discussion, see Machlis 1995). Table 5 illustrates the major scales of protected area management, the key organizational units at each scale, and the management responsibilities associated with these organizations.

At the protected area level, key units of organization include visitor groups, resident populations, park staff, managers of adjacent public lands, and within-park enterprises. At the bioregion level, the protected area is embedded in a wider ecological and social system. Units of concern include local communities, states and provinces, regional offices of park and other natural resource agencies, regional markets, and service economies. At the national level, key units are the national legislatures, central administrations, large non-governmental organizations (NGOs), the media, and other national agencies managing resources. At the realm level, international organizations and other nations' park agencies are central. At the emerging global level, international NGOs, treaty organizations, and world markets become significant organizational units. Each has important management responsibilities, from providing visitor services to strategic planning.

Table 5. Scales of protected area management, key organizational units and management responsibilities

Scale of protected area system	Key organizational units	Management responsibilities
protected area	park staff; resident populations; visitor groups; concessionaires	natural and cultural resource management; identification of threats; visitor services
bioregion	local communities; states and provinces; regional offices; regional service economies	training; monitoring; coordination; policy implementation
national protected area system	national legislatures; central park administration; national NGOs; national travel industries; bilateral NGOs; media organizations	policy formation; funding; evaluation of subordinate managers; acquisition; development strategies
realm	international NGOs; international treaty organizations; other national park agencies	strategic planning; administration of international aid programs; monitoring; training; technical assistance
global system	international NGOs; international travel industry; international treaties; United Nations; world markets	strategic planning; allocation of resources; technical assistance

At each scale, the decision-making process of protected area managers will vary, since different organizational units and political contexts interact. For example, protected area management policies are often manifestations of larger political agendas which may not necessarily mirror local priorities. That is, the management of protected areas is scale-dependent. In addition, each level of management is significantly influenced by the adjacent levels, and all are parts of a nested system of protected area management. Information needs of protected area managers will differ at each scale, though contribute to an overall set of needs. Identifying these specific needs is a critical step in linking social science and protected area management.

4. A Research Agenda for Social Science and Parks

There are numerous social science research issues related to protected areas and requiring scientific study. While needs vary by region and nation, the US experience is a relevant example. The USNPS recently developed a list of critical social science research issues (Machlis 1996). The issues were identified by a questionnaire sent to park managers and scientists, information gathering sessions at several professional meetings, and previous reviews of NPS science needs. The results provide the core of a research agenda for USNPS social science. The issues are organized around a series of critical research questions:

1. Who are park visitors?
2. What are the impacts of visitor use on park resources?
3. What is the relation between parks and surrounding communities and regions?
4. What is the relation between national parks and local, regional, national and international economies?
5. How can threats to parks be mitigated?
6. How effective are agency interpretive, educational and public outreach efforts, and how can they be improved?
7. What organizational and employee issues face the agency?
8. What are the relationships between park management and sustainable development?
9. How can natural and cultural resources management be made more effective?

In Table 6, key initiatives are described for each research question, along with useful sources. Since research needs vary widely throughout the NPS, and many of the issues are relevant at several scales, they are not listed hierarchically or in an order of importance. Certainly, other research questions and applied problems are relevant to parks and protected areas. Yet the key issues described in Table 6 represent a significant agenda for social science. An effective social science program will deliver usable knowledge on these issues.

5. Delivering Usable Knowledge: A Network of Cooperative Protected Area Studies Units

One of the most important challenges facing the world park movement is the integration of scientific information and expertise into park management. How can usable knowledge be delivered to park managers? The US experience provides one potentially useful approach. Since 1970, Cooperative Park Studies Units (CPSUs) have been an effective mechanism for conducting USNPS science, involving the nation's universities in delivering research and scientific expertise to park managers (NRC 1992). Currently, there are eleven CPSUs in the US. They now report to the Biological Research Division of the US Geological Survey, and are currently being expanded to include scientists from several land management agencies, including the USNPS. We propose establishing an international network of similar institutions, Cooperative Protected Area Studies Units.

These research centers and the resulting network are broadly proposed to deliver not just social science, but the full range of biological, physical, and social sciences needed for protected area management. The objectives are to:
1. deliver science and scholarship in all fields of inquiry needed by park managers,
2. provide park managers with consulting, extension and technical assistance,
3. offer professional development opportunities for park agency employees (including workshops, continuing education, specialized training, sabbaticals and graduate degree programs),
4. allow for efficient and timely contracting, conduct and delivery of scientific research,
5. provide park agencies with access to valuable university resources, from laboratories to libraries,
6. have the flexibility to evolve and adapt to meet local, regional and national needs, and
7. serve as units in a global network of CPASUs, providing expertise and scientific information at international levels.

This proposal follows many of the recommendations in several US and international reviews of park-related science, including the US National Research Council's report Science and the National Parks (1992), the IV World Parks Congress (1992), the Ecological Society of America's report on NPS research (Risser and Lubchenco 1992), and the USNPS recent plan for social science, Usable Knowledge: A Plan for Furthering Social Science and the National Parks (Machlis 1996).

5.1 An outline of the concept

A network of cooperative research units would be created to link the full range of university resources to the needs for research and technical assistance of park and protected area managers. The United States, with its large and complex set of national parks, wildlife refuges and other protected areas, has eleven CPASUs. Nations with smaller protected area systems might have one or two CPASUs, depending upon need and resources. The international CPASU network would be coordinated and provided technical support by a coordinating group co-chaired by representatives of participating nations. This leadership group could grow as additional countries became involved. The IUCN's Commission on National Parks and Protected Areas could provide external advice from the scientific community. Communication within the network would be largely digital, using electronic mail, data sharing and the multi-media capabilities of the World Wide Web.

Table 6 Critical Research Questions for the Social Sciences

Research Questions	Key Initiatives for Social Science Research	Useful Sources
1. Who are park visitors?	1) document visitation trends among national/international tourists, and identify different park user groups, 2) identify populations *not* using park system, 3) monitor and analyze visitation data over time, 4) help identify how, where and for what purposes park resources are used, 5) analyze visitor expectations, attitudes and evaluations of park experiences, 6) evaluate and improve methods of reporting visitation statistics, 7) help managers use visitor information available from other agencies.	Clark 1992, Machlis 1989, NPCA 1988, Schroeder 1996
2. What are the impacts of visitor use on park resources?	1) analyze visitor use and distribution patterns, 2) identify critical visitor impacts on natural and cultural resources, 3) describe benefits of visitor use and park experience, and develop plans to increase their provision, 4) develop methods to reduce negative impacts on park resources, and to reduce services and activities incompatible with a park's stated mission, 5) define and describe "overcrowding" of parks, 6) develop, refine and apply "carrying capacity" methods, 7) identify conflicts between park user groups, 8) develop and apply conflict management techniques.	Shelby and Heberlein 1986, West and Brechin 1991, NPS 1994, Krumpe and McCoy 1995
3. What is the relationship between national parks and their surrounding communities and region?	1) clarify role of national parks/protected areas in larger regional mix of recreation opportunities, 2) better understand adjacent community populations, important values and attitudes on parks and protected areas, 3) identify and monitor subsistence/other uses of park resources by local residents, 4) assist managers in forging park-community partnerships, 5) help indigenous peoples assess park-related tourism development, 6) develop/incorporate/refine methods of public participation to balance local and broader regional/national interests, 7) predict impacts of proposed management practices on local residents and visitors, as well as these groups' responses, 8) help managers integrate visitor and community-based views into decision making, 9) assist managers with politically-charged decision making, 10) give input into ecosystem management efforts at park and regional scales, 11) cooperate in partnerships with other government agencies.	West and Brechin 1991, Butler 1993, Metcalfe 1995, Slocombe 1995, Williams and Skabelund 1995, Bernbaum 1996, Cortner 1996
4. What is the relationship between national parks and local, regional, national and international economies?	1) understand economic interactions between parks and adjacent communities, 2) assess local, regional and international economic costs and benefits of parks, 3) develop, improve and apply methods to evaluate ecological benefits of parks, 4) evaluate park entry and user fee systems, 5) predict gains or losses in visitation and evaluate their impact on park management, using regional and international economic indices and forecasts.	Dixon and Sherman 1990, McNeely 1995
5. How can threats to parks be mitigated?	1) evaluate effects of different adjacent land uses on park management, 2) seek understanding of land use changes occurring at the landscape and ecosystem level, 3) predict socioeconomic change (such as migration) that may impact park ecosystems, 4) develop, improve and effectively apply methods of public participation, 5) develop, improve and effectively apply mitigation strategies.	Machlis and Tichnell 1985, Hough 1991, Bryan 1996
6. How effective are agency interpretive, educational and public outreach efforts, and how can they be improved?	1) assess relevance and effectiveness of interpretive programs, media and public contact activities, 2) identify issues and topics that merit interpretive efforts, 3) assess the effectiveness of visitor centers and museums, 4) assist in developing effective communication and environmental education techniques, 5) assess alternative ways to deliver visitor services, 6) describe visitors' willingness to pay for services, 7) conduct surveys and focus groups to examine level of public support for park management among various interest groups and cultures, and explore ways to increase support.	Bishop 1995, Gurung 1995, Lusigi 1995, Munro 1995

7. What organizational and employee issues face the agency?	1) periodically evaluate the "state of the park agency," 2) monitor job satisfaction and understand its influencing factors, 3) help build effective organizational capacities within the agency, 4) develop measures of employee productivity and organizational effectiveness, 5) evaluate field support by central offices, 6) analyze the structure and dynamics of the agency work force, 7) predict socioeconomic trends that will impact the organization and its employees.	NPS 1992 Johnson et al. 1983
8. What are the relationships between park management and sustainable development?	1) conceptualize "sustainability" and devise measures of its various dimensions (e.g., social, cultural, economic), 2) develop techniques for measuring, assessing and monitoring the ecological sustainability of park management practices, 3) explore relationships between local economies and ecological sustainability of park resources, 4) assess the impact of tourism and infrastructure development on ecological sustainability at the park and national level, 6) evaluate experiments with ecosystem co-management regimes where agricultural encroachment, resource extraction or fragmented land ownership patterns threaten park resources	Lele 1991, Costanza 1991, Daly and Cobb 1994, Carpenter 1995, Linn 1995, Metcalfe 1995, Hamilton 1996, Machlis 1989
9. How can natural and cultural resources management be made more effective?	1) identify the human dimensions of natural and cultural resource management, 2) provide critical socioeconomic baseline and trend data for resource management and employee training, 3) provide public input to planning through community surveys, 4) help managers use social science theory, methods and findings in resource management, planning, training and decision making, 5) develop, practice and evaluate ecosystem management	Dillman 1978, Gunderson et al. 1995, Ewert 1996

Key elements of each CPASU would include 1) a host university, 2) partner institutions and agencies, 3) a role and mission statement, 4) a strategic plan and performance goals, and 5) a managers' committee. Each element is described below.

5.1.1 The host university

CPASUs would be largely based at universities. The host university would provide space and basic administrative support, as part of its cooperative agreement. CPASU scientists would both conduct research and act as facilitators in establishing research teams, delivering research and technical assistance on high priority issues to park managers, and engaging the faculty and resources of the host universities. The resident staff would be augmented by university faculty, advanced graduate students, and resource managers and scientists from other domestic and international agencies on rotating assignments and exchanges--providing a critical mass of interdisciplinary skills and tapping the unique benefits of university-based research.

5.1.2 Partner institutions and agencies

Park managers must have access to the best source of scientific inquiry on a given topic. CPASUs would develop partnership arrangements to increase their research and technical assistance capabilities. Partner institutions could also include other national and state agencies, universities, non-governmental organizations (NGOs) and other research organizations. Partner institutions would be linked to the CPASU through formal agreements. New CPASUs could be linked to established 'sister' CPASUs that have experience in dealing with similar park problems--desert ecosystems or marine-based tourism, for example. Park-based scientists could be affiliated with a CPASU through the host university or partner institution. These partnership arrangements would make the faculty, graduate students and facilities of more universities available to park managers, reduce administrative costs, avoid duplication of effort, and encourage cost-sharing among agencies.

5.1.3 Role and mission statement

CPASUs would focus their research agendas to provide research support at an appropriate scale for nearby parks and protected areas, and specialized expertise useful to national and international needs. Each CPASU would prepare a role and mission statement that identifies research, technical assistance, and other services that it is especially qualified to provide. The role and mission statement would be used to guide research and service activities of the CPASU, avoid unnecessary duplication of effort, provide accountability and evaluate performance of the CPASU, and coordinate the CPASUs into a comprehensive network. A CPASU's role and mission statement would evolve as additional agencies became partners to the CPASU, and as the CPASU gained experience in park-related science.

5.1.4 Strategic plans and performance goals

Each CPASU would prepare a strategic plan and performance goals for its research and service activities. The strategic plan would describe the CPASU's ongoing research, anticipated projects, funding needs, and improvements in delivering science and service to park managers and other partners. Strategic plans would provide the broader regional, national, and international context for both large and small-scale research projects. Annual work plans would include performance goals and provide for flexibility. CPASU plans would be used to guide the timely delivery of useful research to managers, develop funding sources, coordinate research activities with other CPASUs, and evaluate CPASU performance.

5.1.5 Managers' committee

CPASUs would create a mechanism for park managers (and those of other participating partner agencies) to provide advice, while maintaining the independence of research and the objectivity of research results. Each CPASU would organize a managers committee, composed of key park managers, university faculty and representatives from partner agencies. The committee would provide advice and guidance to the CPASU, review the annual work plans, and assist in evaluating CPASU performance.

5.2 Benefits of an international CPASU network

The CPASU network would deliver a broad scope of scientific research and technical assistance to protected area management agencies. Research would be conducted at several scales. Including the full range of disciplines used by resource managers (from ecology to sociology) would improve coordination, increase usable knowledge, and minimize administrative costs. The social, biological and physical sciences would be integrated to provide interdisciplinary problem-solving skills and ecosystem-scale research by multi-disciplinary and multi-national teams. Collaborations between scientists, and between CPASUs throughout the international network would be encouraged. 'Sister park' arrangements could be initiated linking parks around the world with similar research needs. Students and faculty would find international collaboration practical and beneficial. With park agency, NGO and university scientists working together within a university environment, the generation, synthesis, and utilization of scientific information will be greatly enhanced.

CPASUs would also provide increased technical assistance to park managers. Managers would have a primary CPASU to draw on for basic technical assistance, training, planning support and other needed services. They would have efficient, timely and cost-effective access to nearby universities (either a host or partner institution). In addition, the CPASU network would provide managers with specialized skills and assistance available from other CPASUs in the international network. Brokering and sharing of CPASU technical assistance would be encouraged; international collaboration would be made easier and more cost-effective.

The CPASU concept must be adapted to reflect national conditions: political systems; educational institutions; park administrations; national laws and policies; civic culture and so forth. Poland is an example. Most scientific research conducted in the national parks is by park scientists, Polish Academy of Sciences and university researchers. Park directors are aided by scientific advisory boards and there is considerable ongoing applied natural research. Applied social science research and technical assistance is relatively uncommon. Some parks, such as Ojców National Park, have emerging partnerships with other European national parks. Hence, several elements of CPASUs are in place; others will need to be modified to meet Polish needs. Yet the key elements, from host universities to managers committees, form the basis of a flexible and practical approach to delivering usable knowledge to protected area managers.

5.3 Funding the CPASU Network

Funding a network of CPASUs is a significant challenge. Costs include support for research scientists and faculty, graduate student stipends, fieldwork expenses, equipment and administrative charges. In the US, CPSU programs range in annual costs from US $75,000 to $400,000. Yet the problem is tractable. CPASUs can initially be established with small staffs, contracting for part-time involvement by university faculty. Equipment can be shared, and administrative costs minimized. Partner agencies, particularly conservation NGOs, international development organizations and private industry can provide support funds or in-kind assistance. Start-up support from philanthropic foundations can be sought. In addition, the costs of coordinating an international network of CPASUs can be minimized through electronic communication and frugal management. International organizations with scientific responsibilities (such as NATO) can provide assistance to the emerging network. Countries with established CPASUs (such as the US) can provide technical support and advice in the development of new CPASUs in other nations (such as Poland, our host).

6. Conclusion

The establishment of CPASUs, and an international network of research stations devoted to protected area management, will significantly expand the capacity of all sciences to deliver usable knowledge. This network will directly benefit the social sciences.

As the 20th century draws to a close, the challenges facing park managers around the world will increase in breadth, complexity and severity. An accurate understanding of the relationship between people and parks is critical to protecting resources unimpaired, providing for public enjoyment, and sustaining human populations dependent upon park resources. Such understanding requires a sound scientific basis. Hence, social science is a necessary partner in protected area management.

6.1 Acknowledgments

Support for this paper was provided by the US National Park Service and the NATO Advanced Research Workshop organizers. Bill Grigsby, Glen Gill, Emilee Ford and Sandy Watson provided valuable assistance.

7. References

Bernbaum, E. (1996). Sacred mountains: Implications for protected area management. *Parks* 6(1):41-48.
Bibelriether H. and Synge H. (1994). Europe. In McNeely J.A., Harrison J., Dingwall P. (eds) *Protecting Nature: Regional Reviews of Protected Areas*, pp. 103-132. IUCN, Cambridge.
Bishop S.G. (1995). Partnerships for ecosystem management and sustainable development: Some biosphere reserve models.In Linn R.M. (ed) *Sustainable Society and Protected Areas*, pp. 5-8. The George Wright Society, Hancock, Michigan.
Bryan H. (1996). The assessment of social impacts. In Ewert A.W. (ed) *Natural resource management: The human dimension*, pp. 45-66. Westview Press, Boulder, Colorado.
Butler R.W. (1993). Pre- and post-impact assessment of tourism development. In Pearce D.G., Butler R.W. (ed.) *Tourism research: Critiques and challenges*, pp. 135-155. Routledge, London and New York.
Carpenter R.A. (1995). Limitations in measuring ecosystem sustainability. In Trzyna T.C. (ed) *A Sustainable World*, pp. 175-97. IUCN, Gland, Switzerland.
Clark R.N. (1992) Alternative strategies for studying recreationists. In Machlis G.E., Field D.R. (eds) *On interpretation: Sociology for interpreters of natural and cultural history*, pp. 24-43. Oregon State University Press, Corvallis, Oregon.
Cortner H. (1996). Public involvement and interaction. In Ewert A.W. (ed) *Natural resource management: The human dimension*, pp. 167-80. Westview Press, Boulder, Colorado.
Costanza R. (ed) (1991). *Ecological economics: The science and management of sustainability.* Columbia University Press, New York.
Daly H.E.and Cobb J.B., Jr (1994). *For the common good (second edition).* Beacon Press, Boston, Massachusetts.
Dillman D. (1978). *Mail and telephone surveys: The total design method.* John Wiley and Sons, New York.
Dixon J.A., Sherman P.B. (1990). *Economics of protected areas: A new look at benefits and costs.* Island Press, Washington D.C.
Ewert A.W., (ed.). (1996). *Natural resource management: The human dimension.* Westview Press, Boulder, Colorado.
Gill, G. (1996) Personal communication. Cooperative Park Studies Unit, University of Idaho, Moscow, Idaho.
Gunderson L.H., Holling C.S., and Light S.S. (eds.) (1995). *Barriers and bridges to the renewal of ecosystems and institutions.* Columbia University Press, New York.

Gurung, C.P. (1995) People and their participation: New approaches to resolving conflicts and promoting cooperation. In McNeely J.A. (ed) *Expanding Partnerships in Conservation*, pp. 223-33. Island Press, Washington, D.C.

Hamilton, L.S. (1996) The role of protected areas in sustainable mountain development. *Parks* 6(1) 2-13.

Hough J. (1991). Social impact assessment: Its role in protected area planning and management. In West P.C. and Brechin S.R (eds) *Resident peoples and national parks: Social dilemmas and strategies in international conservation*, pp. 274-282. University of Arizona Press, Tucson, Arizona.

Johnson D., Field D.R., and Machlis G.E. (1983) *The organization and the employee*. University of Washington, Cooperative Park Studies Unit, National Park Service, Seattle, Washington.

Krumpe E.E. and McCoy L. (1995) Techniques for resolving conflict in natural resource management. In Saunier R.E., Meganck R.A. (eds) *Conservation of biodiversity and the new regional planning*, pp. 67-74. Organization of American States, Washington, D.C.

Lele S.M. (1991). *Sustainable development: A critical review*. World Development 19(6):607-21.

Light S.S., Gunderson L.H., and Holling C.S. (1995). The Everglades: Evolution of management in a turbulent ecosystem. In Gunderson L.H, Holling C.S, Light S.S. (eds) *Barriers and bridges to the renewal of ecosystems and institutions*, pp. 103-68. Columbia University Press, New York.

Lindblom C.E. and Cohen D.K. (1979). *Usable Knowledge: Social science and social problem solving*. Yale University Press, New Haven, Connecticut.

Linn R.M. (ed.) (1995). *Sustainable society and protected areas: Contributed papers of the 8th conference on research and resource management in parks and on public lands*. The George Wright Society, Hancock, Michigan.

Lubczyński L. (1992). *National parks in Poland (annual report 1991)*. Białowieża National Park, Izabelin, Poland.

Lusigi W.J. (1995). How to build local support for protected areas. In McNeely J.A. (ed) *Expanding Partnerships in Conservation*, pp. 19-24. Island Press, Washington, D.C.

Machlis G.E. (1996). *Usable knowledge: A plan for furthering social science and the national parks*. National Park Service, Washington, D.C.

Machlis G.E. McKendry J.E., and Correia M.E. (1996). *A social science plan for South Florida's National Park Service units*. National Park Service, Washington D.C.

Machlis G.E. (1995). Social science and protected area management: The principles of partnership. In McNeely J.A. (ed) *Expanding Partnerships in Conservation*, pp. 45-57. Island Press, Washington, D.C.

Machlis G.E. (1989). Managing parks as human ecosystems.In Altman I. and Zube E.H. (eds) *Public places and spaces*, pp. 255-75, Plenum Publishing Corporation, New York.

Machlis G.E. and D.L. Tichnell (1985). *The state of the world's parks*. Westview Press, Boulder, Colorado.

McNeely, J.A. (1992). The contributions of protected areas to sustaining society. In *IV Congreso Mundial de Parques Nacionales y Areas Protegidas: Plenary Sessions and Symposium Papers*, pp. 1-6 Caracas, Venezuela.

McNeely, J.A. (1995). Partnerships for conservation: An introduction. In McNeely J.A. (ed) *Expanding Partnerships in Conservation* pp. 1-12. Island Press, Washington, D.C.

Metcalfe, S.C. (1995). Communities, parks, and regional planning: A co-management strategy based on the Zimbabwean experience. In McNeely J.A (ed) *Expanding Partnerships in Conservation*, pp. 270-79. Island Press, Washington, D.C.

Mishra, H.R. (1994). South and Southeast Asia. In McNeely J.A., Harrison J., and Dingwall P. (eds) *Protecting Nature: Regional Reviews of Protected Areas*, pp.177-203. IUCN, Cambridge.

Munro D.A. (1995). New partners in conservation: How to expand public support for protected areas. In McNeely J.A. (ed.) *Expanding partnerships in conservation*, pp. 13-18. Island Press, Washington D.C.

National Park Service. (1992). *National parks for the 21st century: The Vail Agenda*. Chelsea Green Publishing Company, Post Mills, Vermont.

National Park Service. (1994). *National Park Service strategic plan: Vision*. Government Printing Office,Washington, D.C.

National Parks and Conservation Association. (1988). *Parks and people: A natural relationship (vol 3). Investing in park futures: A blueprint for tomorrow.* National Parks and Conservation Association, Washington, D.C.

National Research Council. (1992). *Science and National Parks.* National Academy Press, Washington D.C.

Neumann R. and Machlis G.E. (1989). Land-use and threats to parks in the neotropics. *Environmental Conservation* (16) no 1.

Nikol'skii, A.A., Bolshova L.I., and Karaseva S.E. (1994). North Eurasia. In McNeely J.A., Harrison J., and Dingwall P. (eds) *Protecting Nature: Regional Reviews of Protected Areas,* pp. 133-155. IUCN, Cambridge.

Nowicki M. (1992). *Environment in Poland: Issues and solutions.* Ministry of Environmental Protection, Natural Resources and Forestry, Warsaw, Poland.

Olindo P. and Mbaelele M.M. (1994). Sub-Saharan Africa. In McNeely J.A., Harrison J., and Dingwall P. (eds) *Protecting Nature: Regional Reviews of Protected Areas,* pp. 47-71. IUCN, Cambridge.

Regional Environmental Center for Central and Eastern Europe, (1994), *Strategic environmental issues in Central and Eastern Europe (vol 1) Regional Report.* Regional Environmental Center for Central and Eastern Europe, Budapest.

Regional Environmental Center for Central and Eastern Europe. (1994). *Strategic environmental issues in Central and Eastern Europe (vol 2) Environmental needs assessment in ten countries.* Regional Environmental Center for Central and Eastern Europe, Budapest.

Risser P.G. and Lubchenco J. (1992). *Report of a workshop for a National Park Service Ecological Research Program.* NPS Ecological Study Program Workshop Report. Albuquerque, New Mexico.

Schroeder H.W. (1996). Ecology of the heart: Understanding how people experience natural environments. In Ewert A.W. (ed) *Natural resource management: The human dimension,* pp. 13-28. Westview Press. Boulder, Colorado.

Shelby B. and Heberlein T.A. (1986). *Carrying capacity in recreation settings.* Oregon State University Press, Corvallis, Oregon.

Slocombe D.S. (1995). An ecosystem approach to regional planning. In Saunier R.E. and Meganck, R.A. (eds) *Conservation of biodiversity and the new regional planning,* pp. 53-66. Organization of American States, Washington, D.C.

Transport for Leisure Ltd. (1995). *A Sustainable Transport Strategy for Ojcow National Park.* West Yorkshire, UK.

Ugalde A. and Godoy J.C. (1994). Central America. In McNeely J.A., Harrison J., and Dingwall P. (eds) *Protecting Nature: Regional Reviews of Protected Areas,* pp. 301-345. IUCN, Cambridge.

West P.C. and Brechin S.R. (eds.). (1991). *Resident peoples and national parks: Social dilemmas and strategies in international conservation.* The University of Arizona Press, Tucson.

Williams L.T. Jr. and Skabelund L.R. (1995). Bridging the gap between federal, state and local land management agencies: The New River Parkway land management system. In Linn R.M. (ed) *Sustainable society and protected areas,* pp. 293-300. The George Wright Society, Hancock, Michigan.

World Conservation Union. (1992). *Workshop Abstracts. Proceedings of the IVth World Congress on National Parks and Protected Areas.* IUCN, Caracas, Venezuela.

Integration of Information on the Kraków-Czestochowa Jura for Conservation Purposes: Application of CORINE Methodology

Anna Dyduch-Falniowska[1], Malgorzata Makomaska-Juchiewicz[1], Róża Kaźmierczakowa[1], Joanna Perzanowska[1], Katarzyna Zając[1]

[1] Institute for Nature Conservation, Polish Academy of Sciences, Kraków, Poland

Abstract. The Kraków-Częstochowa Jura constitutes a structural and functional entity and is an area of great natural value. It has been identified as a site of European importance in the CORINE system. The individual character of this area is determined by geological conditions, specifically geomorphology, and particularly karstic phenomena. The specificity of the nature of the Jura lies in the presence of very different type of habitats and associated mountain and boreal-mountain species, as well as southern taxa (Pannonic and Mediterranean). The Jura has preserved its natural richness under many centuries of human pressure. To understand this area fully, and formulate conclusions for conservation of the natural heritage of the Jura, it is necessary to collect and arrange all available information on the nature of the area. In this paper the integration of information is proposed according to CORINE methodology.

Keywords. biodiversity conservation, CORINE, Poland, Kraków-Częstochowa Jura

1 Why Integration, Why the Kraków-Częstochowa Jura?

The Kraków-Czestochowa Jura (Figure 1) is of interest because it has preserved its natural richness under many centuries of human use and pressure. To fully understand this phenomenon and formulate conclusions for conservation of the natural heritage of the Jura, it is necessary to collect and arrange all available information on the nature of the area. Truly interdisciplinary research is lacking on the nature of the Jura. Such research would be useful in preparing a general concept of management of this region, which is compatible with the idea of sustainable development.

Data for this paper are part of the data bank of the CORINE (Coordination of Information on the Environment) biotopes project. It is one of many CORINE programmes undertaken in Europe since the eighties. The Kraków-Czestochowa Jura as a whole corresponds with many aspects of CORINE methodology. It has been identified as a complex-site, which contains 29 individual sites. Each of these also meet CORINE criteria (Figure 1).

From the point of view of physical geography the Kraków-Częstochowa Jura should be treated as a geo-complex. It is a large plateau, rising in the west, with a steep rocky edge sloping to the north-east. This plateau is built of Upper Jurassic limestone. The south-eastern area is formed of Triassic and Paleozoic (Devonian, Carboniferous, and Permian) rock, while the central and eastern parts, are Cretaceous (Czeppe 1972). Specific features of the area include limestone monadnocks rising on the plateau, deeply incised valleys and gorges, and numerous caves. The rich and diversified relief is the result of geological processes working for many millions of years after the formation of base rocks of the contemporary Jura.

The Kraków-Częstochowa Jura distinguishes itself among other regions by its very rich vegetation. The Jura contains about 1300 vascular plant species, i.e. nearly 50% of the vascular plant species noted from Poland, over 400 moss species, and 400 lichen species. Approximately 90 plant communities have been described for the area; with about 50 associations identified beyond question.

Figure 1. Distribution of CORINE sites in Kraków-Częstochowa Jura. 1 - sites larger than 100 ha, 2- sites smaller than 100 ha, 3- boundary of complex-site.

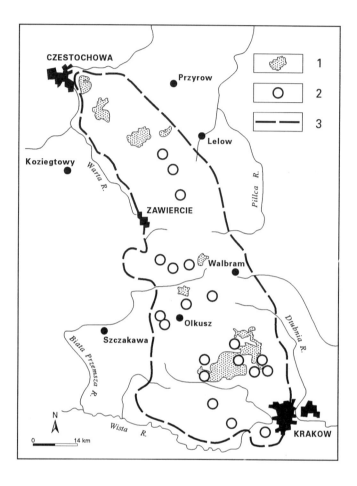

About 70% of the area is occupied by agrocenoses. Forest communities cover only about 20% of the area. The occurrence of the two types of mountain beech forests is noteworthy. In the southern part of Jura on shady and humid northern slopes of ravines there are extrazonal sites of *Dentario glandulosae-Fagetum*, an association characteristic of the lower montane belt of the Western Carpathians. In the central and southern parts of Jura we found patches of *Dentario enneaphyllidis-Fagetum* typical of the lower montane belt of the Sudety Mountains. Rare in Poland and well-developed in Jura is the mountain sycamore forest with the heart's tongue *Phyllitido-Aceretum*, growing in shady and humid places on rocky debris (Medwecka-Kornaœ 1952; Medwecka-Kornaœ and Kornaś 1963). The sunny southern slopes with rocky substratum are covered with the rare in Poland *association Carici-Fagetum convallarietosum*, a thermophilous beech forest (Michalik 1972). In the past on the plateau there were common mixed pine-oak forests. Most of these forests are greatly changed.

Among non-forest communities, the xerothermic vegetation is particularly noteworthy. It is represented in the Jura by natural and semi-natural grasslands developed on rock, grasslands on sand, and thermophilous scrub with hazel. We should mention here, above all, patches of loose natural grasslands on rocky substratum, belonging to the *Festucetum pallentis* association. This develops in two subassociations: *F. p. neckeretosum* in shady places and *F. p. sempervivetosum* in sunny places. Very rare taxa, such as *Thymus praecox*, or taxa which have relic sites in the Jura, such *as Saxifraga aizoon* of arctic-alpine affinity, are involved in this association. Xerothermic and thermophilous communities cover less than 1% of the area of the Jura. However, species in these communities constitute almost one third of the flora of the region. In valley bottoms are found different semi-natural meadow communities and in places, small fragments of peat bog, rush, spring, and water communities. The spring area of Biala River near Olkusz was once a site of the endemic association, *Cochlearietum polonicae* (Kwiatkowska 1957).

The Kraków-Częstochowa Jura flora is dominated by species widely distributed in Europe. Also numerous are taxa whose localities in the Jura are separated from their major ranges. This concerns, above all, relic localities of mountain species, steppe species, and boreal taxa. Two endemic species occur in the area: *Galium cracoviense* growing on limestone tors and *Cochlearia polonica* growing in cold springs on a few replacement sites in the southern part of the area. Both are placed in the Polish Plant Red Data Book (Zarzycki and Kaźmierczakowa 1993). Among other threatened in Poland plant species, there are about 50 which have stations in the Jura.

The character of the invertebrate fauna of the Kraków-Częstochowa Jura is determined by the presence of xerothermic habitats and post-glacial relics, and by connections with the Carpathians. The occurrence of over half of the species of particular invertebrate groups known from Poland, shows the importance of the Jura for the fauna of the country (Szeptycki and Warchalowska-Śliwa 1992). The most characteristic and valuable components of the Jura fauna are xerothermic elements. Two types of xerothermic invertebrate communities occur in the Jura. One group is connected with natural non-forest habitats, which originated as a result of the particular coincidence of topographic, microclimatic and other factors. Another group inhabits open sites arising from deforestation of the area. These two types of habitats differ from one another in soil fauna (Szeptycki and Warchalowska-Śliwa 1992).

Insect groups unrelated to the soil, such as butterflies, hymenopterans, or orthopterans, do not show this differentiation. In a study of the butterflies of Częstochowa Jura, Skalski (1992 a,b) estimated the proportion of lepidopterofauna species confined to dry habitats at 13%. Species now characteristic of this type of habitat in the Jura are *Lysandra argester (Bgstr.), L. coridon (Poda),* and *Meleageria daphnis*.

One may suppose that xerothermic zoocenoses in the Jura have a very dynamic composition. This can be illustrated by the results of studies on the coleopterofauna of the environs of Ojców. Pawlowski et al. (1994) analyzed this group in two periods: 1854-1914 and 1955-1990. In the first period 30 strictly xerothermic species were found, and in the second one, 50. Only 6 species were common for these two periods, i.e. they were noted in the past and at the present time. The authors suggest that these intriguing qualitative differences are connected with the migrations of species from several directions, facilitated by human impact on the uplands of southern Poland.

The majority of the regionally most valuable invertebrate species are connected with beech forests. A characteristic phenomenon is the simultaneous occurrence of western or alpine, and sudetic, and Carpathian species, in the Jura (Szeptycki and Warchalowska-Sliwa 1992). Particularly interesting is the occurrence of some high mountain species in the Jura, such as *Orchesella alticola* which is the representative of *Collembola*. These live in extremely cold places, for example cave holes or on fragments of rocks with such a microclimate. The arrival of these species was connected with a periglacial period, or with some other cold and woodless time.

Worthy of notice also are troglobiontic species of invertebrates. The cockchafers: *Choleva lederiana gracilenta, Catops tristis infernus* and a representative of *Collembola - Onychiurus alborufescens* were found in the Pod Sokkia Góra Cave; *Mesochorutes ojcoviensis (Collembola)* is known from the Nietoperzowa Cave; and the mite *Oribella cavitica*, from the caves of Ojców. The spider *Porrhomma moravicum* is the most numerous troglobionte. The sites of troglobiontes in the Jura are mostly of relic character; they are relics of the last interglacial period (Szelerewicz and Górny 1986).

The distribution of numerous species, valuable from the point of view of biodiversity conservation in the Jura, is an effect of the operation of many geobotanical, microclimatic, topographic and historical factors. It reflects the strong fragmentation of habitats, whose influence on the populations of particular species should be thoroughly analyzed if suitable conditions for conservation of the full biodiversity of Jura are to be created.

A characteristic feature of the mammal fauna of Kraków-Częstochowa Jura is the very rich group of insectivore species (83% of the insectivore fauna of Poland). Almost all known Polish species of bats occur here. Among them are three species characteristic of this area: the seriously threatened lesser horsehoe bat *Rhinolophus hipposideros*, the greater horsehoe bat *Rhinolophus ferrumequinum* and the extremely threatened Geoffroy's bat, *Myotis emarginatus*. Outside of the Kraków-Czestochowa Jura only single localities of the lesser horsehoe bat are known. Geoffroy's bat practically does not occur out of the Jura. The fauna of bats in the Jura is not satisfactorily known. The existing data concern, above all, species which are connected with rocky habitat, and winter in caves or in fortifications (Labocha and Wołoszyn 1994; Postawa et al. 1994). The greater horsehoe bat is sporadically known for the area; the last record dates from 1922 in the Wierna Cave at the CORINE site of Parkowe.

Of the mammal species placed in the Polish Red Data Book of Animals (ed. Glowacinski 1992), 9 species occur in the Jura. Among them there are 6 species of bats (*R. hipposideros, Myotis bechsteini, M. emarginatus, Vespertilio murinus, Eptesicus nilssoni* and *Nyctalus leisleri),* the European beaver (*Castor fiber*), the fat *dormouse (Glis glis*), and the otter (*Lutra lutra*). Most of them are rare species (category R).

The breeding avifauna of the Jura is relatively rich (Walaś and Mielczarek 1994). This is related to the exceptional differentiation of the environment. Birds occur which represent forest habitats, open areas (meadows and fields), scrub and mid-field groups of trees. Some birds are confined to water and marshy habitats, and others are linked to limestone tors (monadnocks). To this last group belong such species as *Apus apus*, or *Falco tinnunculus* which occur here in their natural environment. At the present time these species are known, above all, from towns and villages where they nest in walls, a substitute for their natural rocky habitat.

The reptiles of the Polish Jura include six species with three from the CORINE list. The threatened smooth snake, *Coronella austriaca*, is particularly worthy of notice. The species is recorded at Dolina Prądnika and Góra Zborów. Amphibians are represented by 13 species, including eight from the CORINE list, with *Bufo calamita, B. viridis and Hyla arborea*. More thorough research on the herptofauna has been completed in the northern part of Jura, in the Częstochowa Upland (Kowalewski 1992).

Summing up, the Kraków-Częstochowa Jura is a structured and functional entity - an area of great natural value. It should therefore be treated, independently of its administrative divisions, as one natural area with great inner variation.

2 Impacts on natural values of Human Activity in Prehistoric Times and in the Middle Ages

It is in the Kraków-Częstochowa Jura that the oldest traces of human presence in the territory of contemporary Poland have been found. Human groups appeared in this area about 120,000 years ago when a subarctic climate dominated. A little later, in the period of the last interglacial, small isolated groups of Neanderthal people lived in the Kraków Upland. Their subsistence economy was based on hunting (Kruk 1990). The scale of human impact on the nature of the Jura is thought to have been rather insignificant. People hunted for large animals: mammoths; hairy rhinoceros; cave bears. Small groups of hunters often changed hunting grounds, and traces of their presence were obscured by nature. This stage ended about 40,000 B.C. (Kruk 1990).

Homo sapiens appeared in the Jura in the middle phase of the Würm glaciation. Forest tundra was the preferred area of camping and hunting for bears, mammoths, aurochs, European bison and reindeer. Archeological finds dated from this period are found in the caves: Nietoperzowa and Koziarnia (Jerzmanowice culture), Mamutowa in Wierzchowie (Aurignacian culture), and in the open excavation at Spadzista Street in Kraków. These are sites from the Upper Palaeolithic. This stage of human presence in the Jura ended about 10,000 years ago when the climate became more severe and the Upland turned into arctic tundra (Kruk 1990). Also in this period the impacts on natural values of human activity were likely no greater than the capability of biocenoses to regenerate. Yet some archeological finds suggest intensive hunting pressure in the area which may have led to local extirpations. So the question arises why did humans abandon this area. Did they overexploit the populations of large mammals?

At the beginning of the Neolithic (4,500-1,800 years B.C.) the next stage of human settlement in the Jura began. The economy was based on the extraction and distribution of flint. Agriculture began to develop on the Jura plateau. The first limited groups of farmers most probably settled in small deforested areas, forming isolated aggregations, mainly in the neighbourhood of river valleys. Over time human settlements extended over larger and larger areas. A 50 ha settlement from this period was discovered at Olszanica on the right bank of Rudawa River. The settlers built houses mostly of oak logs; for other purposes hornbeam was used. They cultivated wheat in small fields obtained by clearing of forest by fire. But these fields were quickly abandoned.

A radical change in the method of farming took place in the period of 4,100-3,800 B.C. At that time, larger settlements originated, often on loess hills. They were more sparsely distributed, and suggest more extensive farming. At the beginning of the fourth millenium B.C. the pattern of occupation by settlers again changed. The permanently settled zone moved from the margins of valleys to the plateau. In agriculture more extensive methods of plant cultivation were introduced. The role of stock breeding increased (Godlowska et al. 1995). Information on the methods of husbandry in this period would be helpful in studying the possible consequences of these activities on nature. Undoubtedly, it is the first period when human economic activity brought about the differentiation of habitats and the origin of new regional biocenoses.

At the beginning of the Bronze Age (2,300-1,600 B.C.) a network of permanent settlements was formed; agriculture developed in river valleys, with stock and sheep breeding on deforested heights. Traces of this colonization in the Upland are found in Iwanowice, Szyce, Modlnica and other areas. Fields were most probably cultivated by forest-field rotation. Forest was cleared to obtain a place for plant cultivation, but stumps were left. At first, cereals were sown. After a lapse of time fields were abandoned for 15-20 years. The subsequent successional stages of abandoned fields were used as gathering grounds and as a kind of pasture. After about 20 years young forest was cut down, wood was used as timber, and a cleared area was utilized anew for cultivation purposes (Kadrów 1995). This kind of husbandry probably did not bring persistent negative consequences for the nature of the Jura. Certainly, it contributed to the maintenance of new semi-natural habitats. Conditions were also created for the incoming of species and communities characteristic of open areas.

The Early Bronze Age was followed by the stage of depopulation of the Kraków-Częstochowa Jura. We may suppose that in this period only caves were used as places where groups of people spent nights. Few traces of Lusatia culture have been dated as Bronze and early Iron Age in the southern part of the Jura. This rather long period, from 1,400 B.C. to the arrival of Celts, was characterized by a slower rate of colonization and cultural transformation. The question arises whether the existing earlier open habitats of anthropogenic origin could persist in these conditions. This problem needs further study.

In the 3rd century B.C. Celts appeared in the area of Jura. Traces of their colonization in the La Téne period are concentrated in the southern margins of the Kraków Jura, on both sides of Vistula River. Farther to the north they are relatively scarce (Godłowski 1995). In the La Téne period, i.e. up to the 1st century B.C., the natural consequences of human activity apparently were insignificant. Later on, from the 1st to 5th century A.D. colonization of the Jura intensified. Archeological finds from this period are mainly earthenware, tools and the like. Intense economic activity probably took place throughout the area of Kraków-Częstochowa Jura. At the end of this period intensified use of caves occurred, including some with not very comfortable conditions. Perhaps, they constituted refuges for people during the invasion of the Huns. Archeological finds from this period include different objects of material culture. However, no conclusions on the natural consequences of human activity can be drawn on the basis of this material. The distribution of archeological finds in the Kraków Jura (Godłowski 1995) shows that in this period people chose for settlement, areas that had been inhabited earlier such as river valleys, or adjacent heights. The scale of human impact on the nature of this region was likely rather small in view of the depopulation mentioned earlier.

No traces of early Slavonic occupation have so far been found in the Jura. Traces of two small settlements near Tyniec and three strongholds near Damice, Mników, and Zagorów, have been dated from the tribal period (7th-10th century). It should however, be stressed that the archeological knowledge of this period is not satisfactory. Nevertheless, the opinion seems justified that the Kraków part of Jura was weakly populated in the early Middle Ages and it constituted a base of supplies for main settlement centres (Poleski 1995)

In the late Middle Ages numerous fortifications were raised in the Jura, linking Kraków to Czestochowa. From 1228 when Prince Henry the Bearded built the castle of Skala, through the rest of the epoch of the Piast dynasty to the down-fall of the Polish state at the end of the 17th century, human impact on the nature of the Jura increased systematically. Its scale depended on the needs of castles and fortifications on the Eagle Nests Route. These structures were connected with the development of colonization, industry, and agriculture. A detailed analysis of the changes in nature, which accompanied this development, is not a simple task. Above all, it would be necessary to interpret anew, from the point of view of the nature, the material history of this region. Analyses made by Laberschek (1995) raise a hope that reconstruction of the human-nature relations in the Kraków-Czestochowa Jura will be possible. The later history of human impact on the nature of Jura is the history of mining and metallurgy, accompanied by intense urbanization. Only some parts of Jura have been the subject of historical studies which would enable the reconstruction of human impact on nature. The Prądnik River valley and Olkusz have been studied in this way.

3 Contemporary Threats to the Nature of Jura

The Kraków-Częstochowa Jura is situated among the three large urban-industrial agglomerations of Silesia, Kraków and Częstochowa. There are also numerous villages and small towns, and industrial plants located on the Jura itself. That is why the Jura is greatly affected by human pressure which threatens the existence of many plant and animal species, as well as whole ecosystems. As a result of intense urbanization and the rapid development of systems of roads and power-lines, valuable natural areas decrease. Similar effects are produced by building more and more popular recreational houses, which are especially dangerous for nature as these houses are usually located in the most attractive natural areas. The presence of human settlements is connected with contamination of the environment. The concentration of SO_2 in the air increases due to burning of sulphur rich coal in household stoves. Water and soil are polluted by sewage from farms and untreated municipal sewage (about 78% of the total amount of sewage). Illegal refuse grounds are numerous and dumping of litter directly into rivers and streams is a very common practice.

Many hundred years of colonization of the Jura have caused considerable deforestation and transformation of much of the area into arable land. For example, in the province of Czestochowa, arable land constitutes 60% of the area. Intensive farming is followed by the increased endangerment of the flora and fauna due to water pollution by fertilizers, insecticides and herbicides rinsed from fields and by dumping of refuse from stock-farms. As a result of drainage, hygrophilous species lose their habitats and are scarce in the Jura in any event. Burning of stubble and grassland destroys the invertebrate fauna. Arable soil is less resistant to water erosion, the results of which were observed during this year's (1996) heavy spring and summer rains.

Through clear-cutting and introduction of coniferous plantations in historical times, forest management has reduced floristic composition and transformed the structure of stands. The dying of coniferous trees, particularly vulnerable to industrial emissions, results in uncovering of the forest floor and changes in the structure of the herb-layer. Typical forest species recede and synanthropic and meadow species encroach. The introduction of foreign species, such as *Pinus nigra, Robinia pseudacaccia,* and *Padus serotina* is common. These changes in forest management favour the synanthropization of vegetation and withdrawal of vulnerable native species.

A characteristic feature of the Kraków-Częstochowa Jura, is the occurrence of solution pits and karst phenomena. They cause the escape of water deep into the ground; as a result of this, the network of surface waters is scarce. The natural water deficit has been increased by the deforestation of large areas, river channelization and drainage. Mining has also caused the lowering of the ground water table and formation of depression sinks. Dense population and developing industry bring an excessive uptake of water. Intake from spring water often leads to irretrievable destruction of these features. As a consequence, the Jura suffers from serious water shortage. Permanent drying of habitats contributes to their transformation and the dying off of hygrophilous species.

Industry located at the outskirts of the Jura is very burdensome for the environment. Large sections of rivers carry waters which cannot be directly used because they are polluted by industrial wastes, municipal sewage, and mine waters that are strongly saline and rich in heavy metals. An analysis of the physical and chemical properties of waters in the province of Czestochowa showed that these waters do not meet any of the prevailing water quality standards. Also *Coli* exceeded the acceptable level. Extractive industries (gravel pits, quarries) locally have caused the complete mechanical destruction of vegetation and the upper horizons of soil. Dumps and refuse grounds produce similar results. Particularly dangerous are long-range emissions. Winds bring large amounts of pollutants. Chemical compounds either directly affect living organisms (causing e.g. chlorosis and necrosis of tree leaves), or they accumulate in soil and water, particularly in deep Jurassic valleys. In conditions of still air and high humidity; their effects are visible for some time.

The input of pollutants in the Kraków-Częstochowa Jura was estimated according to the methodology of the CORINAIR programme (one of the CORINE programmes), separately for a group of eight compounds (SO_x, NO_x, NMVOC, CH_4, CO, CO_2, N_2O, NH_3), 11 possible sources of emissions were taken account. The method of data processing allowed for determination of the share of a given source in the emission of particular groups of compounds. For example, the main sources of SO_x emissions are heat and power generating plants (48%), processes of combustion in industry (30%) and local domestic heating (14%). The magnitude of emissions has been estimated for particular provinces, so it is difficult to determine the average level of pollution for the Jura, situated as it is in three provinces, and also because of the great local differentiation of emissions (Table 1) (Pazdan 1995).

The picturesque landscape with its interesting tors and diverse flora and fauna make the Jura exceptionally attractive for tourists. However, the rapid development of tourism, especially motorized tourism, brings with it development pressures for the construction of new roads, parking facilities, and picnic places. Participants in group excursions destroy the forest herb-layer, beat new paths, pick up flowering specimens of plants, litter the area with rubbish, make noise, and startle wild animals. Lovers of rock-climbing leave the paths leading to tors trampled to bare rock. They damage rocky crevices and walls. Speleologists damage the dripstone of caves and startle cave-dwelling bats.

Table 1. Emission of selected groups of pollutants for three provinces in 1990 [in Mg].

Group of pollutants	Katowice	Kraków	Czestochowa
SO_x	662 500	150 000	40,600
No_x	262 500	71 250	22 500
NMVOC	90 000	35 000	27 000
CH_4	2500 000	75 000	62 000
CO	1800 000	700 000	260 000
NH_3	1 650	875	1 100

Apart from the transformations of the environment, caused by human activity, we observe other natural changes, undesirable from the point of view of nature conservation. Preservation of semi-natural habitats, such as xerothermic grasslands, so important for biodiversity conservation, requires human interference. It is necessary to apply historic forms of farming, such as moderate grazing and irregular mowing, otherwise, grasslands left to their fate quickly undergo succession towards scrub and forest. Overshadowing causes a decrease in the number of species because light-loving plant and invertebrate species withdraw from the area.

4 Nature Conservation in the Kraków-Czestochowa Jura

Protected areas in the area of Kraków-Częstochowa Jura include one national park (Ojców - 1,592 ha), 25 nature reserves (c. 1,308 ha in total) and 6 landscape parks (Orle Gniazda, Dlubnianski, Dolinki Krakowskie, Tenczyński, Rudniański and Bielańsko-Tyniecki) covering 102,115 ha in total. Other areas in the Jura are a zone of protected landscape. Thus, the whole area of the Jura theoretically has protection, which means that we are sensible to its great natural value. However, the protective regime of landscape parks - to say nothing of the areas of protected landscape - is rather liberal and practically it does not guarantee the protection of landscape values, threatened habitats, or sites of rare plant and animal species. Another national park (Jura NP) is proposed for the northern part of the Jura (Czestochowa province). This park would include, among others, 4 CORINE sites (Skaly Jurajskie near Olsztyn, Sokole Góry, Parkowe and Góry Gorzkowskie). Identification of such sites of European importance in this area is a strong argument for taking it under protection as a national park which is the highest category of protection in Poland. Most of the CORINE sites selected in the Jura are situated in nature reserves or landscape parks.

5 What We Need to Protect the Biodiversity of Jura: The Distribution of Natural Sites, Fragmentation and Patchiness of Habitats, Ecological Barriers and Corridors

The uniqueness of the Kraków-Częstochowa Jura lies in the fact that despite many hundreds of years of human pressure, very rich flora and fauna have survived there. Some biocenoses have preserved their natural character. Others contain valuable natural objects. That is why the Jura is considered as an area of great natural value. As earlier mentioned, our knowledge of the nature of Jura is incomplete and unsystematic, so this knowledge cannot be the basis for a co-ordinated plan of protection for the whole area. There is relatively rich information on the Kraków Jura (Ojców Plateau) but single reports only for the northern part of the area (e.g. Skały Jurajskie near Olsztyn, Sokole Góry). The results of studies concerned with the fragments of Jura are often extrapolated for the whole area. However, spatial planning and rational conservation of the biodiversity of Jura can not be based on such an approach. Many species have very limited ranges in the Jura. An example is *Galium cracoviense*, a species endemic to Poland, so far known only from limestone tors near Olsztyn. Among the cockhafers, there are many species which have their only sites in the environs of Ojców (Pawlowski et al. 1994). Similar examples may be found in other groups of invertebrates. We know that many species of molluscs have insular distribution in the Jura.

Necessary complements to the faunistic and floristic recognition of the Jura should include the most valuable elements of biotopes (species, plant associations, geomorphological forms). They should be considered as the structural-functional elements of systems to which they belong. Hence, it is necessary to integrate information on them so that this integration can be a basis for spatial planning satisfying the needs of biodiversity conservation. A possible approach to this question is the integration of information according to CORINE methodology. This involves different levels of organization of ecosystems and biocenoses. Within the scope of the CORINE biotopes programme, this integration would comprise the distribution of species important from the point of view of European natural heritage conservation. To meet the requirements of nature conservation on the national, or regional scale, the CORINE data bank should be completed, particularly with field autecological observations.

A next step in the work should be the integration of the distribution of CORINE sites with the land cover map. Analysis of the land cover categories, especially those separating CORINE sites, should be undertaken from the point of view of their functions as ecological barriers or corridors. To make the results of the integration useful in spatial planning on the regional or local scale, further field research should be carried out as well. This will enable the precise identification of ecological barriers and corridors within the geocomplex of the Kraków-Częstochowa Jura. CORINE sites constitute so-called biocentres and it is necessary to ensure contact among them. For example, two sites, Dolina Pradnika and Dolinki Podkrakowskie, are separated by a narrow belt of cultivated fields and settlements. In preparing the land use and management plans for this area, provision should be made for a corridor connecting these two sites, so that the plants and animals can move freely between the sites. Faunistic and floristic studies, much more detailed than those made in the Jura up to now, are needed for this work. It is necessary to take into account the ecological requirements of species and entire biocenoses.

It has been already mentioned that preservation of the natural values of the Jura under conditions of many hundred of years of human pressure is an intriguing challenge. To examine the biogeographic and human conditions influencing the present state of the flora and fauna of the Jura, we are planning a historical study of colonization in selected sites of European and national importance, as well as the effects of this colonization on the present shape of biotopes and biocenoses. Conclusions from this study should be useful in nature conservation. They should be helpful in building up proper relations between local communities and protected areas. It is a well known fact that the success of any undertaking in the field of nature conservation is conditioned by the ecological consciousness of communities. The Kraków-Częstochowa Jura seems to be a particularly appropriate area for comprehensive natural education. That is why it would be advisable to prepare natural and cultural educational or interpretation trails or paths on the basis of selected CORINE sites.

6 References

Czeppe Z. 1972. Rzezba Wyżyny Krakowsko-Wieluńskiej (Relief of the Cracow-Wielun Upland). *Studia Osrodka Dokumentacji Fizjograficznej* 1: 20-30.
Godłowska Z. (ed.) 1992. *Polska Czerwona Księga Zwierząt* (Polish Red Data Book of Animals). PWRiL, Warszawa.
Godłowska E, Kozlowski J.K., Kruk J., Lech J., Rook E. 1995. Młodsza epoka kamienia (The Neolithic). W (In:): *Natura i kultura w krajobrazie Jury.* T.IV. Pradzieje i (Średniowiecze. Praca zbiorowa. Zarząd Zespolu Jurajskich Parków Krajobrazowych w Krakowie, Kraków, p. 49-84.
Godłowski K. 1995. Okres latenski i rzymski (The La Téne and Roman Periods). W (In): *Natura i kultura w krajobrazie Jury.* T.IV. Pradzieje i Œredniowiecze. Praca zbiorowa. Zarzad Zespolu Jurajskich Parków Krajobrazowych w Krakowie, Kraków: 113-136.
Kadrów S. 1995. Poczatki epoki brazu (dzieje iwanowickiego mikroregionu osadniczego) (The beginnings of the Bronze Age). W (In): *Natura i kultura w krajobrazie Jury.* T.IV. Pradzieje i Średniowiecze. Praca zbiorowa. Zarzad Zespolu Jurajskich Parków Krajobrazowych w Krakowie, Kraków: 85-92.
Kowalewski L. 1992. Herpetofauna Wyżyny Częstochowskiej i jej przemiany w ubieglym 20-leciu (Herpetofauna of the Czestochowa Upland and its changes during the last 20 years). *Pradnik, Prace Muz. Szafera* 5: 247-265.
Kruk J. 1990. Osadnictwo pradziejowe na Wyzynie Krakowskiej. W (In): *Jurajskie Parki Krajobrazowe województwa krakowskiego.* Informator krajoznawczy. Ed. J. Partyka. Zarzad Zespolu Jurajskich Parków Krajobrazowych Województwa Krakowskiego, Wydawnictwo Karpaty, Kraków: 21-22.
Kwiatkowska A. 1957. Rozmieszczenie warzuchy polskiej (*Cochlearia polonica* E. Fruhlich) w okolicy Olkusza (Distribution of *Cochlearia polonica* E. Fruhlich in the environs of Olkusz). *Fragm. Flor. Geobot.* 3: 11-15.
Laberschek J. 1995. Osadnictwo sredniowieczne w swietle zródel pisanych (Mediaeval settlement in the light of written sources). W (In): *Natura i kultura w krajobrazie Jury.* T.IV. Pradzieje i Sredniowiecze. Praca zbiorowa. Zarzad Zespolu Jurajskich Parków Krajobrazowych w Krakowie, Kraków: 163-178.
Labocha M., Woloszyn B. W. 1994. Dekady spisu nietoperzy na Wyzynie Krakowskiej (Winter bat censuses in the Krakowska Upland). W (In): *Zimowe spisy nietoperzy w Polsce 1988-1992. Wyniki i ocena skutecznosci* (Results of the winter bat census in Poland: 1988-1992). Ed. B.W. Woloszyn. Publikacje Centrum Informacji Chiropterologicznej ISEZ PAN Kraków: 104-122.
Medwecka-Kornaœ, A. 1952. Zespoly lesne Jury Krakowskiej (Forest Associations of the Jurassic Region near Kraków (Cracow). *Ochr. Przyr.* 20: 133-236.
Medwecka-Kornaœ, A, Kornas J. 1963. Mapa zbiorowisk roœlinnych Ojcowskiego Parku Narodowego (Vegetation map of the Ojców National Park). *Ochr. Przyr.* 29: 17-87.
Michalik S. 1972. Cieplolubne lasy bukowe na Wyzynie Krakowsko-Czestochowskiej (Thermophilous beech forest Carici-Fagetum /Moor 1952/ emend. Hartmann, Jahn /1967/ in the Kraków-Czestochowa Upland). *Fragm. Flor. Geobot.* 18: 215-225.
Pawlowski J., Mazur M., Mlynarski J.K., Stebnicka Z., Szeptycki A., Szymczakowski W. 1994. *Chrzaszcze (Coleoptera) Ojcowskiego Parku Narodowego i terenów oœciennych* [Beetles (Coleoptera) of Ojców National Park and its environ]. Ojcowski Park Narodowy, Prace Muz. Szafera, Ojców.
Pazdan W. 1995. CORINAIR'90. ATMOTERM s.c., Opole.
Poleski J. 1995. Wczesne Œredniowiecze (The early Middle Age). W, *In Natura i kultura w krajobrazie Jury.* T.IV. Pradzieje i Sredniowiecze. Praca zbiorowa. Zarzad Zespolu Jurajskich Parków Krajobrazowych w Krakowie, Kraków: 137-150.

Postawa T., Wegiel A., Zygmunt J. 1994. Dekady spisu nietoperzy na Wyzynie Czestochowskiej (Winter bat censuses in the Kraków Upland). W (In): *Zimowe spisy nietoperzy w Polsce 1988-1992. Wyniki i ocena skutecznosci* (Results of the winter bat census in Poland: 1988-1992). Ed. B.W. Woloszyn. Publikacje Centrum Informacji Chiropterologicznej ISEZ PAN Kraków: 130-148.

Skalski A.W. 1992a. Zmiany fauny motyli dziennych Wyzyny Czestochowskiej (Changes in butterfly fauna of the Czestochowa Upland). *Pradnik, Prace Muz. Szafera* 5: 191-222.

Skalski A.W. 1992b. Rozsiedlenie motyli dziennych (*Lepidoptera: Papilionoidea+Hesperioidea*) na Wyzynie Czestochowskiej [Distribution of butterflies (*Lepidoptera: Papilionoidea+Hesperioidea*) on the Czestochowa Upland]. *Ziemia Czestochowska* 18: 179-192.

Szelerewicz M., Górny A. 1986. *Jaskinie Wyzyny Krakowsko-Wielunskiej*. [Caves of the Kraków-Wielun Jura] Wydawn. PTTK Kraj, Warszawa-Kraków.

Szeptycki A., Warchalowska-Sliwa E. 1992. Charakterystyka fauny Wyzyny Krakowsko-Czestochowskiej (Characteristic of the fauna of the Kraków-Czestochowa Upland). *Pradnik, Prace Muz. Szafera* 5: 149-159.

Walas K., Mielczarek P. (eds.) 1994. *Atlas ptaków legowych Malopolski 1985-1991* (The atlas of breeding birds in Malopolska 1985-1991). Biologia Silesiae, Wroclaw.

Zarzycki K., Kazmierczkowa R. (eds.). 1993. *Polska Czerwona Ksiega Roslin* (Polish Plant Red Data Book). Instytut Botaniki PAN, Kraków.

The Potential Role Of Biosphere Reserves In Piloting Effective Co-Operative Management Systems For Heritage, Landscape and Nature Conservation

Henry Baumgartl[1]

[1] Karkonosze Foundation
PO Box 2298
NL-1620 EG Hoorn
THE NETHERLANDS

Abstract. Contributing to the sustainability of our biosphere through effective conservation of landscape and biological diversity is a complex effort, involving a wide variety of scientific, professional, social, cultural and economic considerations. For conservation to work it is, therefore, essential to find common ground and to be absolutely certain everyone understands each other without the constraints or potential threats of, for instance, unfamiliar languages, concepts or professional jargon. In spite of many recent attempts to create the right conditions for interdisciplinary groups to work together productively, the present state of landscape and biological diversity, and any realistic prospects for the near future, do not appear encouraging at all. Some very strong action needs to be taken soon to improve communication and to develop effective information/knowledge exchange mechanisms. The Biosphere Reserve concept, as developed since the early 1970s in the UNESCO Man and the Biosphere (MAB) programme, is well suited to provide a solid base for the development of such mechanisms. This is especially so in Central and Eastern European Countries (CEECs), where the MAB programme has been received very well and innovation is a must, rather than a may be, in all layers of society.

Keywords. Biosphere reserves, co-operation, communication, Internet, landscape and biological diversity, nature conservation, science, Europe

1 Introduction

Contributing to the sustainability of our biosphere through effective conservation of landscape and biological diversity is a complex effort, involving a wide variety of scientific, professional, social, cultural and economic considerations. For conservation to work it is, therefore, essential to find common ground and to be absolutely certain everyone understands each other without the constraints or potential threats of, for instance, unfamiliar languages, concepts or professional jargon. In spite of many recent attempts to create the right conditions for such interdisciplinary groups to work together productively, any realistic prospects for the near future do not appear encouraging at all. Some very strong action needs to be taken soon to improve communication and to develop effective information/knowledge exchange mechanisms. The Biosphere Reserve concept, as developed since the early 1970s in the UNESCO Man and the Biosphere (MAB) programme, is well suited to provide a solid base for the development of such mechanisms. This is especially so in Central and Eastern European Countries (CEECs), where the MAB programme has been received very well and innovation is a must, rather than a may be, in all layers of society.

Conceptually biosphere reserves are to fulfil three complementary functions. Namely, a conservation function, to preserve genetic resources, species, ecosystems and landscapes, a development function, to foster sustainable economic and human development, and a logistic function to support demonstration projects, environmental education and training, and research and monitoring related to local, national and global issues of conservation and sustainable development.

Based on experience from the beginning of the Programme in 1971 until the early 1990s, MAB is now entering a new phase, focusing on the following elements:

a. The development and full use of the existing network of sites, identified as biosphere reserves, of which 328 exist in 82 countries as of June 1995. Biosphere reserves are areas of terrestrial and coastal/marine ecosystems, where, through appropriate zoning schemes and management mechanisms, the conservation of ecosystems and their biodiversity is combined with the sustainable use of natural resources for the benefit of local communities, including relevant research, monitoring, education and training activities. Biosphere Reserves represent a major tool for implementing the concerns of Agenda 21, the Convention on Biological Diversity and other international agreements. The future of the World Network of Biosphere Reserves will be guided by the Strategy and Statutory Framework for the world network drawn up at the International Conference on Biosphere Reserves held in Seville, Spain, in March 1995.
b. Continuing efforts to reconcile conservation and sustainable use of biological diversity with socio-economic development and maintenance of cultural values; these efforts are needed at the ecosystem and landscape levels, and should cover different geographical units such as catchment basins, land-water interfaces and urban-rural systems in different parts of the world.
c. Building up human and institutional capacities, including communication networks using modern technologies, to help countries address complex, cross-sectoral issues of environment and development.

This new phase of the MAB programme will be conducted in close co-operation with the appropriate partners such as UNEP, FAO, ICSU, IUCN, ISSC and relevant programmes such as Diversitas (IUBS, SCOPE, UNESCO, ICSU, IUMS), Ecotechnie (UNESCO, Cousteau Foundation), People and Plants (UNESCO, WWF, Kew Botanic Gardens), the Global Terrestrial Observing System (UNESCO, UNEP, FAO, WMO and ISCU), as well as other relevant UNESCO activities. The above activities concern mainly the UNESCO - Division of Ecological Sciences (SC/ECO) (Source UNESCO - MAB - WWW pages at http://www.unesco.org/mab/).

Since 1976, with the first Biosphere Reserves, it has been the intent of the MAB programme to protect nature and promote good stewardship for human use of the biosphere without degrading or destroying its natural resources. At first, the MAB concept challenged established points of view, and the Biosphere Reserves were little more than a vision identified and approved as such, but rarely fully functional and frequently lacking even a skeleton management system to support implementation of the integrative and essentially interdisciplinary feature of the MAB concept.

Much has changed since then, and the goals, as well as the methods, proposed by MAB are now to be found on almost all scientific and political agendas. At the turn of the 1990s, the biosphere reserve concept has been given a precious opportunity to substantiate its promise as the changes in Central Europe leave political and economic voids to be filled with new, and hopefully better, systems.

2 Central Europe

Central European countries may well pride themselves on their extensive areas with high biodiversity values. However, these values exist largely as a result of political, economic, and military considerations that are no longer valid, and it is one of the most important challenges of our time to provide the framework that will safeguard these values for future generations. A strong asset in the realization of this complex task is the many highly trained conservation professionals and scientists in Central Europe. They are the critical factor in any serious effort to develop and implement biosphere reserves as practical realities in the Central European countries.

3 An Integrated Conservation Management Effort In Central Europe

The Global Environment Facility (GEF) programme as managed by the World Bank, has responded to this potential by initiating a major effort to strengthen selected Biosphere Reserves (BRs) in Central Europe. Since 1991, the scope and substance, as well as the number of projects, countries, and additional sponsors have steadily increased to cover a wide variety of elements of biosphere reserve management.

Important in all Central European GEF and associated projects, are close connections through committees, management meetings, workshops, training sessions, consultation, and reporting. Continuously improving co-ordination through effective exchange of information and ideas has created a strong base for co-operation at all levels, as well as recognition of the need to expand external involvement in biodiversity protection by closely involving local Government Organizations (GOs), Non-Government Organizations (NGOs), and other groups (e.g. schools) in the process. Another positive effect of the co-ordinated approach has been the commitment to share attained skills with others through the GEF framework. To translate experiences to others with often different backgrounds has generated a widening perspective for everyone involved.

The GEF also provided essential funding to strengthen local research in Biosphere Reserves (BRs). Over 250 scientists received support from GEF Central European Biodiversity Protection projects. Many also attended priority scientific meetings where selected subjects have been summarised and made available to a wider audience.

UNESCO-MAB and GEF initiated a joint effort to train BR people in the use of facilities provided by the Internet. The first workshop, with people attending from Czech Republic, Slovakia and Poland, was organised in 1995 in Warsaw, and was received very well. Although the participants had very different backgrounds, they all agreed that the Internet would provide them with a tool to work as an active participant in an international network of peers, and help considerably in effectively implementing the interdisciplinary context of biosphere reserve management.

The apparent results of the foregoing approach have attracted further support from:

- the MacArthur Foundation of the United States, supplementing funds made available through the Slovak GEF biodiversity protection project, to establish a Foundation for the Protection of Biological Diversity in the trilateral biosphere reserve (Poland, Slovakia, Ukraine) of the Eastern Carpathian Mountains.
- the European Union, and its regional programme for the Environmental Action Plan for Central and Eastern Europe, to directly support three projects of biosphere reserves, with trans-national co-operation in place or under development, along the southern border of Poland. These projects deal with institutional strengthening and some urgently needed investment to prevent the irreversible loss of biodiversity values in these areas.

A particularly interesting output of one of these projects was the establishment of a bilateral conservation management structure in the Karkonosze / Krkonose Bilateral Biosphere Reserve (Czech Republic - Poland). The establishment process may well be considered a practical model for similar activities elsewhere. All three projects included a general requirement to ensure active involvement of local stakeholders (i.e. workshops) and dissemination of experience to professional peers.

At the conclusion of these projects, the EU provided some additional funding to organise a workshop (June 1996) with the specific purpose of generating an Action Plan to be presented to the European Commission, based on the experience generated by the effort and with recommendations for further strengthening the integrated management of these valuable areas.

- A number of small grants have also been provided from various sources (e.g. the Dutch Ministry of Housing, Spatial Management and the Environment, Environmental Partnership for Central Europe) to support NGOs directly involved in the training effort, but not covered by the larger projects. This includes, for instance, a grant to employ a person full time for the Polish office of the Karkonosze / Krkonose Bilateral Biosphere Reserve, and a similar grant forthcoming in the Czech Republic.

4 Conclusion

Without the conceptual framework of biosphere reserves, the foregoing co-operative efforts would have been impossible. It is, therefore, somewhat surprising to note that any resistance to the efforts came from members of the UNESCO-MAB community, and were mainly concerned with interpretive issues over the use of words.

The communication issue is the most important one now as much of the knowledge needed for better management is available amongst the various disciplines and professions. Financial resources are also often available to support the realization of viable biodiversity protection objectives. The communication problem in the MAB community may be largely attributable to an apparently deeply rooted inability to communicate effectively, often as a result of adherence to the definitional detail required by scientists in their work but unrealistic and unnecessary in interdisciplinary and integrative work. The consequence is difficulty in linking biological, ecological, social and economic ideas and information for conservation of landscape and biological diversity as well as preservation of cultural heritage on a bioregional scale.

Research to find viable methods, techniques and performance indicators for more effective communication and information exchange is an urgent need for the immediate future.

5 Reference

UNESCO - MAB - www pages at http://www.unesco.org/mab/

The Berezinski Biosphere Reserve in Belarus: Is Ecotourism a Tool to Support Conservation in the Reserve?

Sylvie Blangy[1]

[1] 123 Rue de la Carrierasse
34090 Montpellier, France

Abstract. The Berezinski Reserve has been a protected area for the past 70 years. In 1979 it was designated an UNESCO Biosphere Reserve. The Reserve is located 100 km north of Minsk, the capital of Belarus. It is located in the European-Siberian region of the Palearctic between the watersheds of the Black and Baltic Seas in the southern Taiga zone. It stretches over 120,000 ha. A rural population of 1500 is spread out in the buffer zone in eight small communities and in one large settlement located in the centre which hosts the administrative buildings and headquarters of the Reserve. The inhabitants live from traditional agriculture of Kolkhose type, involving cattle raising, hay making and potatoes and rye growing. The Reserve has not experienced a high level of tourist visits. School children, hunters and officials were the main visitors during the Soviet period. It was only in 1994 that the Reserve first hosted a group of naturalists from France. Berezinski Biosphere Reserve can aspire to becoming a new destination in Eastern Europe for Western naturalists. The potential is high. But the natural resources are threatened. The Reserve is under high hunting pressure and there is a sense of urgency.

Keywords. Berezinski Biosphere Reserve, Belarus, biosphere reserves, ecotourism, nature conservation, community, international aid and assistance, research, exchanges

1 Introduction

The Berezinski Reserve has been a protected area for the past 70 years. In 1979, it was designated an UNESCO World Biosphere Reserve. It was given the European Diploma in 1994. The Reserve is located in Eastern Europe 100 km north of Minsk, the capital of the Republic of Belarus. It is located in the European-Siberian region of the Palearctic between the watersheds of the Black and Baltic seas in the southern taiga zone extending over 120,000 ha. The Berezina river runs for 110 km across the Reserve and is supplied by small tributaries, large lakes and marsh areas. Forests cover 80% of the area. Some of it is old growth pine forest or black alder and pubescent birch marshy woods. There are natural woodless mires, considered as unique natural complexes within Eastern Europe.

A rural population of 1500 people is spread out in the buffer zone in 8 small communities and in one large settlement located in the centre which hosts the administrative buildings and the headquarters of the Reserve. The inhabitants live from traditional agriculture of the Kolkhose type, involving cattle raising, hay making, and potatoes and rye growing.

The Reserve is home to 50 species of mammals, 230 species of birds, as well as 10 amphibian, five reptile, and 34 fish species. It contains endemic populations of about 600 beaver, 800 elk, 120 otter, 40 bears, 30 lynx, 20 wolves and 40 badgers. About 12 pairs of osprey, 25 pairs of black stork, 8 pairs of short toed eagle, 35 pairs of grey crane, and 20 pairs of owl breed in the Reserve. Permanent breeding grounds of the golden and the white tailed eagle and of the eagle owl still exist. Species of the temperate zone such as the peregrine falcon, the black and willow grouse, the three toed woodpecker, and the golden plover also breed in the Reserve. Seven-hundred and eighty species of plants are found in the Reserve, many of which are rare in Belarus.

This density of wild plant and animal life makes the Reserve very attractive for naturalists from Western Europe. The wolf in particular is one of the strongest motivations for visiting the Reserve.

Berezinski can be described as one of the most exciting and exceptional wildlife destinations on the doorstep of Western Europe because it remains largely undiscovered and under-reported.

Until 1994, the Reserve had not experienced a high level of visitation. School children, hunters and officials were the main visitors during the Soviet period. It was only in 1994 that the Reserve first hosted a group of naturalists from France. This experience was very positive and the Reserve decided to extend this form of tourism. The Reserve managers were very motivated to develop a sustainable way of managing the natural resources.

The main potentials of Berezinski are the following:

- Its location in Central Europe and its easy access, although the visa procedure needs to be improved
- The variety of large mammals which are rare or have disappeared in Western Europe, the diversity of landscapes (wetland and marsh areas)
- A high sense of hospitality within the staff as well as among local villagers
- A high capacity for integrating tools and concepts and a strong motivation for sustainable development and tourism in the management team
- Vivid cultural traditions such as singing, dancing, embroidery, cooking, local products
- The human presence with 8 traditional villages, wooden houses, sauna and empty rooms and houses available for guests (bed and breakfast)
- The existing cottages and dachas devoted to very special guests (officials, hunters) which can be made available to foreign naturalists
- The diverse environmental education infrastructures (museum, conference center, video cassette, interpretive signs, board walk, watching tower, hides...) and the awareness campaign undertaken with visiting school children.

The following can be considered as drawbacks:

- The management of wildlife in general is questionable
- Hunting, poaching, berry picking, and fishing put high pressure on the natural resources
- The population of wolves is very limited within the Reserve (20 individuals). This population could be higher (cf expertise on wolf). Wolves are still hunted in Belarus
- The impact of hunting on wildlife behaviour is high. As a result wildlife is elusive and observations are difficult. The staff feel bound to track the wildlife for successful observations. There is a lack of adequate wildlife observation protocol, for example watching without disturbing and a minimum impact observation policy
- The lack of applied scientific research on management and the lack of strong scientific collaboration with foreign laboratories
- The lack of funding and an appropriate strategy for conservation in Belarus
- The inadequacy of accommodation for naturalists
- The lack of a bilingual naturalist guide
- The infancy of the tourism industry in general in Belarus and the lack of private initiatives
- The absence of inbound operators with conservation in mind

2 Technical Assistance and Scientific Cooperation

The potential of the area remained unknown until the French Natural Regional Park of the Vosges du Nord in 1991 initiated a scientific cooperation program with the Reserve. With the financial support of the French Embassy and the Council of Europe, and with the assistance of the French Park, a few expert exchanges and consultancies were then carried out by a journalist, a photographer, a wolf specialist, an ecotourism consultant, a wildlife tour operator, and a botanist.

A Berezinski delegation has been hosted twice in France by the Vosges du Nord Natural Regional Park. The ornithologist, the botanist, the director, and the tourism officer of the Reserve, as well as a scientist from Minsk University have had the opportunity to visit other Biosphere Reserves and meet biologists in France. Part of the same delegation has been offered the chance to attend several training programs on Environmental Education and on Interpretation and Tourism in England, Spain and Slovakia.

2.1 Results

The expert visits have been followed by several reports widely distributed in the Reserve and several contacts with French and British wildlife tour operators.

- Four tour operators have programmed seven trips in 1996, mainly wildlife watching trips and a couple of participatory packages (a bird census and atlas up-dating trip). Donations have been collected from visitors and converted into binoculars, bird books, telescopes, torches
- Two French journalists and photographers have published two articles in Environmental Magazines which has brought much attention to the Reserve.
- Hunting tourism is being slowly replaced by wildlife tourism
- Short visits of French researchers have brought some benefits to the Reserve through discussions and exchanges. A Belarus student is now being hosted by the French Park, completing a Ph.D. on a comparative botanic study. The wolf specialist has brought some knowledge and new tips on how to census big mammals.
- A foundation and non-government organizations have been created for the Reserve.

2.2 Problems Encountered

Belarus, as a newly independent country with no energy resources, is in great economic difficulties. The system of protected areas is suffering from lack of funds and from delays in wage payments. Other sources of funding like international aid, sponsorship and private investments are very welcome, but they are scarce. No financial support has been found for 1997 for infrastructure and new means of transportation. Belarus is not eligible for European Programs such as Phare, Tacis, GEF, Ecos-Ouverture, UNESCO, or Ford Foundation.

The 10th anniversary of Chernobyl and the recent political situation have brought negative publicity to Belarus as a tourism destination. Some of the trips have been cancelled due to the risk of radioactivity and political instability. The income from wildlife watching groups has not been properly invested into conservation and tourism infrastructures. The field of ecotourism is becoming a source of conflict among the staff. The tourism department has not been set up yet and the staff needs training.

The status of the wolf remains problematic. It is still persecuted and this persecution is based on ideas lacking a solid scientific foundation. It is unlikely that large game populations are adversely affected by wolf populations. The Reserve can accommodate and sustain a significant resident wolf population and the wolf can be a foundation element in promoting Berezinski Reserve.

2.3 Projects for 1997

The Council of Europe and the Natural Regional Park of Vosges du Nord remain the main support of the Berezinski Reserve. They plan to finance and encourage several actions and consultancies for:

- an evaluation of the 1996 tourism program, followed by a draft of an ecotourism strategy, a management plan, and some guidelines for visitors and tour operators.
- the setting up of a credible tourism department at the Reserve headquarters, consisting of a energetic and carefully appointed team. The representatives of this department should undertake a wildlife tour overseas and an appropriate training program for the tourism division should be designed (packaging, quotation, marketing).
- development tools to measure visitor impacts on natural resources
- the development of a community based tourism package and some combined tour packages between Belarusian Reserves

- the creation of a ground operator in Minsk who works with conservation in mind
- the design and production of various promotion and information tools like a map for visitors, a presentation brochure, a newly designed logo and a package of educational materials for children with translation into English and French
- a promotional week, offered to selected wildlife tour companies that would each send a representative and test the packages
- the organization of a conference on ecotourism for Eastern and Central European Countries with field trips and case studies in the various reserves and a trade show between Western and Eastern tour operators hosted by the Reserve.
- the development of collaboration, and the extension of the research program to other countries and other fields (host foreign visiting scientists for long periods of time and send Belarusians to foreign laboratories). A wolf specialist from Canada (radio tracking) has been contacted. A Belarus ornithologist will be hosted in France for 6 months next year.
- the greatest possible protection to the wolf, recognizing it as the most important asset of Berezinski Reserve; a special scientific cooperation program is planned to evaluate the impact of the wolf on game predation
- collaboration with Bielowiesa Pushka, and other reserves of Belarus (Berezinski can be a leader in the conservation and ecotourism program in Belarus) and collaboration with neighbouring countries like Poland on promoting nature tourism in Eastern Europe
- exchanges on sustainable management with other reserves and parks in Europe and in other continents (American and Australian...)
- development of exchanges with other towns for educational, cultural, artistic, and environmental purposes with Western countries
- fundraising and partnerships with sponsors, environmental investors and subscribers.

3 Final Observations

The scientist Jean-Claude Génot, wildlife and conservation officer at the Vosges du Nord Natural Regional Park, has played a major role in the development of sustainable tourism within the Berezinski Reserve. His involvement requires a lot of dedication and his part time job as the Berezinski representative is been contested by the French Park authorities. The management of the natural resources needs to be seriously improved. Applied research is needed in wolf and wildlife management.

In terms of sustainable tourism, the short-term assistance of the two experts from the Council of Europe has been very useful, but is not sufficient. Other sources of funding and assistance than those of the Council of Europe are needed for the Reserve. It is still too early to let them go their own way.

Canada-Hungary National Parks Project 1992-1995

Tom Kovacs[1]

[1] Natural Resources Conservation, Parks Canada
25 Eddy Street, Hull, Quebec, Canada K1A 0M5

Abstract. The Canada-Hungary National Parks Project was set up in 1992 in response to a request from the Hungarian National Authority for Nature Conservation, for Canadian assistance in developing the fledgling Hungarian national parks system. The Project was supported financially by Canadian External Affairs under its past aid program for Central and Eastern Europe. The project was divided into phases: which included a familiarization visit to Canadian national parks by senior Hungarian officials; the provision of Canadian documents; technical training; professional advice; and extended on-site training opportunities in Canadian national parks. The project was concluded in 1995. Parks Canada acted as executing agency under contract with External Affairs. A fifth phase of the Project, lying outside of the contract requirements, will be conducted by Parks Canada to evaluate the project outcome.

The purpose of this paper is to describe the Project, as an example of international cooperation, and to offer recommendations for further action and research.

Keywords. Hungary, national parks, grassland, exchange, planning, training, research

1 Discussion

1.1 Project Objectives

By 1992, Parks Canada, the federal program responsible for Canada's national parks, national historic sites and heritage canals, had developed considerable international experience, but not in managing multi-year projects involving Central and Eastern Europe. When External Affairs approached Parks Canada to design and carry out a national park project in Hungary, Parks Canada entered unfamiliar territory. The Project was considered as a pilot with other possible future projects to follow in Central and Eastern Europe. Since the early 1990's, the European geopolitical landscape has changed considerably, as has the level and focus of Canadian foreign aid. As a consequence, no similar projects have been sponsored since by Canada in Central and Eastern Europe.

Being the first project of its kind, there were no previous examples to fall back on for design purposes. The Project was tailored to meet the needs stated by Hungarian officials and the needs perceived by me as the project manager. The mutually agreed upon objectives for the Project were:
- the provision (and in some cases the translation) of key Canadian documents on the conservation of protected areas. For example, the National Parks Act, Canada's Green Plan, park policies, strategic plans, park management plans, visitor services plan, and natural resources management plans;
- the transfer of GIS technology;
- the provision of training in Hungary by visiting Canadians and,
- the provision of on-site training to Hungarian officials in Canada.

Given the relative lack of management and operational experience in the young and rapidly expanding Hungarian national parks system, the provision of training became the centrepiece of the project. It was offered both in the formal (as in the case of the GIS course offered in Canada) and informal sense. Informal training consisted of visiting, observing, and experiencing Canadian national parks and working in them for extended periods. Specific aspects of training were to include management practices, warden training, marketing, ecotourism, fundraising, park management planning, hospitality industry, service excellence, and environmental impact assessment. In the end, not all the initially identified training needs were met. This was due in part to changing priorities over the life of the Project, lack of resources and of in-house Parks Canada expertise in certain areas, and the limitations imposed by park operational requirements.

1.2 Project Structure and History

The Project was carried out in four phases:
- Phase 1 consisted of project scoping, needs identification and initial project design. It was completed in 1992;
- Phase 2 was a familiarization tour of some Canadian national parks offered to senior Hungarian national park officials to allow further project definition. This phase was also accomplished during 1992;
- Phase 3 consisted (in part) of the provision of technical advice in Hungary by recently retired Parks Canada officials under contract with Parks Canada. The four ex-officials are members of the Second Century Conservation Club, an organization of past Parks Canada employees with vast and varied national parks experience and expertise. In 1993 the team spent six weeks in Hungary observing, providing advice, holding meetings, making presentations and submitting a written report to the Hungarian authority. The second part of this phase was the GIS training given to three Hungarian national parks staff at Carleton University in Ottawa. Canada also gave Hungary some SPANS software to permit the application of the TYDAC GIS technology in Hungary.
- Phase 4 consisted of the management training program. It was designed to give Hungarian officials extended on-site management experience in Canadian national parks. Four management training participants spent a total of 15 months in various locations. The training programs were individually designed to meet specific requirements. Detailed management internship outlines were prepared for each participant to meet the stated needs as well as performance expectations. The participants had a hand in shaping their own individual objectives; as a result they were highly motivated and worked hard to achieve them.

1.3 Project Results

During the four years of the Project all its major objectives were met. Since 1992, a large volume of information has been transferred to Hungary, consisting mostly of documents on Canadian national parks. Other avenues of information transfer were the activities of the Second Century Conservation Club, the GIS training at Carleton University, the numerous meetings with the Project participants and follow-up contacts. The process of information exchange continues past the end of the Project.

The major result of the Project is the experience gained by Hungarian participants in the operation and management of Canadian national parks. Since there is no real substitute for first hand experience, it would be difficult to overstate the value of working and living in a national park setting. In fact, this is the kind of experience Parks Canada values in its own employees. The combined total of 15 months by four Hungarian officers represents a significant learning opportunity which should have a multiplier effect, given their anticipated influence on the Hungarian national parks scene. The participants are either in leadership positions now or have the potential for them in the future.

While two of the management program participants visited and observed parks as their primary means of learning, two others carried out actual work and prepared two written reports. It was the latter arrangement that proved to be most beneficial to all concerned, i.e. to both the training interns and their hosts. It should also be noted that, since the Hungarian participants did not receive a Canadian salary, their work can be considered as a significant in-kind contribution to the Project by Hungary.

An important product of the four year project is the sustained personal working relationships that have been established between Hungarian and Canadian colleagues which will last long past the Project and enable the continuation of information exchange. This is of lasting mutual benefit, as evidenced by some new initiatives which have surfaced since the end of the Project. The proposed collaboration on a grasslands national parks workshop in Hungary for 1997 is an example.

Although the Project's focus was on national parks, other benefits accrued, including exposure to Canada, a multi-cultural country, with democratic institutions and governance. This exposure was recognized to be of added value by project participants. This is not to suggest here, nor was it suggested at anytime throughout the Project, that the 'Canadian way' was anything other than an example of one of the many ways of doing things. At no point was it implied that the Canadian way was the right way or that it was even transferable. I hold the view that it was up to the Hungarian participants to discern what might be useful for their own home setting.

Lastly, there was the largely unforeseen but important benefit of mutual learning. Not surprisingly, the information flow took place in both directions and Canadian participants have also learned in the process. As well, the Project allowed us to build on our project management experience and, since it has been well received in Hungary, it will enhance Parks Canada's international reputation. Clearly, there were additional important benefits to both parties.

2 Recommendations for Future Action and Research

2.1 General Recommendation

It was indicated earlier that, for several reasons, some of the lesser objectives were not met. Training for wardens, ecotourism, marketing, fund raising, the hospitality industry, park management planning, and environmental impact assessment did not take place. Some of these training needs remain valid and should be addressed in the future. Given that the Canadian Government is unlikely to provide further assistance at this time for protected area development in Central and Eastern Europe, future action and research needs will have to be met from other funding sources. This could be put to test, however, by one of the Central or Eastern European countries initiating a formal request to the Canadian Government for funding. Clearly, the dominant sources of support for research have to be European based. By sponsoring this workshop and acting on its recommendations, NATO, for example, can become a significant contributor to the development of protected areas, heritage conservation, tourism and sustainable development in Cooperating Partner countries. This would be a highly appropriate rounding out of its existing support for environmental research.

It is recommended, therefore, that:
i) NATO include in its funding program an on-going, long term commitment to support research and activities related to the development and management of protected areas in the Cooperating Partner countries;
ii) NATO establish a Visiting Fellowship Program to foster the transfer of knowledge and information, professionalism, cooperation and partnership through staff exchanges between NATO and Cooperating Partner countries.

2.2 Recommendations Respecting Hungary

The Project had no formal requirement to look into the research needs of Hungary. As a result, the following recommendations are incidental observations indicating possible voids in research activity. The recommendations should be validated by Hungarian experts. Bearing this is mind:

1. It is recommended that a comprehensive research needs analysis be undertaken by the Hungarian authority for the national parks program to place future research effort on a solid footing by identifying a research strategy with priorities. In addition to the comprehensive study, certain specific areas have been noted for research.

The Hungarian national parks authority is strong in the natural sciences which serve the organization well. This strength is reflected in staff attitudes that still tend to favour a view of national parks as nature preserves, rather than places that also provide education and enjoyment of appropriate activities to visitors. There is now a conscious effort to promote a wider view and use of parks through the distribution of information brochures, the building of visitor centres and infrastructure. Nevertheless, a more fundamental effort is required to broaden the national park constituency in Hungary, based on social science. In my view this is a crucial task facing the program, for without broad based public support the Hungarian national parks movement is unlikely to flourish in the long term.

2. It is recommended, therefore, that research be undertaken by the Hungarian authority into: the potential role of social science in managing national parks; increasing public awareness of their value; and offering educational and enjoyable park visitor programs.

Reference was made earlier to certain training requirements identified but not met during the conduct of the project. In addition, there are other training requirements that should be undertaken in order to meet the challenges of a changing and growing organization. These include training in public relations, information and communications, visitor reception, nature interpretation and a host of social science areas dealing with visitor impacts, levels of acceptable change and risk assessment, to name a few.

3. It is recommended that research be conducted by the Hungarian authority to assess the training and educational requirements of the Hungarian national parks staff in order to ensure that the right mix of staff capabilities exist within the organization to meet present and future program needs.

During the course of the Project, I have seen little evidence of an integrated approach toward the examination and understanding of a greater ecosystem perspective in managing Hungarian national parks. Yet such a perspective is especially important in a small, densely populated country with intensive agriculture. A shift toward ecosystem based management would require research into broader ecosystems, ecological monitoring, regional integration, sustainability, land management practices, local history and culture. Foremost, the approach would require natural resources inventories on which to base all future work. The ecosystem approach to park management is necessary to mitigate cross boundary influences, to lessen the island biogeography effect, to increase local awareness and support for parks, and to merge local and conservation interests. The net effect would be greater effectiveness in protection and increased societal benefit.

4. It is recommended that research be undertaken toward the national implementation of the ecosystem based approach to managing Hungarian national parks.

Hortobagy is the first and most widely known Hungarian national park. It is also a grasslands national park, although other national parks in Hungary also have grasslands components. Lake Ferto national park, for example, contains the westernmost grasslands that occur across Asia and Eastern and Central Europe. Since grasslands are considered cultural landscapes, resulting in part from grazing and hunting activities, they have unique cultural, economic and conservation requirements. These requirements need to be researched thematically on an international scale in order to reach a better understanding of the particular management practices of these important areas.

5. It is recommended, therefore, that a thematic study be conducted by the Hungarian authority toward the greater understanding of the management requirements of grassland areas internationally.

In conclusion, I offer these recommendations for future funding considerations by NATO and within the Central and Eastern European context, with the cautionary note that they represent incidental observations and, as such, require validation.

A Sustainability Appraisal of the Work of the Yorkshire Dales National Park Authority

David Haffey[1]

[1] Countrywise Consultants, Eastland Banks House, Dipton Mill Road, Hexham, Northumberland, NE46 1RY, United Kingdom.

Abstract. An appraisal of the work of the Yorkshire Dales National Park Authority assessed the extent to which its plans, policies and programmes conformed to sustainability principles. The methodology established a 'sustainable position' for each of the Authority's principal functions in order to characterise a sustainable approach to the management of the Park: this also provided a basis for measuring past performance and recommending changes in future policy and practice. New approaches and techniques are described which may be relevant to the management of other National Parks and protected areas.

Keywords. National Parks, sustainable development, sustainability, appraisal, management policy, management practice

1 Sustainability and English National Parks

1.1 Introduction

Following the Rio conference of 1992, the concept of sustainable development has come to be accepted as the principle that should guide our use of the world's resources. Since then, many governments, including the British government (H.M. Government, 1994), have committed themselves to a more sustainable pattern of development and have adopted policies and action plans to achieve this objective. In practice, progress has been relatively slow, not least because of the range and complexity of the issues to be addressed and the social and economic implications of imposing constraints or limitations upon development.

In 1994, the Countryside Commission, the British government agency that is charged with conserving the natural beauty of the English countryside, recognised that the English National Park system provided an ideal opportunity to test the application of sustainable development principles to the use and management of a discrete area of natural landscape. This view stemmed from the fact that these Parks are not only 'protected areas' that are to be conserved for the nation but they are also working, living, managed landscapes that provide a livelihood for their indigenous communities. As such, these Parks offered an opportunity to develop an approach that could possibly act as a model for the sustainable management of other parts of the country.

1.2 Sustainability Appraisal

To investigate this issue in greater depth, the Countryside Commission and the Yorkshire Dales National Park Authority contracted the Countrywise environmental consultancy, together with the Baker Associates planning consultancy, to undertake a 'sustainability appraisal' of the work of the Park Authority. The aims of this study were to appraise the plans, policies and programmes of the Park Authority, as prepared and executed in accordance with its statutory function, and to determine the extent to which they conformed to sustainability principles. In addition, the study was required to make recommendations as to the changes that should be made to improve the Authority's performance in this area.

This appraisal developed new approaches and techniques which may be relevant to other National Parks that are concerned to ensure that their own policies and actions accord with sustainability principles. It is hoped that this paper might provide a useful guide to the procedures that could be adopted in carrying out such an appraisal. Further details are available in the full report made to the Countryside Commission and the Yorkshire Dales National Park Authority (Haffey and Baker, 1995). Before describing the methodology and conclusions of the appraisal, it may be helpful to outline the background to National Parks in England and to describe the relevant features of the Yorkshire Dales National Park.

2 The Yorkshire Dales National Park

2.1 Background

Compared to many European countries, Great Britain was slow to develop a network of National Parks. Public pressure for their creation was born in the early part of the century and stemmed from a concern to preserve beautiful countryside and to secure or retain access to mountains and moorlands. Following several decades of debate and discussion, legislation was passed in 1949, allowing the creation of National Parks with the statutory purposes of '...preserving and enhancing the natural beauty of the areas....and promoting their enjoyment by the public.' In 1995, these purposes were amended with the aim of making the Parks better able to respond to the challenges and opportunities that now face them.

2.2 Landscape and History of the Yorkshire Dales

The Yorkshire Dales National Park was designated in 1954 and covers an area of 1761 square kilometres in the Pennine Hills of northern England. The Dales have a distinctive landscape of open grass moorlands and impressive limestone crags that are dissected by many deep glaciated valleys. The present day landscape has been largely created by the hand of humans. Over the centuries a succession of settlers have left their mark on the land by clearing the original cover of broadleaved woodland, building settlements and roads, cultivating crops and introducing sheep grazing to the hills and fells, and later, building the dry stone walls and field barns that today so epitomise the Yorkshire Dales.

This landscape retains many features of natural and cultural importance. The natural history interest of the Dales centres on the remnant semi-natural broadleaved woodlands, traditionally managed species rich hay meadows, heather moorlands, limestone grasslands and areas of limestone pavement. These habitats support a wide diversity of flora and fauna and encompass many sites of regional, national, or international importance

Centuries of human occupation have also left their mark on the landscape in the great wealth of features of archaeological and historical significance. The remains of burial mounds, megalithic monuments, hut circles and enclosures testify to the early patterns of human occupation. In more recent times, the vernacular architecture and traditional building styles of the villages and settlements and the distinctive pattern of stone field boundaries have created an irreplaceable record of its social and cultural history.

2.3 Economic and Social Factors

The local economy of the National Park is based primarily upon agriculture. Farming takes place on over 95% of the land and it provides a livelihood for the majority of local people, either directly, or indirectly through related employment. Most of the higher land of the hills and fells is rough grassland, which is managed primarily for sheep rearing, with the production of hay and silage for winter fodder being a priority on the more fertile enclosed fields of the valley bottoms.

In recent decades, tourism has become increasingly important to the local economy. Since the designation of the Park there has been a dramatic increase in recreational use such that an estimated eight million people now visit the area each year. This has led to the development of a wide range of infrastructure and services to meet the needs and expectations of an increasingly mobile, affluent and discerning population of day visitors and tourists. The majority of visitors to the National Park come to sightsee, to go for short walks and to enjoy the peace, solitude and natural beauty of the Dales countryside. Others come for more active pursuits such as hiking, angling, cycling, rock climbing, potholing or hang gliding.

2.4 Management of the Yorkshire Dales National Park

The administration of the National Park is the responsibility of the Yorkshire Dales National Park Authority, a committee comprising elected members of local authorities within the region and representatives of the Secretary of State for the Environment. The role of the Authority is to conserve and enhance the natural beauty of the Park and to promote its enjoyment by the public.

The enacting legislation of 1949, which was based more upon discretionary powers than statutory duty, did not foresee the pressures that have now come to bear upon National Parks, nor did it endow the Park Authorities with the powers necessary to deal with these pressures. In particular, there are very real limitations on the ability of the Dales Park Authority to control the management of land since over 97% of the Park is in private ownership. The pursuit of National Park purposes is therefore very dependent on the ability of the Authority to influence the actions of private owners and occupiers of land. The legislation passed in 1995 to amend the purposes of National Parks and the duties and composition of National Park Authorities is unlikely to radically alter this situation.

Pressures upon the resources of the Yorkshire Dales National Park arise from many quarters and, directly or indirectly, affect the Park's natural and cultural heritage, the beauty and character of its landscape and its qualities of wilderness and solitude; these pressures can be categorised into three main types:
- development proposals, such as housing and quarrying activities;
- changes in types and patterns of farming and other forms of land use;
- recreation and visitor use, especially in relation to footpath erosion, traffic congestion and loss of wilderness qualities.

Over the years the Park Authority has developed policies and work programmes which seek to achieve its statutory objectives and the aim of this study was to assess the extent to which the Park Authority's intentions and actions measure up to the principles of sustainability. It is recognised, however, that at the time when existing policies and programmes were framed, the concept of sustainability, as it is now understood, was not an explicit consideration.

3 The Appraisal Methodology

3.1 The Scope of the Study

To establish a baseline for the study, the appraisal used the definition of 'sustainability' originally adopted by the Countryside Commission in a paper on 'Sustainability and the English Countryside' (1993); namely that 'human use and enjoyment of the world's natural and cultural resources should not, in overall terms diminish or destroy them'. The appraisal process was concerned primarily with an assessment of the Authority's three main policy documents, the 'National Park Plan' (1984), the 'Yorkshire Dales Local Plan' (1993), and the annual 'Bid and Functional Strategy' (1993). In addition, supporting information on the Authority's performance and programmes of action was gathered through detailed discussions with staff and the examination of relevant publications, reports, Committee papers, case files, work programmes and budgetary information.

To simplify the appraisal procedure, the work of the Authority was divided into five 'functions', corresponding to those used in the Authority's 'Functional Strategy', namely;
- Conservation
- Planning
- Information and Interpretation
- Recreation
- Support to the Local Community

3.2 Definition of a 'Sustainable Position'

At the heart of the appraisal process was the establishment and definition of a 'sustainable position' to characterise a sustainable approach to each of these five functions. These positions were then used as a reference point against which the Authority's performance in each function could be judged. This 'sustainable position' was also identified as an objective which the Authority should seek to achieve and which should henceforth govern the nature and direction of all future work. It was recognized that, in practice, sustainability would be very difficult, if not impossible to achieve but it was argued that this should not diminish its importance as an objective to aspire to and one which should be a major determinant of future policy. Since the definition of a 'sustainable position' for each function is such a key element in this process, it is perhaps useful to give one example, namely, a 'sustainable position' for the Authority's 'conservation' function

A requirement of the definition of 'sustainability' adopted in this appraisal is that the Park Authority should protect the Park's stock of natural and cultural resources from loss or damage arising from the use and enjoyment of the area. Certain elements of this stock can be identified as 'critical environmental capital', a term which refers to those features that are acknowledged to be of value, whose loss or damage would be very serious and which, once damaged or destroyed, could not be readily replaced (English Nature, 1994). Central to a sustainable position for the Park Authority with respect to its conservation function is a commitment to the protection of this critical environmental capital. In a report to the Countryside Commission, which formed part of the background research for the Commission's Position Statement, Grove-White, Phillips and Toogood (1993) concluded that the implication of an area's inclusion as part of the nation's critical capital would be that it would be treated for all practical purposes as inviolable.

The definition of stock as critical environmental capital will require that only minimal, if any, change is acceptable. However, it is implicit in this approach that it is neither practical nor realistic to preserve everything and that the protection of environmental stock is an active process in which change is managed rather than stopped. Priorities also evolve as the value of natural, historic, or cultural features is periodically reassessed.

A 'sustainable position' for conservation was therefore defined as:
- no further loss or degradation of defined critical environmental capital including designated conservation sites and defined tracts of landscape, wilderness areas, wildlife habitats, geological features, and cultural and historic resources;
- enhancement of critical environmental capital wherever possible;
- maintenance of overall levels of non-critical environmental stock and the acceptance of change, within defined limits, with losses preferably compensated by gains elsewhere;
- acknowledgement of background environmental change arising from natural and anthropogenic influences;
- an overriding commitment to the retention of the character, culture and qualities of the Yorkshire Dales.

3.3 Aims of the Appraisal Process

The appraisal process established a comparable 'sustainable position' for each of the Authority's other four functions and used these to carry out a comprehensive assessment of the Authority's performance. The aims of this process were:
- to comment upon the Authority's policies insofar as they were relevant or contributed to the defined sustainable position for each function;
- to identify areas for additional policies or programmes to improve performance;
- to suggest modifications that would improve the effectiveness of existing policies;
- to address the constraints inhibiting a more sustainable approach, in terms of other objectives which conflict with that approach, or in terms of external influences such as legislation and national policy.

4 Conclusions of the Study

4.1 General Points

In terms of the detailed aspects of the Park Authority's policy and practice, the study produced a range of conclusions and made specific recommendations with respect to each function. In a more general context, the overall conclusion of the study was that the Park Authority had managed its responsibilities in an effective manner, with the result that the essential character and quality of the Park had been safeguarded successfully. In part, this was felt to be due to the natural linkage that exists between the principles of conservation and of sustainability which meant that many matters of importance to the pursuit of sustainability had already been well served by the work of the Authority in fulfilling its statutory purposes. However, the requirement to adopt an explicit concern for sustainability meant that in future the Authority had to look beyond these statutory responsibilities and have regard to a number of other considerations.

First, a concern for the sustainability of the Park established a more stringent and complex set of criteria by which the performance of the Authority could be judged. It brought wider issues into play, exposing conflicting trends in the use and management of the Park, and imposing a longer time horizon over which any attrition or loss of critical resources had to be considered.

Second, sustainability encompasses matters that are fundamental to the maintenance of the Park's environmental capital but which had then been little influenced by the work of the Authority. For example, the Authority had, and still has, little influence over factors that affect air and water quality, soil condition, and the diversity of flora and fauna.

Third, the pressures upon the Park had changed significantly, both in their character and intensity, over the forty-five years since the purposes of the National Parks were first defined. In the Yorkshire Dales, as in other National Parks, conflicts between the two statutory purposes have given increasing cause for concern as the public has acquired greater mobility and as countryside recreation has become a more popular leisure pursuit.

Fourth, the Park was seen not to exist in isolation. Its condition was influenced by external events which lay beyond the control of the Park Authority and possibly also national government. Pressures for new development within the Park were also affected by socioeconomic factors that were governed by much wider trends. Conversely, the presence of the National Park and the policies pursued by the Authority could affect the surrounding land and its dependent communities. Constraints upon development within the Park could lead to pressures for demand to be accommodated in the vicinity of, but outside, the Park boundary. The effect of the National Park as a popular leisure destination was also to generate large volumes of traffic over a relatively wide area.

4.2 Recommendations to the Yorkshire Dales National Park Authority

The study concluded that the National Park Authority should make a policy decision to adhere to sustainability principles in the pursuit of its statutory functions. It was recognised that this would be a bold and onerous commitment but it was felt that it was necessary if, in future, the Park was to be managed on a sustainable basis. This decision required the Authority to review all aspects of its current operations, to evaluate the options, to redefine its objectives and to set out a clear strategy for achieving them. In particular, it was recommended that action should be taken to address the following issues:
- conservation - the identification and protection of the Park's critical environmental capital;
- planning - the establishment of planning policies which could effectively protect the Park from unwanted development, especially in relation to new housing;
- information - the promotion and dissemination of a new 'message' for the National Park centred on the importance of sustainable management;
- recreation - the achievement of levels and patterns of recreational use that were consistent with the Park's capacity to sustain them;
- the development of policies and programmes that reduced public dependency on, and the environmental impact of, private car use.

It was argued that the adoption of policies and programmes that sought to achieve these objectives would set the Authority on a new course that would better address the present and anticipated future pressures upon its resources and thereby more closely meet the needs of the Park and its dependent communities.

The main impediment to the pursuit, and possible achievement of these objectives was considered not to be a lack of will or commitment on the part of the Authority but the limited powers and resources bestowed upon it by central government. It was also recognised that the Authority's ability to adopt and implement appropriate policies would be deflected or constrained by external, political, economic and social influences. The National Park could not be managed in vacuum; it is wholly dependent upon communities outside the Park for many of the goods, services and facilities that are necessary for everyday life. Sustainability could therefore not be approached in isolation from external influences and all future steps towards this goal had to recognise that policies for the Park needed to be integrated with policies for the sustainable use and development of the environment as a whole.

5 Lessons for the Future

This study concluded that National Parks, as nationally important landscapes, ought to stand at the forefront of any steps towards achieving national sustainability. After all, if National Parks cannot be managed according to sustainability principles, then there can be little cause for optimism that the rest of the countryside, which is often less sympathetically managed, will be adequately safeguarded for future generations.

It could be argued that, by adopting a rigorous approach to sustainability, National Parks provide an ideal opportunity to demonstrate how sustainability might be pursued, and possibly achieved, in the management of a nation's natural and cultural heritage. This view is consistent with the conclusions of the National Parks Review Panel, appointed by the British government in 1992, that the concept of sustainable development is consistent with objectives of National Parks and that these Parks......'could therefore serve as one demonstration of, or testbed for, the nation's commitment to sustainable use' (Edwards, 1991): a viewpoint that is equally valid for National Parks and 'protected areas' in other parts of the world.

Whilst the theoretical and philosophical basis of this approach is well established, the mechanisms by which it should be applied to the practical management of a defined and statutorily protected area are far less clear and have received comparatively little attention. In particular, it is essential that research is now undertaken to develop and refine the operational procedures necessary for translating sustainability concepts and principles into action on the ground. The formulation of policy is not the difficult part; it is the implementation of policy. If National Parks are to provide a useful model for the sustainable management of other areas, we need to gain a better understanding of the mechanisms for, and implications of putting such policies into practice so that we can not only tell others how it should be done but also show them how it should be done.

6 References

H.M. Government. (1994). *Sustainable Development: The UK Strategy*. HMSO. London, UK.

Haffey, D. and Baker, J. (1995). *Sustainability Appraisal of the Plans, Policies and Programmes of the Yorkshire Dales National Park Authority*. Internal report to the Countryside Commission, Cheltenham, U.K. by Countrywise Consultants (Hexam, U.K.) and Baker Associates, Bristol, U.K.

Countryside Commission. (1993). *Sustainability and the English Countryside*. CCP 432, Countryside Commission, Cheltenham.

Edwards, R. (1991). *Fit for the Future: Report of the National Parks Review Panel*. CCP 334. Countryside Commission. Cheltenham.

English Nature. (1994). *Sustainability in Practice*. English Nature. Peterborough.

Grove-White, R., Phillips, A , and Toogood, M. (1993). *Sustainability and the English Countryside*. Centre for the Study of Environmental Change, Lancaster University. Lancaster, U.K.Yorkshire Dales National Park Committee. (1984). *National Park Plan: First Review*. North Yorkshire County Council. National Park Office, Leyburn, North Yorkshire, U.K.

Yorkshire Dales National Park Committee. (1993). *Yorkshire Dales Local Plan: Deposit Copy*. North Yorkshire County Council. National Park Office, Leyburn, North Yorkshire, U.K.

Yorkshire Dales National Park Committee. (1993). *1995/96 Bid and Functional Strategy* North Yorkshire County Council. National Park Office, Leyburn, North Yorkshire, U.K.

Forest Management and Planning and Local Populations: Assessing the Case of Niepołomice Forest, nr. Kraków, Poland

Jerzy Sawicki[1], Rafal Serafin[2], Bogumila Kuklik[3], Katarzyna Terlecka[4], Tomasz Terlecki[4]

[1] Chair, National Parks Unit of the Polish Ecological Club, Kraków, Poland
[2] Director, Polish Environmental Partnership Foundation, Kraków, Poland
[3] Independent Consultant, Bialystok, Poland
[4] Polish Ecological Club, Kraków, Poland

Keywords. Assessment, forest management, conservation, tourism, institutions, planning natural values, public education, property rights, parks and protected areas, Poland

1 Introduction

Forest issues in Poland are undoubtedly a hot topic. The Forestry Act in force now for five years has proven inadequate, especially with respect to the management and control of non-state forests. Forest issues are currently being addressed in a highly centralized way at the level of the Ministry of Environmental Protection, Natural Resources and Forestry through its Forestry Department and the Directorate-General of State Forests. Yet there is growing evidence of jurisdictional conflicts between authorities responsible for forests and those responsible for nature conservation. An additional influence in forestry issues stems from the growing aspirations of local governments to exert and extend the scope of their jurisdictions. One thing is evident, namely that the economic, legal and institutional issues with respect to Polish forests are clearly not being addressed systematically. There is little doubt that this situation requires initiation of various analytical and other studies, as well as development and implementation of innovative legal and institutional reforms.

The National Parks Unit of the National Board of the Polish Ecological Club has taken the position that forests constitute one of the most important elements of our nation's ecosystems and so must be given considerable attention. This conviction coincides with growing international concern for forest conservation in the wake of the World Environmental Summit in Rio de Janeiro, among international organizations, such as the UN, as well as local and transnational NGOs. Poland's efforts to prepare for European Union membership have also drawn attention to environmental quality issues, including forest issues. In this context, the National Parks Unit of the National Board of the Polish Ecological Club undertook to prepare a case study because of forest management in Poland that serves to illustrate the linkages between forests and local populations. The Niepołomice Forest near Kraków was selected for the purposes of or assessment as it illustrates many issues that are typical in Polish forest management situations. The case study is presented in the spirit of making more basic information about forest management issues, more widely available, for the purposes of improving understanding and decision-making with respect to forests.

2 The Legal and Institutional Context

The focus of this paper is the Niepołomice Forest (Figure 1) which is analyzed and assessed from the perspectives of natural values and resource management in relation to local populations. Legal and policy issues of Polish forestry are briefly reviewed as these provide the context for the case study. In assessments of progress towards Sustainable Forest Management (SFM) and adherence in Poland to the forest principles adopted at the Rio Summit, it is vital to understand something of the challenges posed by the economic, legal and institutional reforms in forestry underway in Poland. To understand the Niepołomice Forest case, it is important to gain an appreciation of the contemporary legal and institutional context of Polish forest management and planning.

Figure 1. Proposed Niepolomice Landscape Park

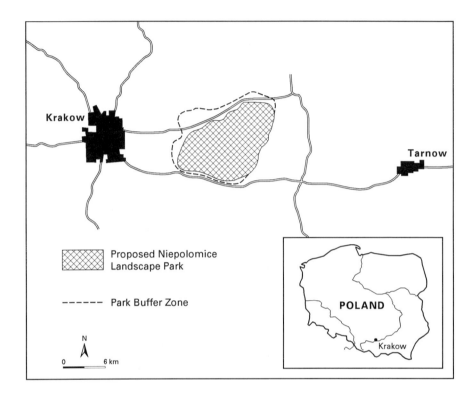

Two statutes relate directly to forestry in Poland. These are the Law on Forestry and the Law on Nature Protection, dating from 1991. Forestry-related issues, however, are also dealt with in at least twenty other statutes, such as the Law on Land Use Planning, Law on Agricultural Land, Law on Hunting, Law on Environmental Protection. All issues relating to conservation broadly defined fall within the remit of the Ministry of Environmental Protection, Natural Resources and Forestry (*Ministerstwo Ochrony Srodowiska, Zasobów Naturalnych i Leśnictwa*) where the most relevant departments are the Department of Forestry (*Departament Leśnictwa*) and the Department of Nature Protection (*Departament Ochrony Przyrody*). The General Directorate of State Forests (*Dyrekcja Generalna Lasów Panstwowych*) which has responsibility for national forest policy and planning is part of the Department of Forestry. The Directorate works through Regional Forestry Directorates (*Regionalne Dyrekcje Lasów*) which have responsibility for several voivodships. These in turn, oversee more local Forest Offices (*Nadleśnictwa*) which are responsible for larger forest systems and for directing activities through Forest Stations (*Leśnictwa*), the lowest level organizational unit in the Polish Forest System.

The total forest area of Poland extends over approximately 27% of the country (1996 figures), of which 18% are forest lands which are not state-owned. The Law on Forests in its current form - which has existed now for five years - contains almost no provision for mandatory control of non-state owned forest land. The practical outcome of this is that extensive logging is taking place on most privately-owned forest lands motivated by the desire for short-term profits. There are several examples of clear-cuts, but more typically logging is associated with no replanting or afforestation programmes. Only a very few voivodes (i.e. state governors - local representatives of central government) have exercised their right under the Law on Forestry to empower the relevant Regional Forestry Directorates with control of private forest lands.

The Department of Nature Protection has responsibility for protected areas. These include 20 national parks which fall within the remit of the National Park Board *(Krajowy Zarzad Parków Narodowych)*. The Department also has responsibility along with relevant local state governors or Voivodes *(wojewoda)* for landscape parks of which there are over a hundred - as well as nature reserves. Each voivodship has a Conservation Officer *(Wojewódzki Konserwator Przyrody)* who is responsible to the National Conservation Officer *(Krajowy Konserwator Przyrody)*. At the level of landscape parks, there are jurisdictional conflicts among their directorates and those of the forestry administration which often have responsibility in the same geographical area. It has so far proven impossible at the national level - in one Ministry - to work out effective and coordinated responsibilities between the Departments of Forestry and Nature Conservation.

The environmental threat to Polish forests results, above all, from air pollution from local, as well as distant industrial areas. The outcome of this is that to all intents and purposes there are no completely healthy large forest complexes left in Poland. Forest damage can be observed almost everywhere to a larger or smaller degree, associated most probably with acid precipitation or new kinds of disease outbreaks which are often only slowly coming to be studied, such as the consequences of genetic changes in insects.

A second threat to Polish forests is related to direct human impact resulting from on the one hand, seeing forests as simply timber stockpiles which leads to logging pressures, and on the other hand, from stresses associated with growth of human settlement which lead to recreation and tourism impacts, as well as other human pressures. Perhaps the most visible manifestation of this is uncontrolled building development in many of Poland's forested areas.

It is important to underscore that the General Directorate of State Forests has for years been active in conservation of large forest ecosystems, quite independently of the Department of Nature Conservation. Indeed, following assessments of forest damage, over 50% of Polish forests have been designated for special protection through management practices that are to ensure their ecological functions. This means: special provisions for ensuring forest health; a preference for natural afforestation, limiting regulation of water use; promoting species most adapted to the given habitat; clear-cutting and other extensive logging practices; and limiting exploitation of resin and stumpwood. In the past few years, foresters have been developing a new approach to forest conservation -- the so called Development Forest Complexes *(Promocyjne Kompleksy Leśne)*. There are currently five of these designated in Poland, including all the large forest complexes such as the Bialowieza Forest, the Knyszynski Forest and the mountain forests of Beskid Żywiecki and Silesia. Under this designation, a special type of conservation-oriented forest management is to be practiced with the goal of preserving the most valued forest ecosystems through severely limiting logging and other commercial production-oriented activities. It is important to note that the concept of the Development Forest Complex as a conservation mechanism is not recognized in the Law on Nature Conservation. This example shows how forest and nature conservation authorities can often be in competition with one another.

3 Forest Management and Planning in National Environmental Policy and the Basis for Forest Policy

To understand the specifics of the Niepołomice Forest situation, it is important to gain an appreciation of how national environmental policy relates to forest management, planning and conservation. The importance of forest conservation as a long term undertaking was recognized in the National Environmental Policy adopted in 1992, in the Implementation Programme for National Environmental Policy to the year 2000 (1994) and in the Principles for National Forest Policy adopted in 1996. All these documents acknowledge that forest resources must be considered as the basis for ensuring overall environmental quality. The goals of forest resource management and planning are therefore stated as follows:

- increasing the biological resilience of forests and forest ecosystems;
- conserving genetic resources and wild plants and animals, and also increasing forest cover especially in important water recharge areas;
- enhancing and restoring degraded forest habitats, and improving the species mix of existing forest stands;
- improving in concert the nature conservation and economic production functions of forests;
- undertaking action aiming to enhance the biological productivity and resilience of forest margins.

Achieving these goals requires an increase in forest cover of up to 30%, especially in the context of reforesting marginal agricultural lands.

A special Forest Conservation Education Centre is to be created along with a strategic national programme highlighting a conservation-oriented forest management model. This new approach to forest management and planning requires that forest policy be more clearly linked to broader socio-economic issues, through for example:
- using public and private resources to finance forest conservation in state and other types of forests;
- developing legal mechanisms and economic incentives for encouraging private owners to participate in the conservation of forest resources;
- developing planting and afforestation programmes on agricultural lands for the purposes of forest production;
- systematic monitoring of forest quality that takes into account a wide range of environmental indicators at the national, regional and local levels.

Design and implementation of forest policy aiming to promote conservation functions alongside more conventional production functions, requires also that special attention be given to ensuring that tourism and recreational use of forest lands is sustainable. Action must be taken not just in planning for tourism and recreation in forested areas, but also to strike an appropriate balance between conservation needs and public access to and use of forest lands. This challenge cannot be left solely in the hands of State Forestry authorities, as for the purposes of tourism and recreational use, various tourism-related organizations, NGOs and local governments must be treated as important stakeholders.

These latest trends in forest policy at the national level require that forest authorities must progressively reorient their activities through for example:
- developing and implementing a programme for a Forest Conservation Education Centre;
- including local governments and other organizations with a stake in forest management in a multi-stakeholder approach to forest management and planning that is characteristic of the conservation-oriented model of forest management.

By presenting the Niepołomice Forest as a case study, the aim is to assess the current state of forest management and planning.

4 The Geographical and Historical Context of the Niepolomice Forest

The Niepołomice Forest provides the focus for this case study as it constitutes a large forest complex with high natural values located close to the industrial centres of Kraków, Tarnów and Bochnia. The Niepołomice Forest lies in the southern part of the Vistula river valley known as the Sandomierz Lowland at the confluence of the Vistula and Raba rivers. The terrain is relatively flat and the climate is typical of much of Central Europe. At present settlement does not intrude into the forest interior, but is limited to three local municipalities (*gminas*) which contain approximately 20 settlements.

Historically, the Niepołomice Forest was part of the great forest complex which linked the Carpathian Mountains to Central Poland. It is first mentioned in the 13th century, but in the years that followed it was subjected to intense pressure and was almost lost altogether along with other Polish forest complexes. But luckily in the 16th century, the Niepołomice Forest was granted royal protection as it was recognized as the most important source of food for the King's table in Wawel Castle in Kraków. Even by this time, however, the large fauna of the forest had been almost extirpated to the extent that bears, bison and aurochs were imported for hunting purposes from Białowieża Forest. In the partition period, the Forest became the property of the Austrian Emperor and it remained largely intact despite being logged for the purposes of the salt mines in Wieliczka and Bochnia. Significant damage was inflicted in the First and Second World Wars. Nonetheless, the Niepołomice Forest remains today largely intact and within the jurisdiction of only one Forestry Office located in Niepołomice. What is more, the greater part of the forest remains state property. Private forest lands constitute only about 10% of the forest complex and the management practiced on these lands does not constitute a significant problem or threat. It is also worth noting, that a site extending over several hectares within the forest is managed as a bison breeding station. The area is closed off and constitutes something of a gene bank for the bison population. The station contains several bison and is managed by Ojców National Park on the basis of a lease agreement with the General Directorate of State Forests.

5 Natural Values of the Niepołomice Forest

Four types of forest ecosystem are represented across the 110 km^2 of the Niepołomice Forest. Biodiversity typical of the vegetation of the whole Carpathian region can be found here including:

- multi-species lowland forests associated with *Carpinion betuli* (including stands of *Tilio-Carpinetum stachyetosum, Tilio-Carpinetum tipicum, Tilio-Carpinetum caricetosumm pilosae*)
- fertile wetland forests associated with *Alno-Padion* (including stands of *Circaeo-Altnetum, Fraxino-Ulmetum*)
- marshy alder forests and other vegetation associated with *Alnetea glutinosae* (including stands of *Carici Elongatae-Alnetum, Salici pentandro-cinareae*)
- mixed forests on acid soils (including stands of *Pino-Quercetum, Molinio-Pinetum, Gacinio uliginosi Pinetum*)

Especially valued from a natural point of view are the hornbeam forests in the northern and southern parts of the Forest which have been designated as nature reserves (Lipówka, Kolo and Dębina). These are old-growth forest stands. Here and throughout the Forest, many individual trees have been designated as nature monuments (as per the Law on Nature Conservation). These are the remains of the extensive old-growth forest that dominated during the Middle Ages. Without question, the Niepołomice Forest can also be seen as a gene bank containing many tree species and their genotypes, as well as the whole range of temperate forest ecosystems.

The flora is also quite distinctive. Of particular interest are the stands of lowland birch (*Betula humilis*) which are found in the Polanie Błoto area. Of further interest are stands of Atlantic vegetation which are found here at the eastern and southern limits of their range. Royal Fern (*Osmunda regalis*) an unusual subatlantic fern, also is found here. Although there are still many wetland species, their abundance and variety is much less than in the past. Significant plants include: *Andromeda polifolia, Oxycoccus quadripetalus, Eriphorum vaginatum, Ledum palustra*. There are also interesting mountain floral species, such as: *Alchemilla glabra, Alnus incana, Arabis halleri, Veronica montana*.

Many of the species found in the Forest are listed as protected in Poland. Approximately thirty protected species occur in significant stands, including a variety of rare mushroom species in the northern part of the Forest -- many of them to be found only here in Poland. It is worth noting that mushrooms and lichens are extremely sensitive to air pollutants and the Niepołomice Forest has been subjected to intense air pollution impacts arising from the large steel complex at Nowa Huta located only a few kilometres to the west. There is little doubt that air pollution has caused wide spread changes in the Forest flora, especially with regard to mushrooms and lichens. But the nature and extent of changes in the vegetation has yet to be seriously studied.

The distinctive character of the vegetation in the Niepołomice Forest has resulted in designation of the area as a special geobotanical region on Polish botanical maps.

Aside from forest ecosystems, field or grassland ecosystems are important in the Forest. Many of these retain a semi-natural character. Several examples of marshlands also provide habitat for many rare species, such as: *Cirsium rivulare, Polygonum bistorta, Lychnis flos-cuculi*. The more open field ecosystems include: buttercup and other associations (*Ranunculus acer, Lychnisflos cuculi*), bent grass associations (*Carici-Agrostidetum caninae*); and, polygonaceous associations (*Cirsio Polygonetum*). All of them possess a rich flora, to the extent that in some northern parts of the Forest there are as many as 60 different species per 100 m. The largest impacts on species richness and diversity are on the forest edges where agriculture and forestry-related activities have been most intense, especially with respect to the use of chemical agents. Wetland vegetation is well represented in the Forest, especially with respect to raised and lowland bog plants.

Niepołomice Forest is one of the few remaining places in Poland where fauna typical of extensive old growth forest can still be found. The large diversity of forest biotopes, coupled with extensive areas of forested lands and nearby field, wetland and water habitats translate also into high biodiversity. Of value in this respect are several ecological corridors along the valleys of the Wisla and Raba rivers that connect the Niepolomice Forest to the mountain ecosystems of the Carpathians and the forest lands of the Sandomierz lowlands.

Well conserved forest and field ecosystems provide habitats for many rare and endangered species at the European scale. These include for example the Lesser Spotted Eagle (*Aquila pomarina*), redshank (*Tringa ochropus*), Short-Eared Owl (*Asio flammeus*), Tawny Owl (*Strix uralensis*), and Penduline Tit (*Remiz pendulis*). A high density of most mammal species occurs in Poland, including deer *(Capredus capreolus)*, *Muscardinus afellanarius*, and *Clethrionomys glareolus*. Old trees provide habitats for a rich insect and reptile fauna. In terms of breeding birds, there are in parts of the forest over 10 pairs per hectare - one of the highest levels in Poland.

Five nature reserves extend over an area of over 100 ha and protect the most valued forest stands. Four of them are forest reserves: Lipówka, Kolo, Dlugosz Królewski and Debina, whereas the Wielisko Kobyle reserve is an old Vistula riverbed. Additional reserves are planned, although plans are also under way to designate the whole Niepołomice Forest area as a landscape park - a designation that provides somewhat less rigorous protection than a national park.

A voivode (state governor - local representative of central government) is responsible for designation, and the case for creating one in the Niepołomice Forest based on the extensive available documentation, is currently being analyzed by a special group. Moreover the Voivodship Commission for Nature Conservation (*Wojewódzka Komisja Ochrony Przyrody*) lent its support to the landscape park proposal in 1995 by passing an unanimous resolution on the matter. There would appear to be little question that all planned additional protection measures can contribute to reducing human pressure on the natural ecosystems. But additional protective measures translate also into limitations on economic and other activities that can practiced in the area under protection, as well as in surrounding buffer zones - thereby impacting local populations which have already expressed their concerns publicly. However, it is those people who are linked with forestry that provide the most open opposition to the proposed landscape park designation. The conflict between forest and nature conservation authorities in this case results directly from incomplete and inadequate division of jurisdictional responsibility and a lack of clear goals and operational directives. There is no need for conflict as both forest and nature conservation authorities agree on the need for improving the effectiveness of nature conservation. Yet the designation of the area as a landscape park will do little to resolve conflicts and inconsistencies between the two authorities.

6 Forest Management in the Niepołomice Forest

A single Forest Office headquarters in Niepołomice is responsible for forest management across the whole area. The Forest Office comprises eight more local forest offices with a ninth specializing in game management. Over 90% of the forest is state owned, with less than 10% in private hands of which 10% is owned by local governments (*gminas*). Private forest lots are scattered at about 200 sites and are largely characterized by low quality forest serving mostly as sources of fuel wood (Table 1).

Table 1. Percentage Forest Cover by Species Type

Species	%
Pine	65.6
Larch	1.0
Spruce	0.4
Beech	0.9
Oak	17.9
Ash	1.4
Hornbeam	0.4
Birch	1.4
Alder	10.6
Poplar	0.3
Linden	0.1

The presence of other species, such as fir is minimal. Average timber production is approximately 106 m³/ha and the average age of tree stands is 42 years. Harvesting takes place on the basis of detailed management plans which are implemented in annual increments to reach production targets measured in m³. The total annual production target is currently 34,988 m³. This includes group felling of 17,557 m³ and selective felling of 17,441 m³ in an area totalling 778 ha. Clearcutting is not generally practiced with the exception of monocultural plantations or for sanitary purposes. In principle, each logged area is reseeded and afforested. Annual revenues from forestry amount to 500,000 zlotys and are in part used for afforestation. Timber from the Niepołomice Forest is processed by local pulp mills located no more than 100 km distant.

Over the past few years, the Niepołomice Forest Authority has been negotiating with the Agricultural Agency for State-owned Agricultural Land *(Agencja Rolna Skarbu Pants)* over a proposal to transfer 350 ha of agricultural lands left over from the Iglopol State Farm to its jurisdiction for afforestation. The lands have little value for agricultural purposes and their afforestation would not only increase the contiguous area of the Niepołomice Forest, but also contribute greatly to improving the hydrological conditions of the whole region. The lack of a decision in this matter provides a clear example of how the activities of the Agricultural Agency are not consistent with national environmental policy. Nationally, the Agricultural Agency is responsible for over 300,000 ha of agricultural lands that have been earmarked for afforestation. Over the next five years the General Directorate of State Forests is to take into its jurisdiction 100,000 ha. The main criterion for transfer is the potential for afforestation. As of today, only 20,000 ha have been transferred. The slowness of the transfer process stems from the formal requirement that the change of use must be approved in local land use plans and the fact that the costs of the whole operation must be borne by the General Directorate of State Forests.

With the foregoing in mind, there is an urgent need to streamline the transfer and rezoning procedures. Indeed, it may eventually turn out to be simpler to recommend changes in the land use plan, and then following field analysis, transfer the rezoned lands to local municipalities. But it is worth noting that most local municipalities have yet to review their land use plans under the new legislation that came into force in 1995. The result of this situation is that lands earmarked for afforestation are lying fallow. Perhaps it is in the interest of all concerned with increasing forest cover to ensure transfer of lands is to State Forests as quickly as possible

Aside from commercial logging, tree harvesting is related to prophylactic and sanitary activities. The basis for all forest management activities is the forest operational plan *(operat leśny)* which is approved by the General Directorate of State Forests. The long term goal is to systematically increase the beach-oak component of the forest. Unfortunately, activities in this direction cannot be well coordinated with the renaturalization of the Niepołomice Forest due to changes in recent years in the hydrology of the area. Employees include professional staff: forest inspectors; wardens; forest officers - 25 persons in all; administrative staff - 14 persons; and forest labourers - 18 persons. Work related to larger endeavours is contracted out to private companies. The Niepołomice Forest Office has living quarters for approximately 70. An important part of the Forest Office's responsibilities relates also to game management. Two game management districts are found in the Niepołomice Forest as well as a

Regional Game Breeding Station (*Ośrodek Hodowli Zwierzyny Lownej*). The principal game is deer, wild pig or boar, and stag with frequent hunting trips organized for foreign visitors mainly from Switzerland and Germany. There is also a lively trade in hunting trophies. The annual revenues from hunting are approximately 70,000 zlotys.

Over the past few years, the natural resources of the Forest have been regenerating -- mainly as a result of the reduction of industrial air pollution. The closure of the nearby Skawina Aluminium Works following public pressure by environmental groups was very important in this regard. Indeed, over the past three years a regeneration of mushrooms and cranberry has been observed. In turn, this has led to a return of human foraging in the forest which is resulting in renewed pressures on sensitive ecosystems and impacts on mammal populations.

One of the main challenges facing the Niepołomice Forest District is the problem of insect pests and fungal diseases. In 1995, old forest suffered badly from leaf pests, especially pine webworm and leaf moth. Webworm threatened over 40 ha of pine forest and in 1996 helicopters will be used to fight new outbreaks. Leaf moth is primarily a pest that attacks oak and in 1995 threatened over 300 ha. No special action was taken, however, as oak regenerate themselves on their own. Fungal outbreaks affected four of the Forest's principal species and required some preventive action to hinder wider outbreaks. Aside from insect pests and other infections, the Forest also suffers other damage related to physical impacts, such as snowstorms and human impacts related to human action, including air pollution or alteration of drainage patterns. Oak dieback is usually attributed to industrial pollution impacts and estimates suggest that over 200 ha of the Forest are seriously affected. In sum, it is fair to say that forest management activities as currently practiced, are in large measure focused on monitoring threats and taking action to prevent or reduce them.

7 Human Settlement Around the Niepołomice Forest

The Niepołomice Forest is spread across three local municipalities (*gminy*): Niepołomice, Drwina and Kłaj. Together, the three gminy cover an area of 287 km^2 with a total population of 36,684 of which 19,946 live in the urban centre of Niepołomice. Although the average population density is 128 persons/km^2, most areas are thinly populated on account of the large percentage of forested land in the area. Overall, the area is characterized by population growth (approx. 6.1% annually), mostly due to immigration to the urban and rural municipalities of Niepołomice. The surrounding rural areas are, however, characterized by outmigration and net population loss. This situation is explained to a large degree by the proximity of large urban centres. The age structure presents itself as follow: 0-18 years - - 3.3%, 18-59 years -- 52%, 60 and over 17.7%.

The Niepołomice Forest is surrounded in large part by agricultural land owned as individual private holdings with an average size of about 3 ha. The proportion of the local population employed in forestry is very low, although there can be seasonal highs depending on the needs of the Forest District Office. The vast majority of local people have two jobs: cultivation of a farm holding and working in a non-farm job away from place of residence which typically offers few job opportunities (underdeveloped services, commerce, industry). This situation is typical of rural areas of southern Poland lying in the vicinity of large urban and industrial centres. The system of dual employment described above is a direct result of the communist era of central planned industrial development.

The local municipalities surrounding the Niepołomice Forest were empowered in March, 1990 when the Polish Parliament passed a new Local Government Act. Local governance is often seen as one of the basics of democracy. Yet in the Polish case it is the greater political context that dictated the nature and extent of that local governance. In the Partition Period, local governments were bastions of Polish national culture and identity. In the post-war communist period in contrast, local governments were strictly controlled by the United Peoples Workers Party (PZPR) -- the Polish Communist Party -- which implemented and assured Soviet policy and influence. In this context, self-governance and autonomy of local government were completely destroyed and the National Committees (*Rady Narodowe*) that supposedly represented local interests were a complete and utter fiction.

By embarking on a radical path of reform in 1990, Poland has striven to undertake far-reaching administrative and institutional reforms that despite a wide range of difficulties, have opened the way for genuine local self-governance. The development of local governance is yet at an early stage. The legal and jurisdictional frameworks have only indicated general principles and many practical procedures and operational responsibilities have yet to be defined more precisely.

A more recent step aimed at granting local governments greater autonomy and self-governance is rooted in the Land Use Planning Act of 1994. Local land use plans comprise two basic elements. The first is an analysis of constraints and the general directions of land use policy. This element provides the basis for all policy and planning activities undertaken by or promoted by the local municipality. The land use plan proper -- the second element -- has primarily a regulatory function based on zoning land uses and the principles of land use activities that meet prevailing legal and administrative criteria. In the case of areas owned by State Forests, local land use plans must be in compliance with forest management plans prepared by State Forest authorities.

In sum, the reforms begun in 1989 initiated a completely new basis for establishing relationships between local governments and other institutions in interacting with them. With the foregoing in mind, in what ways do the local municipalities associated with the Niepołomice Forest use their new found powers to promote local development?

The municipality of Niepołomice stands out among the gminy of southern Poland as one that is well governed and has been able to create an attractive climate for foreign investment (Coca-Cola has built one of its biggest bottling plants here). As a result, the gmina is able to generate substantial revenues which are reinvested in the further development of urban infrastructure (water, sewage and telephone systems), supporting small business, and improving the attractiveness of the area for tourist purposes (e.g. construction of a recreation park, renovation of the Castle), and ensuring also that as many of the local people as possible have their place of work in the area. There is little question that this very positive situation is linked closely to the municipality's geographical location close to the Kraków urban area and within reach of rail, road and even Kraków city buses.

In contrast, the municipality of Drwina located to the north of the Forest, furthest away from Kraków, has few of the advantages of Niepołomice. The situation is different yet again in the case of the municipality of Kłaj. In contrast to Niepołomice, the main source of revenue in the local economy is from agriculture and small scale manufacturing. Only in Niepołomice is infrastructure development under way with respect to sewer lines, gasification and water lines. Over 60% of Kłaj is forest and due to the conservation-oriented management practiced within it, forest-production related revenues are minimal. With this in mind, the local municipality sees its future in tourism and recreation development. The analysis completed for the local land use plan recognizes the importance of forest resources for recreation development. Although the analysis for the land use plan is an internal document for the local government, the lack of consultation and coordination with the Niepołomice Forest Office in its preparation is testimony to the fact that relations between the local government and the Office are not what they should be. Moreover, minor conflicts over land ownership between the local government and the Office contribute to unnecessary tensions arising from the fact that legal proceedings replace communication.

In sum, the situation in all three local governments associated with the Niepołomice Forest shows that effectiveness in promoting local economic development is directly related to the degree of cooperation achieved with the Niepołomice Forest Office.

8 Local Populations and the Forest

The Niepołomice Forest District Office has overall responsibility for the Niepołomice Forest. The state forest administration contributes only in a minor way to the budgets of local governments, mostly through a forest tax that is levied by them. The District is responsible for all management in the State Forests and provides also professional services to owners of private forests with respect to the preparation of forest management and afforestation plans. All three local municipalities are primarily agricultural in character, even though forests cover as much as 50% of the area. Non-state forest lands which are currently experiencing at the national scale, serious problems in assuring appropriate forest management practices, make up only 846 ha of mostly very young stands of low commercial value. In this case they are well managed by the Niepołomice Forest Office on the basis of special management agreements.

It is worth noting at this point, that the regulations arising out of the Law on Forestry provide opportunity for private land owners to secure tax benefits in return for reforesting land. Indeed, the current legal regulations encourage reforestation of lower quality soils (formally, class IV, V and VI soils). This is especially significant with respect to class IV soils which are not relieved of agricultural land tax, but are relieved of all taxes if they are afforested for a period of 40 years. Moreover, local governments can receive funding for preparing forest management plans and seedlings can be provided from the Voivodeship Fund for Environmental Protection.

In terms of local policy and planning, the Niepołomice Forest is seen as a potential source of tourism-related revenues. Indeed, development plans of the local governments associated with the Forest are focused on promoting the natural values of the Forest. This widespread interest in making use of the Niepołomice Forest for weekend recreation and other tourist-related activities demands analysis of the human impact on the natural values of the forest. The need for estimating carrying capacities in this way is recognized in the Principles of National Forest Policy. In practice this means undertaking assessments of potential impacts on specific areas, such as the Niepolomice Forest which are seen as opportunities for development. The local municipalities associated with the Forest have already undertaken infrastructure development and begin efforts to obtain investment related to tourism and recreation. It is worth noting here, that presently the area is little developed in terms of recreation and tourism facilities. Hotel and other tourist accommodation simply does not exist in any significant quantity or standard. In this context, assessments can provide the basis for balancing tourism pressure against the needs of conservation.

In undertaking the research being reported upon here, the Polish Ecological Club held a meeting to assess the extent and effectiveness of cooperation between the local governments and the Niepołomice Forest Office. Especially interesting were the views expressed by the Municipality of Niepołomice on the tourist and recreational development of the Niepołomice Forest and its surrounding areas. The Municipality was of the view that the brunt of recreation and tourism development should take place outside the Forest proper. A tourism development strategy highlighting the attractiveness of the Niepołomice Forest should draw attention to the other recreational and tourism opportunities in the municipality, such as lakes, sports facilities and heritage resources. The strategy promoted by the Municipality of Niepołomice is consistent with the accessibility programme developed by the Niepołomice Forest Authority. In this situation, there is clear opportunity for limiting access to the forest to ensure conservation effectiveness, while also promoting recreation and tourism in other parts of the municipality.

Cooperation between the Municipality of Niepołomice and the Forest Authority is promising for the future, although each party must yet clearly acknowledge the scope of its jurisdiction and the need for coordinating implementation of their respective statutory responsibilities. The remaining two municipalities of Drwina and Kłaj did not participate actively in expressing their views during the course of this research. Their lower level of involvement can be explained most likely by their largely reactive stance in local and regional issues.

Currently, 2,886 ha are designated for recreation and tourism purposes by the Niepołomice Forest Office, concentrated mostly in the western part of the Forest. The Office has prepared car parking, camping and other tourism related facilities as part of overall forest management. These measures by themselves, however, are not sufficient to ensure effective protection in nature reserves, wildlife habitat or forest regeneration areas. Tourism development of the Niepołomice Forest that assures effective nature conservation demands much closer cooperation and coordination among the Forest management authorities, local offices of central government, and local government. This is largely absent at present. As a case in point, the existing system of six tourist trails is largely autonomous, developed and managed by the National Tourist and Countryside Association (PTTK).

There is little doubt that sustainable tourism development is one of the most pressing challenges for ensuring effective forest management in the area. Yet there is no integrated approach at present to tourism development. An integrated approach would appear essential prior to undertaking any significant tourism promotion and related infrastructure development. The current lack of tourist trails and weakly developed accommodation and other tourism-related facilities mean that weekend and day visits dominate tourism patterns. The presence of large numbers of recreationists and mushroom pickers in certain periods leads also to difficulties in ensuring appropriate behaviour which in turn, lead to fire and other hazards. Recognizing the prospect of growing impact from recreation and tourism, the Niepołomice Forest Office has developed a recreation management plan for its forested areas in the municipality of Niepołomice. A new 300 car carpark is now being built, along with another camping area and educational and bicycle trails. There are also plans to develop a petting zoo for hooved animals and an enclave for viewing bison. The motivation for these activities lies in creating areas which are attractive for recreationists and so will lessen pressure on other parts of the forest which are important wildlife and rare plant habitats.

According to the Niepołomice Forest Office, when compared to other forested areas, poaching of game animals and illegal tree cutting are not of significant proportions. This is due in part to the relatively small percentage of privately-owned land and an effective forest management system based on well-equipped forest rangers. Also there is not the economic pressure for illegal treecutting and poaching here that exists in other parts of Poland. The municipalities associated with the Forest are not centres of wood related industry nor are there local craft traditions based on forest resources. Moreover, almost all the local settlement is now supplied by natural gas, so there is little demand locally for fuel wood. In addition, many of those living in the three municipalities work in the urban-industrial centres of Kraków, Bochnia or Wieliczka and so are not dependent on their livelihood on the forest and its resources. Also it is worth noting that the development strategies of the local municipalities recognize the importance of protecting the natural values of the Forest. For this reason, there is an evident preference for promoting environmentally-friendly and ecologically sensitive industry (especially in the Drwina municipality). All three municipalities are committed to developing comprehensive sewer systems, to building local sewage treatment plants and to dealing effectively with the problem of waste disposal.

There is also the question of whether the environmental awareness of residents of the Forest municipalities is sufficient to justify basing a development programme on conservation of the Forest's natural resources. Experience to date suggests that much can be done in this area, as appropriate strategy development and subsequent implementation requires the active involvement of local people. Indeed, the Forest Conservation Education Centre programme proposed as part of the National Forest Policy to the year 2000 should most certainly be based on pilot projects implemented in forest complexes, such as the Niepołomice Forest. Such activities should lead to involving people in practical forest conservation activities in partnership with State Forest staff.

Issues of environmental education have been largely left to the Forest Office by the local municipalities, even though the Office has no resources allocated for developing and promoting environmental education activities. The Office in turn, considers environmental education activities a low priority when compared to the main focus of forest management and planning. Nonetheless, Office staff participate actively in working with schools, providing an information service to local farmers, and co-organizing forest clean-ups with youth groups. There is no active environmental NGO in the area, although there is a proposal to develop a local branch of the Centre for Environmental Education in Kraków that is being supported by the municipality of Niepolomice.

9 Conclusions

Forest management and conservation in the Niepołomice Forest case suggests several conclusions that would appear to have wider applicability in Poland.

9.1 Management and Property Rights Issues

The choice of the Niepołomice Forest for this case study was a deliberate one in the sense that it is a typical forest complex in many ways. Indeed, it provides an exemplary case of forest conservation, planning and management. The Forest lies close to large urban-industrial areas, surrounded by well-managed farmland, with less than 10% in private ownership. The greater part of the forest resources are managed by the state forest administration -- in this case the Forest Office in Niepołomice. The large variety of commercial and conservation-related operations currently practiced in the Forest suggest that a well-coordinated management system with clear jurisdictional responsibilities is essential. An important point in this regard is the articulation of a vision of the future state of the Forest, and the preparation and implementation of management and planning activities to realize that vision through, for example systematic replanting and afforestation activities aimed at changing the species mix in the forest.

The question arises as to whether the various forest management activities currently in place would be sustained and improved with privatization and reprivatization of the forest - perhaps to several different owners. The experience of the past five years with the Law on Forestry which exempts private forest lands from any effective regulation, suggests that this situation would be much less favourable from a sustainable forestry standpoint. There would likely be much greater commercial pressure on short term profit from logging and much less emphasis on replanting, afforestation and species-mix management. Recognizing the Niepołomice Forest to be a large ecosystem of growing importance in terms of ecological value at the larger regional and national scale, leads to the suggestion that forest complexes such as the Niepołomice Forest should be placed under the jurisdiction of a single authority with a clear mandate for conservation. Unfortunately, many of the proposals for restructuring and privatizing forest management and forest related industry appear not to be well-thought out from a longer term sustainable forestry perspective.

Key issues related to assessing the implications of different restructuring and privatization scenarios would appear to warrant further analysis, especially with regard to legal issues, property rights, forest management responsibility, and interaction with the State Agency for Agricultural Land (*Agencja Rolna Skarbu Pants*).

In the Principles of National Forest Policy (1996), the chapter dealing with private ownership of forests proposes that private owners create voluntary associations. An important opportunity in this regard lies in promoting nature conservation in areas not formally under protection, in partnership with foresters and private owners of appropriate lands. This field could also be an important area of activity for non-governmental organizations active in promoting nature conservation. Some initial moves in this regard are being made through a project now being implemented by the Institute for Sustainable Development in Warsaw and the Environmental Partnership for Central Europe (EPCE). The project is striving to identify opportunities for nature conservation outside of protected areas. Developing a locally acceptable and effective formula for promoting nature conservation outside protected areas that is effective will require legal and economic reforms that will provide appropriate incentives. The approach would seem a promising one as it addresses directly many of the challenges that have posed difficulties for the State Agency for Agricultural Land, the State Forest Authority and others.

With regard to the jurisdiction of local government, state administration and forestry authorities, any situation that involves several institutions with jurisdiction over resources can lead to conflicts. In the case of a large forest complex, such as the Niepołomice Forest, where local populations and proximity to large urban areas mean that human impacts on natural values cannot be easily avoided, effective coordination is essential between various central and local government authorities for the purposes of conservation.

Current legislation, especially the Law on Land Use Planning, does outline jurisdictional responsibility but provides no procedure and little incentive for effective coordination among different authorities. Indeed, each of the three local municipalities in the Niepołomice Forest has its own development vision and is working somewhat independently to promote investment, tourism and recreation infrastructure and the local economy. All major investment plans and infrastructure development proposed by the local municipalities, and indeed by the Forest Office, should be subjected to assessment and wide-ranging consultation. Putting an end to fragmented developments with little pre-planning or sensitivity to the larger context is a vital prerequisite to ensuring sustainability of the Niepołomice Forest ecosystem. Given the large number of land use planning and nature conservation specialists in nearby Kraków, the Niepołomice situation could become an important focus and precedent for planning and management innovation.

Another issue worth considering for a moment is the proposal to create a Niepołomice Forest Landscape Park. The consequence of such a designation would be the appearance of yet one more authority with management responsibility. It is well to remember that in all of Poland, there is at present not a single example of effective coordination among landscape park, local municipalities and forest authorities. With this in mind, any institutional innovation in the existing management system must be linked to redefinition of jurisdictional responsibilities of the different agencies, especially with regard to conservation of landscape values. Potential for jurisdictional conflict between the Forest Office and the Directorate of the proposed Landscape Park with respect to conservation of landscape values can be avoided. This is because responsibilities of the Landscape Park directorate extend beyond forest issues into a broader land use planning framework which also includes conservation of a variety of other heritage and historic values. The aim of the landscape park is to conserve the character of the landscape as a whole. In the specific case of the Niepołomice Forest, it is essential nevertheless to ensure that jurisdictional responsibilities are clearly spelled out at the designation stage.

9.2 Tourism Issues

Tourism and recreation-related impacts on the Niepołomice Forest will continue to grow and so management activities will require closer cooperation between local municipalities and State Forestry authorities. Joint management activities would need to distinguish between different types of tourism and recreation, striving to promote hiking, biking and riding activities in ways that minimize impacts on natural values. Planning that coordinates activities and marketing of the three local municipalities is essential. Moreover, these activities need to take into account the natural and cultural heritage of the area both in terms of assuring appropriate protection, and promoting heritage as the principal attraction for tourists and recreationists.

Tourism promotion and development in the Niepołomice Forest area have been recognized by all three local governments as an important objective. Yet preparation of an appropriate tourism infrastructure development plan will require considerable coordination, if it is to conserve the heritage values of the area and not add to destructive pressures. The plan needs not only to provide for heritage conservation, but also promote a strategy that will increase the attractiveness of the area as a whole in an increasingly competitive tourist and recreation market in and around the Kraków area. Indeed, the heritage values of the area could provide a basis for distinguishing the Niepołomice area as an important tourist destination. Promoting the area in this way, however, requires considerable investment in research and publications of various kinds.

9.3 Issues of Public Education

Although environmental awareness is growing in Poland, it is still inadequate. In situations such as those of Niepołomice Forest, for conservation to be effective, local populations must also be active participants. Environmental education relating to forest conservation and management targeted at both younger and older persons must be an essential component of forest management and planning. This means promoting awareness of the values and significance of Niepołomice Forest and other forest complexes within the school curriculum, as well as through outside school activities. An important opportunity offered by the Niepołomice Forest situation is the development of educational and self-learning programmes that build on local natural and cultural heritage values as a basis. Educational authorities could play an important role in encouraging teachers to use local heritage values as a basis for designing geography, history, art and even language teaching. This approach could also lead to involving parents in educational and learning activities through the development of related educational materials. The Niepołomice Forest Office should also develop and implement its own educational programmes which would actively involve forest staff and help build a capacity for contact with the public. On account of its interdisciplinary demands, developing effective educational activities in this field would require close cooperation with local governments, educational, and environmental groups. Indeed, it may even prove appropriate that an environmental group, such as the Polish Ecological Club, play a key facilitating role in the development and implementation of the programme.

10 Final Remarks

This case study was prepared as a contribution to promoting a wider understanding and implementation of the ecological basis for forest management that is consistent with the Declaration of the Environmental Summit in Rio de Janeiro, the Conference of Environmental Ministers held in Helsinki and IUCN's Parks for Life programme. This study in fact represents a step in this direction.

International agreements, such as the Convention on Biological Diversity and the Forest Principles of Agenda 21, as well as initiatives such as the Intergovernmental Panel on Forests (IPF) are currently seeking to promote Sustainable Forest Management (SFM) in various parts of the world. These agreements and initiatives should provide a framework and basis for assessing the current status of forest management effectiveness and progress towards sustainable forest management in forest complexes, such as the Niepołomice Forest. In conjunction with other national and international NGOs (such as those involved in the IPF), monitoring and assessments of progress towards sustainable forest management could be carried out for presentation to citizens, local authorities, national governments and international groups. This study in fact represents a step in this direction.

The Agenda 21 document adopted at the Rio Summit contains a chapter dedicated to the principles of forest management. No preference is expressed as to ownership questions, thereby leaving the issue open as to whether sustainable forest management is best undertaken in the context of public or private ownership. Indeed, Agenda 21 recommendations are clearly to be implemented by governments in ways that take into account local conditions and circumstances. It is worth noting that among others, the following principles are listed as essential elements of effective sustainable forest management:

- extending the involvement of the private sector (this by no means the same as extending private ownership), trade unions, rural and other local communities as well as environmental and other non-governmental groups in forest-related activities and in promoting public information and education activities in the country as a whole;
- promoting forest-related issues through public educational programmes with the aim of raising public understanding and awareness as to the role of forests in society and the different ways of managing them;
- guaranteeing public involvement in the development and implementation of forest-related management programmes in order to ensure relevance to local needs and circumstances;
- promoting forest use and management in ways that are consistent with the principles of sustainable social and economic development.

Agenda 21 draws attention to the need to strengthen national and subnational institutions responsible for forest management and development by promoting linkages among other local, regional and international organizations involved in forest-related activities.

11 Acknowledgements

The study was financially supported by BothEnds from the Netherlands and was based on a review and analysis of statistical and other information available on the Niepolomice Forest. These included documentation made available to the authors on the proposed Niepolomice Landscape Park by the Kraków Voivodship Environment Department, and other materials made available by the Regional State Forest Directorate, the Niepolomice Forest Office and the three local municipalities of Niepolomice, Drwina and Klan. Moreover, the authors were able to meet and interview representatives of the institutions listed above to verify and discuss issues arising from the study. The authors would like to acknowledge the help of the staff of all the institutions mentioned above. In particular, the authors would like to thank Dr. Alfred Król, director of the Regional State Forest Directorate and Dr. Bozena Kotonska of the Kraków Voivodship Environment Department. We extend also special thanks to Mr. Stanislaw Kracik, the Mayor of Niepolomice and a Member of Parliament for the Kraków region for his extensive written responses to questions posed during the course of the study and to Mr. Sennik, Director of the Niepolomice State Forest Office for his time, support and involvement in the study.

12 References

Chiechanowicz, J. (ed.) 1994. Ochrona środowiska: zbiór przepisów (Environmental protection: a compendium of regulations). Gdańsk: Wydawnictwo Prawnicze LEX (in Polish).
Głowaciński, Z. and S. Michalik. 1996. Kotlina Sandomierska (Sandomierz lowland). Warsaw: Wiedza Powszechna (in Polish).
IUCN. 1994. Parks for Life: action for protected areas of Europe. Polish summary published by Polish Ecological Club, 1994.
IUCN. 1994. Guidelines for protected area management categories. Gland: IUCN.
Jacyna, I. 1978. Las nie obroni się sam (Forests won't protect themselves). Warsaw: Krajowa Agencja Wydawnicza (in Polish).
Kachniarz, T. and Z. Niewiadomski. 1995. Nowe podstawy prawne zagospodarowania przestrzennego (New legal foundations of land use planning), Warsaw: Instytut Gospodarki Przestrzennej i Komunalnej (in Polish).
Kiełczewski, B. and J. Wiśniewski. 1982. Las w środowisku życia człowieka (The forest in human society) Warsaw: Państwowe Wydawnictwo Rolnicze i Leśne (in Polish).
Lasy Państwowe. 1993-96. Biuletyn informacyjny (information bulletin). General Directorate of State Forests. Warsaw (in Polish).
Leśkowa, A. 1981. Zaczęło się od świętego gaju (It started with the sacred grove). Warsaw: Krajowa Agencja Wydawnicza (in Polish).
Ministerstwo Ochrony Środowiska, Zasobów Naturalnych i Leśnictwa 1992. Polityka ekologiczna państwa (National environmental policy). Warsaw: Ministry of Environment, Natural Resources and Forestry (in Polish).
Ministerstwo Ochrony Środowiska, Zasobów Naturalnych i Leśnictwa 1995. Trwały rozwój lasów w Polsce - stan i zamierzenie (Sustainable forest development in Poland). Warsaw: Ministry of Environment, Natural Resources and Forestry (in Polish).
Ministerstwo Ochrony Środowiska, Zasobów Naturalnych i Leśnictwa 1996. Założenia polityki leśnej państwa (Principles of national forest policy), Ministry of Environment, Natural Resources and Forestry, Warsaw (in Polish).
Radecki, W. 1991. Samorząd terytorialny i ochrona środowiska (Local government and environmental protection). Warsaw: Foundation for Local Democracy Development (in Polish).
Sommer, J. 1993. Prawo ochrony środowiska w Polsce - poradnik praktyczny (Environmental protection law in Poland - a practical manual). Wrocław: Towarzystwo Naukowe Ochrony Środowiska (in Polish).
Szujecki, A. 1992. Czy lasy muszą zginąć? (Must forests die?) Warsaw: Wiedza Powszechna (in Polish).
United Nations. 1994. Forest principles - non-legally binding authoritative statement of principles for a global concensus on the management, conservation and sustainable development of all types of forest. Agenda 21.
Wićko, E. 1979. Gospodarstwo leśne i przemysł drzewny w Polsce (Forest management and timber industry in Poland), Warsaw: Państwowe Wydawnictwo Rolnicze (in Polish).
Wojewódzki Urząd Statystyczny w Krakowie 1995. Rocznik Statystyczny Województwa Krakowskiego (Statistical yearbook of the Kraków Voivodship). Kraków: Wojewódzki Urząd Statystyczny w Krakowie (in Polish).
Wydział Ochrony Środowiska UW 1994. Informacja o stanie środowiska w województwie krakowskim w roku 1993. Kraków: Biblioteka Ochrony Środowiska (in Polish).
Zakład Badań Środowiska Rolniczego i Leśnego PAN 1991. Strategia ochrony żywych zasobów przyrody w Polsce (strategy for conservation of biological resources in Poland). Poznań: Zakład Badań Środowiska Rolniczego i Leśnego PAN (in Polish).

Glossary of Key Terms

1. *State Forests* (Lasy Państwowe), known also as the State Forest Management Administration are organized in three tiers: Directorate-General of State Forests, Regional Directorates of State Forests and District Forest Offices. The Director-General of State Forests is responsible for forest quality.
2. *Municipality or Local Government* (Samorząd terytorialny) refers to a self-governing community, together with a geographical jurisdiction. Local government is an institution that is distinct and separated out from the state administration on the basis of national legislation with the purpose of autonomous implementation of defined public activities. In Poland, local or municipal government is known as the gmina which are both urban (cities) and rural. An urban gmina can include also surrounding rural areas and in some cases is conjoined as gmina i miasto (gmina and city) to include both rural and urban territories.
3. *Conservation Forests* (Lasy ochronne) are forest lands which are designated for a special purpose or use. These uses can include: conservation of land and soils; river banks and springs; pollution-damaged forests; seedling forests; wildlife habitat; protected species; forests significant from a natural or scientific point of view; national security; forest buffer zones within urban administrations with populations of over 50,000; buffer zones around sanatoria and spas; as well as forest lands in mountain areas near the treeline. The Minister of Environment, Natural Resources and Forestry designates conservation forest lands.
4. *Development Forest Complexes* (Leśne kompleksy promocyjne) are forest areas where conservation or ecologically oriented management practices are to be promoted with the purpose of integrating in a sustainable manner forest production and conservation of ecological values and processes.
5. *Local land use plan* (Miejscowy plan zagospodarowania przestrzennego) covers the whole or part of a gmina or an association of gminas. The plan designates as necessary: borders; zoning and principles of land uses, with special reference to public activities; and conservation goals. Principles of land use include: use of lands for municipal infrastructure and public utilities; building permits and permission; types of use; designating building lots for commercial and residential purposes; approval of environmental protection; as well as securing effective management of natural resources and protection of agricultural and forest soils. The plan also allows for temporary or interim land uses, places limits on the geographical extent of development, and identifies degraded lands for restoration. Preparation of the plan is one of the obligations of the gmina. The Council of the gmina is responsible for initiating the preparation of relevant plans, unless such plans are mandatory. On approval of the plan by Council, the plan becomes local law.
6. *Forest management* (Gospodarka leśna) refers to the whole range of activities related to the maintenance of forests, including conservation, planting, sanitary cutting, species management, as well as forest production, including timber, wildlife and other forest products (plants, bark, etc.). Forest management activities are undertaken according to forest management plans which identify tasks related to conservation, maintenance and new planting, as well as timber and wildlife production. Forest management is practiced in a strategic and systematic fashion in the main part only in state forests.
7. *Forest management plan* (Plan urządzania lasu) identifies priorities (timber production, afforestation, sanitary practices, wildlife management), as well as technical and infrastructure needs. Plans are prepared for a ten year period. For non-state forests an abridged plan is prepared. The Minister of Environment, Natural Resources and Forestry approves plans for state forest, whereas the voivode approves plans for non-state forests. Local land use plans must conform with approved forest management plans.
8. *Nature reserve* (Rezerwat przyrody)is a designated area where ecosystems are largely natural and largely undisturbed by human activities. The nature reserves are vital for the survival of valuable species of plants and animals or for the maintenance of important geologic features and processes. A buffer zone can be designated in the area surrounding the reserve. The Minister of Environment, Natural Resources and Forestry designates nature reserves. Following designation a management plan must be prepared which must be approved by the Minister.

9. *National park* (Park narodowy) is an area not less than 1000 ha which is characterized by significant scientific, social, cultural and educational values and so designated for protection. Nature in a national park is subject to protection in toto. A buffer zone around the park provides additional protection. Access to a national park is subject to special controls, including payment, hiking only on marked trails or with recognized guides etc. National parks are created on the basis of a Cabinet Decree; the statute of a national park is established by the Minister of Environment, Natural Resources and Forestry. A nature conservation plan is prepared for the national park and the surrounding buffer zone and must be approved by the Minister of Environment. Land use plans must be in conformity with the national park nature conservation plan. A director is responsible for management of the national park and is advised by a Scientific Advisory Committee.
10. *Landscape park* (Park krajobrazowy) is an area designated for protection on account of its natural, historical and cultural values. The goal of a landscape park is the preservation and promotion of these values in the context of human use and development. A buffer zone can be designated around the park for additional protection against external pressures. The Voivode or governor designates landscape parks. A conservation management plan is prepared for the landscape park and its surrounding buffer zone. Local land use plans must be in conformity with the conservation management plan.

Glossary prepared on the basis of : T. Kachniarz and Z. Niewadomski. Nowe podstawy prawne zagospodarowania przestrzennego (New legal foundations for land use planning) and Ministry of Environment, Natural Resources and Forestry. Sustainable forest development in Poland.

PART 3
Learning from the Experience of Others

Monitoring, assessment and regular reporting is necessary not only on what is happening in our home areas, but also in other places so that we can learn from the experience of others. The papers in Part 3 are offered in this spirit and also because they throw light on the papers and ideas presented in Parts 1 and 2. Highlights for Part 3 include:

- Science, regulation and negotiation can be used effectively in combination, in resolving multiparty conflicts in protected areas, for example skiing in Pilsko Mountain, Poland.

- Ecological restoration can be combined with interpretation and tourism programs in promising ways as for example in Hungary.

- Conflicts over the balance between economic development and conservation - for example on new private lands in formerly communist Central Europe - can be reduced considerably by strategic planning built on international conventions and programs and financial and technical support, as for example in Bulgaria.

- Opportunities for bioregional and other ecological and collaborative approaches exist in the mountainous lands of Europe, for example in the Carpathian region of Poland, Ukraine and nearby countries.

- Natural, economic and social aspects of protected areas need to be linked and planned more interactively and effectively as for example in Estonia.

- Historical and cultural factors can have considerable effect on planning, managing, and deciding upon protected areas, especially in newly independent nations such as Lithuania, where national parks now are established and run largely on historic and ethnic grounds.

- New technical and research methods and techniques such as computer-based Geographical Information Systems (GIS), Remote Sensing and Risk Analysis can be used in cross-disciplinary ways to provide the information needed to deal with longstanding issues such as wetland conservation, as for example in Turkey.

- Areas outside of North America and Europe are experiencing loss of regionally and internationally important natural values as well as the capacity to support local and national heritage conservation and sustainable development; ways need to be found to offer technical assistance and support in often challenging circumstances, as for example in Libya.

Conflict Between Skiers and Conservationists and an Example of its Solution: The Pilsko Mountain Case Study (Polish Carpathians)

Adam Lajczak[1], Stefan Michalik[1] and Zbigniew Witkowski[1]

[1] Institute of Nature Conservation, Polish Academy of Sciences,
Lubicz 46, 31-512, Kraków, Poland

Abstract. The top part of Pilsko Mountain (Polish West Carpathians) is distinguished by its outstanding natural values. Mountain peat-bogs are a unique phenomenon at these altitudes. Plant species include *Allium schoenoprasum* at its only known site in the Polish Carpathians. Among the animals are the endemic rodent, *Pitymys tatricus* and the high montane bumblee-bee, *Bombus pyraneus*. Both are known from only a few localities in the Polish Carpathians. In the areas utilized by skiers and tourists there are signs of degradation. Forest habitats have been fragmented. Erosion has increased considerably. Encroachment threatens animal species. To stop further environmental deterioration and to protect the local flora and fauna, areas of particular natural value were proposed as nature reserves, while those already degraded were to be restored. The proposals were discussed and implemented. The land owners agreed on the establishment of nature reserves in the area of their properties, while the ski lift owners were to pay the costs of vegetation and soil restoration. On the other hand the NGOs and ecologists accepted the use of part of the mountain as a ski sport area.

Key words. Pilsko Mountain, natural values, skiing, tourism, threats to nature, habitat deterioration, rehabilitation solutions, biodiversity, endemic species, conflict, negotiation

1 Introduction

From the outset, ski-lift development on the top part of Pilsko Mountain aroused concern among those engaged in nature conservation. Yet their concerns and opposition was ignored and ski-lifts were constructed. One of the lifts was even built without the required permits and documentation. About the same time a landscape park was designated in the area of the Beskid Zywiecki Mountains which included Pilsko Mountain - in spite of formal by some local governments. Both these events caused prolonged and escalating conflicts between conservationists and developers. This paper summarizes the results of a research project conducted in the Pilsko Mountain area in the years 1993-1994 (Witkowski 1996). The investigations aimed to assess the natural values of the area and the degree to which they are endangered and subject to environmental degradation. The practical motivation of the study was the need to find methods of protecting the nature of Pilsko Mountain amid growing degradation pressures and to develop ways of resolving conflicts between conservation and development.

2 Study Area

The study focussed on the top part of Pilsko Mountain in the Beskid Zywiecki Mountains, a massif lying on the border between Poland and Slovakia. The Polish part includes about 25 percent of the mountain top on the north-east side. Large bog-springs and peat-bogs are located in a broad plateau area, mostly in two mountain meadows: Hala Cebulowa and Hala Miziowa (Michalik 1992a and b). Above them, are two other meadows: Hala Pod Kopcem and Hala Slowikowa, which are markedly changed by skiing (Figure 1).

Figure 1. Location of planned protected areas on the top part of Pilsko Mt.; 1. Dwarf pine and upper montane spruce forest reserve, 2- mountain pasture (2a-2b) and forest (2c) reserve, 3 and 4 – peat bog nature monuments, 5 – mountain pasture protected as "area of ecological stability' (a new category of protected area in Poland).

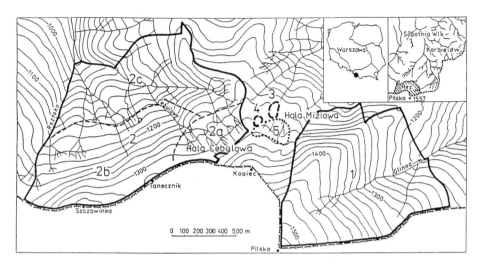

The lower parts of meadows are surrounded by spruce stands representing the upper mountane forest belt *(Plagiothecio-Piceetum)*. Higher up is a community transitional between spruce forest and dwarf-pine *(Pado-Sorbetum)*. At the top of Kopiec Mountain are dwarf-pine thickets *(Pinetum mughi carpaticum)*. The meadows are overgrown with different plant communities. The lower areas are dominated by a Gladiolo-Agrostietum deschampsietosum association. Higher up the dominant community is alpine Vaccinium heath *(Vaccinietum myrtilli)* (Michalik 1996a).

The top part of Pilsko receives an annual average of 1300 mm of precipitation. Favourable snow conditions allow skiing for over 100 days every year (Lajczak 1996a; b; c). The top of Pilski Mountain is intensely used by tourists and skiers. Tourist facilities include a shelter-hostel and another under construction. Both are on the Hala Miziowa meadow. There is also a station belonging to the Volunteer Mountain Rescue Service. Two ski-lifts on the Hala Miziowa lead to the top of Kopiec and to the Hala Slowikowa meadow. Skiers utilize vast forest glades connected by ski runs located on former mountain pastures.

3 Natural Values of the Top Part of Pilsko Mountain

3.1 Inanimate Nature

The massif of Pilsko Mountain (1557 m a.s.l.) is the third highest in the Polish Carpathians. The massif is associated with severe climate and distinctly marked zonation of plant formations (Lajczak 1996a). In this respect the area is unique in the Carpathians and Poland. An elevation of the order of 1250-1350 m a.s.l. is a geographical and climatic rarity at the regional scale (Table 1).

Table 1. Unique Characteristics of the Nature and Landscape of the Pilsko Massif

Unique feature	Range of uniqueness
Altitude above sea level	in the scale of Carpathians
Highmontane climate	in the scale of Carpathians
Zonation of plant formations	in the scale of Carpathians
Geomorphology of the area of Hala Miziowa (landslide niche and moraine)	in the scale of Carpathians
Peat-bogs of considerable thickness in the Hala Miziowa and Hala Cebulowa meadows	in the scale of Carpathians
Cryoplanation terraces around the peak	in the scale of Carpathians
Alpine plant formations (upper forest limit and dwarf-pine belt)	in the scale of Carpathians
Richness of mountain and mountain-boreal species of plants and animals	in the scale of Europe, Poland, Carpathians

An interesting configuration of the massif is found in the neighborhood of Hala Miziowa. Geomorphologists suspect that the niche above this meadow is the remnant of a large landslide reworked by glaciation in the Pleistocene. The amphitheatre-like rampart within Hala Miziowa is a moraine. After glacier recession a small lake was formed in this place and later on, it was transformed into a mountain peat-bog (Lajczak 1996a). Apart from its natural value, this peat-bog constitutes a record of changes in the climate and nature of the area and is a unique phenomenon at the regional scale. A detailed palynological analysis will allow for reconstruction of the evolution of the peat-bog environment. It is the only site in the Polish Beskidy Mountains where an approximately 4.5 m thick peat-bog may turn out to contain the history of changes caused by climate and humans in the environment of a mountain slope at an altitude of 1250-1500 m. Cryoplanation terraces surrounding the peak also contribute to the distinctness of this part of Mount Pilsko. The great concentration and size of peat-bogs is a feature distinguishing the region of Hala Cebulowa and Hala Miziowa meadows at the scale of Carpathians (Table 1).

3.2 Animate Nature

Unique at the national scale is the transitional zone between the mountain spruce forest and the dwarf-pine belt (Table 1). It is worthy of mention that plant communities in this most western massif of the Carpathians are similar in their character to comparable formations in the Sudety Mountains (Michalik 1992a; 1996a). The flora of the investigated part of the Pilsko Mountain study area comprise 80 mountain species, including 18 alpine and 27 subalpine. Some of them (*Salix herbacea*, *Empetrum hermaphroditum*, *Euphrasia picta*, *Melampyrum sylvaticum ssp. carpaticum*, *Sweertia perennis ssp. alpestris*, and *Allium schoenoprasum*) are rare in Poland and in the Carpathians. These species are most numerous in the Hala Miziowa and Hala Cebulowa meadows, in small bog-springs above the Hala Miziowa and in the dwarf-pine belt near the top.

The greatest rarity among the mammals is the Tatra pine vole (*Pitymys tatricus*), which is a West Carpathian endemic species and of regional importance in this area. The species inhabits a transitional zone between forest and dwarf-pine. Other rarities at the national scale include some other species of small mammals (*Sorex alpinus, S. minutus*, and *Sicista betulina*) and birds (*Anthus spinoletta, Turdus torquatus*, and *Loxia curvirostra*). Among the insects are two interesting species of bumble-bees, high-montane *Bombus pyrenaeus*, which occurs in only three other sites in Poland, and rare, boreal-mountain *B. jonellus* (Adamski 1996; Faber 1996).

These valuable natural elements tend to be generally distributed over the study area but have other places of concentration. These places are the dwarf-pine belt and the zone of transition between mountain spruce forest and dwarf-pine as well as near the top part of the massif. These areas are the habitat of the Tatra pine vole *Pitymys tatricus*. The other area distinguishable by the separateness and richness of its flora and fauna is the Hala Cebulowa meadow. The peat-bogs on Hala Miziowa are sites deserving special protection because of their exceptional geomorphology and because they record historic changes in the area. The peat bog is rare and quite rich in flora. Smaller areas supporting threatened plant and animal species are located in the lower part of the Hala pod Kopcem meadow.

4 Ski Facilities on the Top Part of Pilsko Mountain as a New Element in the History of the Utilization of this Area

4.1 Area and Shape of Pilsko Meadows and Glades and Links Between Them

The meadows situated near the top of Pilsko had a definite shape. They were horizontally wide and vertically narrow. The meadows were connected with each other and with villages by narrow navel strings of roads. These allowed for limited migration of species. The introduction of skiing has changed this traditional spatial structure. Skiers favour strongly inclined slopes. So former pastures have been stretched vertically and narrowed horizontally. Old narrow roads have been changed into ski runs and routes for ski-lifts. These links are broad. They radically change the microclimatic conditions, the rate of the flow of waters and the chances of migration of species.

4.2 Area, Age Structure, Species Composition and Health of Stands

Before humans colonized these areas, their lower parts up to about 1200-1250 m a.s.l., were covered by stands dominated by beech and fir (Kawecki, 1939, Ralski, 1930). That is why the forests with a greater share of these species are considered as better preserved than stands dominated by spruce. Sheep farming introduced little change in this area. Fluctuations in the intensity of this activity led to some instability and to a mosaic of patches of heaths, shrubs and young trees (Michalik 1990).

An important factor influencing the health and vitality of stands is air pollution. The results of analyses (Orzel 1993; Bandola-Ciolczyk and Kurzynski 1996) clearly indicate a rapid decrease in the rate of growth of trees in the upper montane belt of Pilsko Mountain This process started towards the end of the fifties and intensified in the later part of the eighties. During the last two decades the incremental growth of the diameter of trees has decreased by about one third (Orzel 1993).

Damages by skiing are concentrated, almost exclusively, in the immediate vicinity of ski runs and affect all age classes. Shrubs and saplings have their tops cut. Young trees lose branches. Older trees have the bark injured (Bandola-Ciolczyk and Kurzynski 1996). These damages, however, are not so harmful as indirect ones. Creation of broad ski run corridors has exposed the forest face, which has resulted in dangerous changes in the local microclimate and wind strength. A visible consequence of this is the dying of forest along the ski run on the top part of Skrzyczne (Orzel 1993). The authors of a report on the forests of Pilsko (Bandola-Ciolczyk and Kurzynski, 1996) suggest that this process begins on the top part of the massif .

4.3 Permanent Human Investments

Until the present century no built structures were located in the top area of Pilsko massif, except for shepherds' sheds built of local material (Kurzynski et al. 1996). With the development of mass tourism the first tourist shelter was erected in this area. As a result of the development of skiing, a system of permanent ski-lifts was installed. Another structure is the building belonging to the Volunteer Mountain Rescue Service. All these structures need power and water supply, which has brought further installations. All these objects and activities produce waste and garbage. These by-products encourage synanthropic plant (Michalik 1992a) and animal species (Adamski 1996; Faber 1996) to invade the investigated area.

4.4 Landscape

Changes in the landscape caused by sheep farming led to the formation of a coarse-grained spatial distribution of large forested areas, meadows and glades. Skiing tends to fragment the forest area around ski runs and to form numerous small 'streams' of ski traffic (Mielnicka 1996). This activity changes the previous coarse-grained spatial distribution of forest and meadow patches into a fine-grained one (Hansson 1992). The activity causes local fragmentation of the forest ecosystem, which results in the replacement of typically forest and shade-loving plant and animal species for light-loving, common ones (Fabiszewski et al. 1993; Michalik 1996a).

4.5 Soil, Erosion and Waters

The direct influence of skiing on soils greatly varies depending on soil characteristics and resistance (Langer 1996). The greatest changes occur in podzolic soils in the dwarf pine belt. They are concentrated in close proximity to the upper stations of ski lifts. The area of endangered soils has been estimated here at 0.4 ha. Within this area significant changes have affected c. 0.13 ha. The remaining area of soils is unchanged by skiing (0.12 ha), or partly changed (0.15 ha). Degradation of soils involves the mechanical stripping of the humus layer of podzolic soils. A decrease in the thickness of this layer in the most heavily changed area was over 7 cm on the average, as compared with soils in a control area.

The area of endangered brown soils in the Hala Miziowa meadow has been estimated at c. 0.30 ha. Mechanical transformation of soil is less important there; although there are substantial changes in the chemistry of soils. Soils affected by tourists had a much greater specific weight and lower ability to absorb rainwater (Lajczak 1996b and c).

An analysis of the course of the process of erosion in the areas utilized by skiers indicates that skiing is one of the three factors responsible for the intensification of erosion in the study area. The two others were attributed to hiking and sheep grazing on the basis of botanical survey and research on the displacement of soils and solifluction. It has been found that the erosion caused by skiing covers about 30% of the eroded area and erosion caused by hiking about 70% (Michalik 1996a, Lajczak 1996b and c). The effect of sheep grazing has not been calculated but is thought to be less than 1%.

The effect of tourist traffic is still greater when we take into account the overall mass of eroded soil. In this case tourists are responsible for nearly 90% of the process and skiers for only 10%. The contribution of sheep is estimated at less than 1%. However, the effect of skiing is indirectly greater because it leads to the formation of large corridors stimulating pedestrian traffic.

4.6 Vegetation and Plant Populations

The most important changes in the vegetation of the top part of Pilsko Mountain took place in the early period of colonization of this area by Wallachian shepherds. At this time, as a result of sheep grazing, new plant communities developed and became consolidated, and the local flora was enriched (see Michalik 1992b). It has been estimated that semi-natural non-forest communities support about

50% of the flora of mountain areas. Certain unique species, such as *Allium schoenoprasum* have extended their ranges and increased in numbers, most probably because of sheep farming. Sheep farming also caused spreading of a large group of alpine species in the forest belt (Kurzynski et al. 1996).

No significant effect of skiing has been found on unique plant communities, or on rare and endangered plant species. The areas where these communities and species occur only slightly overlap areas used for skiing. However, skiing effectively damages vegetation on convex and strongly inclined slopes. Areas affected by skiing are characterized by a thinned sod, and a lower number, or even the disappearance, of flowering plant species. Nevertheless, it should be stressed that these processes mostly affect areas covered by alpine Vaccinium heaths (*Vaccinietum myrtylli*) and poor sward (Michalik 1996a).

4.7 Fauna

The introduction of sheep grazing in the higher parts of mountains caused many negative changes in the local fauna. One of the negative changes was the introduction of exotic ubiquitous forms which can successfully eliminate autochtonic species, as shown for example by snails (Dyduch-Falniowska 1990). Our observations indicate that this process has begun on the Hala Miziowa meadow where ubiquitous synanthropic mammal (Adamski 1996) and bird species (Faber 1996) have successfully competed with native ones for food and territories. A positive effect of sheep grazing is the enrichment of local species connected with open areas as well as the creation of new habitats for mountain bird species that need areas clear of trees and shrubs for nesting (Faber 1996). It has been found however, that species richness and numbers of the local fauna of mammals, birds, and selected groups of invertebrates are much lower in ski areas (Adamski 1996; Faber 1996). These are poor sites for pollinating (Apoidea and butterflies) and for herbivorous insects, especially at higher altitudes.

5 Conclusion: The Necessity to Regulate Use of the Pilsko Mountain Massif

The conflicts associated with skiing on the massif of Pilsko Mountain need to be addressed in a systematic and coordinated way, so as to ensure that the area can be permanently used on a long term basis without increasing degradation of nature and environment. The problems should be tackled by:
1. Solving questions connected with nature conservation.
2. Undertaking restoration works
3. Organizational, administrative and legal regulation of the ways of using the top part of Pilsko Mountain (Krzan et al. 1996).

The areas of great natural value should have absolute conservation priority. Here nature reserves and other strict forms of protection should be created. Relevant proposals were presented in the paper by Michalik (1996b). With the agreement of the owners of the areas, the preservation of the natural values of Pilsko Mountain seem to be relatively easy, because the most valuable structures and areas are not directly threatened by skiing and because the existing organizational, legal and technical solutions seem to guarantee the success of conservation measures (Michalik 1996b). In the conservation management plan for Zywiec Landscape Park which is to be prepared in the near future, the ski area of Pilsko Mountain will be treated separately as in the case of the area of Kasprowy Wierch Mountain in the Tatra National Park (Skawinski 1993).

Preservation of areas where environmental degradation has been proceeding seems to be more difficult. Two potential solutions to this problem have been discussed. The first is the lowering of the upper station of the longer ski lift and reduction of the capacity of both ski lifts; together with a limited programme for the restoration of the environment. The second solution involves introduction of a wider programme for the restoration of the environment at the expense of the ski lifts owners. This has been viewed as a less controversial but more expensive solution.

Both proposals included preservation of tourist trails and their separation from ski runs; protection and resodding of places undergoing erosion; the construction of artificial barriers, transverse to the main direction of winds; and the prohibition of the use of snow grooming equipment in the upper parts of ski runs near ski lifts (Krzan et al. 1996).

After several meetings and discussions the ski lift owners, the landholders and the administration of the Zywiec Landscape Park accepted the second proposal as follows:

a) The land owners accepted the establishment of the planned protection measures - two nature reserves, one nature monument and two ecological stable areas - on their property,
b) The soil and vegetation was to be rehabilitated at the expense of the ski-lift owners,
c) The primary agreement was for a two year period,
d) Any rejection of the program would render the agreement invalid and result in the termination of the ski lift licenses.

Concerned scientists proposed a systematic approach regulation of the management of the Pilsko massif. This would involve determining the natural, technical, spatial, architectural, tourist and skiing conditions for optimal use of the area by skiers in the Pilsko Mountain ski resort. The results would be included in the statutory development and management plans for the area. They would also be a basis for further investments and modernization of facilities, installations, tourist trails and ski runs (Barnes 1995; Tsuyuzaki 1994). Only when this is achieved can we speak of an integrated or systems management approach based upon environmental criteria and permanent use without degradation of the natural values of the area.

6 References

Adamski P. msc. 1996. Small mammals of the top area of Pilski Mountain and the estimation of the impact of the tourism on teriofauna (in Polish). *Studia Naturae* (in press).

Bandola-Ciolczyk E. msc. 1996. The contamination of spruce and dwarf pine needles by sulphus and heavy metals in the top area of Pilsko Mt (in Polish). *Studia Naturae* (in press).

Bandola-Ciolczyk E., Kurzynski J. msc. 1996. Indices of the vitality and mechanical damages of upper montane spruce forest and dwarf pine on ski trails and control plots in top area of Pilsko Mt (in Polish). *Studia Naturae* (in press).

Barnes S. 1995. Balancing act - Oregon ski resort uses Geospatial Computer Technologies to achieve environmentally sensitive, economically viable development. *Geo Info System* 5, 1: 27-35.

Bzowski M., Dziewolski J. 1973. Damages caused by foehn in the spring of 1968 in forests of the Tatra National Park (in Polish). *Ochr. Przyr.* 38: 115-154.

Dyduch-Falniowska A. 1990. Molluscs of the Tatra Mts mountain meadows (in Polish). *Studia Naturae* A, 34: 145-162.

Faber M. msc. 1996. Birds of the top area of Pilsko Mt and the estimation of tourism impact on the avifauna. *Studia Naturae* (in press).

Fabiszewski J., Wojtun B., Zolnierz L., Matula J., Sobierajski Z. 1993. Changes in abundance of the ground cover plants of the upper montane spruce forest in Sudety Mts in the consecutive degradation phases (in Polish). In: *Karkonoskie badania ekologiczne I Konferencja*, Wojnowice, 3-4 grudnia 1992, Ofic. Wydawn. Inst. Ekologii, Dziekanów Leuny: 77-85.

Hansson L. 1992. *Ecological principles of nature conservation, application in temperate and boreal environments*. Elsevier Appl. Sci., London and New York.

Inglot S. 1986. *Remarks on the history of Polish country and farming* (in Polish). Ludowa, Spoldz. Wydawn., Warszawa 448 str.

Kawecki W. 1939. *Forest of the Zywiec district, their present and future* (in Polish). Pr. Rolno-Leune PAU 35.

Krzan Z., Lajczak A., Michalik S., Skawinski P., Witkowski Z. msc. 1996. Project of recultivation and rearrangement of skiing and hiking on the top area of Pilsko (in Polish). *Studia Naturae* (in press).

Kurzynski J., Lajczak A., Michalik S., Mielnicka B., Witkowski Z. msc. 1996. Outline of the history and forms of exploration of the Pilsko Mt (Polish West Carpathians) (in Polish). *Studia Naturae* (in press).

Langer M. msc. 1996. Impact of skiing on the physico-chemical characteristics of the top area of Pilsko Mt (in Polish). *Studia Naturae* (in press).

Lajczak A. msc. 1996a. Abiotic environment of the Pilsko Mt with special regard to its top area (in Polish). *Studia Naturae* (in press).

Lajczak A. msc. 1996b. Chosen elements of the weather on the northern Pilsko slope with special regard to the Hala Miziowa glade (in Polish). *Studia Naturae* (in press).

Lajczak A. msc. 1996c. Impact of skiing on the soil erosion of the top area of Pilsko Mt (in Polish). *Studia Naturae* (in press).

Michalik S. 1990. Vegetation succession in a mountain glade in Gorce National Park during 20 years, as a result of pasturage abandonment (in Polish). *Pr¹dnik* 2: 137-148.

Michalik S. 1992a. Dangers and problems of active protection of biocenoses of the subalpine glades of Gorce National Park (in Polish). *Parki Narod., Rezerw. Przyr.* 11, 4: 25-37.

Michalik S. 1992b. Vegetation of the Pilsko nature reserve in the Beskid Zywiecki Mountains (Western Carpathians) (in Polish). *Ochr. Przyr.* 50, 2: 53-74.

Michalik S. msc. 1996a. Influence of skiing and hiking on the vegetation of the top area of Pilsko Mt (in Polish). *Studia Naturae* (in press).

Michalik S. msc. 1996b. Projected net of the protected territories in the top area of Pilsko Mt (in Polish). *Studia Naturae* (in press).

Mielnicka B. msc. 1996. Skiing and hiking on the top area of Pilsko Mt (in Polish). *Studia Naturae* (in press).

Orzel S. 1993. Assessment of dynamics of diameter increment in the mountain spruce stands exemplified by selected areas in the forests of Silesian and Zywiec Beskids (in Polish). *Acta agr. et Silv.* 31: 3-15.

Ralski E. 1930. Mountain pastures and meadows of the Pilski Mt in the West Beskidy Mts (in Polish). *Pr. Roln.-Leune PAU* nr 1: 1-156.

Siemionów A. 1984. Wadowice country (the landscape-touristic monopgraphy) (in Polish). Kom. Turyst. Górsk. Oddz. PTTK Ziemia Wadowicka w Wadowicach, Wadowice 628 pp.

Skawinski P. 1993. Kasprowy Wierch and Goryczkowa Valley: Human impact on nature in the Tatra Mountains (in Polish). In: *Ochrona Tatr w obliczu zagrozen* (red. W. Cichocki). Muz. Tatrz. im. dr T. Chalubinskiego, Tatrz. P.N., Wyd. Muz. Tatrz.: 197-226.

Research, Conservation, Restoration and Eco-tourism in National Parks: Experience from Hungary

Mihaly Vegh[1] and Szilvia Gori[1]

[1] Hortobágy National Park (HNP),
H-4024 Debrecen, Sumen U2, Debrecen, Hungary

Abstract. In Central Europe (CE) as well as Hungary, the changes in the political and economic systems are very much affecting the system of protected areas. The responsible management authorities have to face difficult situations and have to give more precise and adequate answers for those people who are within or around different protected areas. International conservation organizations are continuously developing their requirement systems. These demands are pushing the CE countries to develop their conservation systems further. Hungarian nature conservation started to deal with this task at the beginning of the 90s. Programmes were launched basically using domestic funds but also with significant external financial support. Each of the programs was contributed to creating a new systematic approach. The fundamental element was a research program set up according to the IUCN requirements, called: the basic natural status survey. Based on the results, especially in National Parks, a zonation system was developed. Parallel with the zonation system a PHARE project was carried out which made proposals for introduction of eco-tourism according to the prescriptions laid down in the zonation system. Related management plans and restoration projects were also set forth. Both of these activities have strong research needs. Also today when the economic aspect of conservation activity seems to be of vital importance, the introduction of eco-tourism into national parks systems stimulates research into a field which was not present in the classical conservation methodologies. This field is the socio-economic area and notably the economic valuation of protected areas. If we take into account a basic principle of the conservation, namely the necessity of the participation of local people in conservation activity - especially in densely populated Europe - then eco-tourism projects together with their research needs, are a sound tool for enabling parks to give better responses to the pressures upon them.

Keywords. National parks, protected areas, conservation, management plans, zoning, economic evaluation, eco-tourism, public participation, privatization, Hungary

1 Introduction

At the beginning of the 90s, as a result of the changes in the political and economic structure of the country, nature conservation had to face new and basic challenges. One of the most important factors was the restructuring of the land ownership system. According to a set of new laws the process of compensation, privatization and the transition of the former cooperative farms was started. In order to be able to preserve the natural heritage of the country, the nature conservation authority had to be well prepared because in many cases it had to give very precise answers about areas and natural values. On the other hand, international organizations were continuously refining their requirement systems and meeting these demands was also a very important factor driving nature conservation. In the case of national parks, a very well defined request was addressed to Hungary to establish a scientifically based zonation system.

Due to our traditional lack of roads and infrastructure, a lot of natural values were preserved; values which had already disappeared from the other parts of Europe. The official nature conservation body realized this fact and launched numerous programmes aimed at assuring continued protection. The majority of the projects was financed from domestic resources but a significant international financial contribution was also involved, mainly from the European Union's PHARE Program. Each of the projects was part of a new systematic approach to conservation practice.

The fundamental component of the process was the so called basic nature status survey. In order to have a unified countrywide system, the whole scientific establishment - research institutes, universities, consultants, national parks, NGOs - was involved in a consultation procedure. As a result of this, a jointly accepted system was worked out. This research method contained all those elements which later made it possible to apply GIS (Geographical Information System).

A priority list was also set up for the protected areas system. In the first place were the national parks; thus the majority of research efforts were undertaken in these areas. The second priority was landscape protection areas and nature reserves where work still under way. During the course of the research in the national parks, it was recognized that work should also be done in the socio-economic field because only in this way can a really adequate system be set up. Such work for the national parks was hitherto totally unknown, so field research projects were implemented to find possible ways of giving better answers to new demands.

With significant external financial help, zonation and eco-tourism projects were carried out in the three largest Hungarian national parks. These projects were very closely linked with the results from the basic nature status survey process. The result was the setting up of the zonation system. At the same time consultants were carrying out data collection on the land ownership situation and its trends, economic developments in the respective areas, tourism facilities, and possibilities for eco-tourism development. Combining all these data resulted in a refinement of the zonation system originally established purely on natural grounds. The IUCN zoning prescriptions were also adapted to the specific conditions of the respective parks (Table 1). If we now look at the final result it can be seen that it is a relatively complicated one. But if we take into consideration other aspects than nature conservation it can be easily understood. This exercise proved to be very useful and has generated further work which is leading to a management plan.

A very user friendly tool has been developed in a form of a chart, which takes into account the biological aspects of different zones as well as the land use forms. The tool gives a precise description about where, when and what can be done (Table 2). This chart makes it possible to determine: which zone/subzones have to be managed especially for conservation purposes; which are available for mass tourism; which can sustain eco tourism; and which need restoration.

Parallel with the ongoing political, economic and ownership changes, nature conservation had to face serious financial restrictions too. New ways had to be found to generate revenues. The zonation and eco-tourism projects were an excellent opportunity to elaborate projects and plans for the introduction of revenue generating activity and also to make suggestions for higher levels of financial and legislative management. Such management can help the work of nature conservation very much. A good example is the Canadian revenue generation system, which is also in an introductory phase. This system allows the park authorities to keep much of their planned revenue and use the funds in servicing the National Park system, in establishing new parks, producing educational materials, or promoting introduction of a revenue generation system to all those parks that can provide adequate facilities for visitors.

The Hungarian national parks at this stage unfortunately cannot provide the variety of services that the Canadian system can, but a certain level can be reached. Eco-tourism projects for the respective national parks have been used to identify the possibilities where the first steps can be taken. A few projects have been elaborated for the Hortobágy National Park (HNP) (Figure 1). We will give some information about the results which already have been achieved.

Table 1. Description and legal status of IUCN zones in the Hortobágy National Park

ZONE	SUB-ZONES	DESCRIPTION, LANDUSE & FUNCTION	LANDUSE RIGHTS	OWNERSHIP LEGAL STATUS
A	-	Relatively large natural areas without any type of economic use in order to protect species and biotopes of special concern and/or biogeographically typical ones	NATIONAL PARK Authority	State-owned, managed by the National Park
B	B1	Natural or semi-natural biotopes surrounding areas of ZONE A, with strictly controlled and limited agricultural and forestry activities with the priority of nature conservation in order to protect species and biotopes of special concern and/or biogeographically typical ones	NATIONAL PARK Authority	State-owned, managed by the National Park
	B2	Biotopes developed as a consequence of human activities (fish-ponds, depositories, tree-plantations, etc.) with special, limited land-use forms, with the priority of nature conservation in order to protect species of special concern	NATIONAL PARK Authority	State-owned, managed by the National Park
	B3	Extensively used ploughed areas with some valuable nesting and feeding bird species of special concern, controlled and managed in order to provide best conditions for these species, some of them are planned to change into grasslands	NATIONAL PARK Authority	State-owned, managed by the National Park and private owners, Nature Conservation Law, authoritive power
	B4	Buildings or other settlements with their relatively small (less, than 10 ha) surrounding areas	NATIONAL PARK Authority and others	
C	C1	Extensively used grasslands and tree-plantations inside the present protected area of the national park, where economic use is allowed but restricted, serving as a buffer between natural and cultural areas	NATIONAL PARK Authority	State-owned, managed by the National Park
	C2	Ploughed fields and relatively larger settlements (with not more than 150 inhabitants) inside the present protected area of the national park, where economic use is allowed but restricted, serving as a buffer between natural and cultural areas	NATIONAL PARK Authority	State-owned, managed by the National Park
	C3	Intensively used grasslands, tree plantations, forests and ploughed fields outside the present protected area of the national park, serving as a buffer between natural and cultural areas, but ploughed fields also as nesting and feeding sites for valuable bird species and also for landscape protection	Others (private owners, cooperative farms)	Nature Conservation Law, regulatig power, local and regional planning law
	C4	Intensively used ploughed fields (and also some settlements or their parts) for (restricted) economic use, outside the present protected area of the national park which could be potential nesting and feeding sites of valuable bird species and also have a landscape protection function	Others (private owners, cooperative farms)	Nature Conservation Law, regulatig power, local and regional planning law

Table 2. General regulation concept of different land forms in the IUCN-zones of the Hortobágy National Park

ZONE	Ploughing	Grazing	Mowing	Reed cutting	Forestry activity	Fishing	Hunting	Linear establ.	Other establ.	Use of chemicals	Water management
A	Not allowed	Not allowed	Not allowed	Not allowed	Only to promote the renewal of the forests by indigenous tree species	Not allowed	Not allowed	Not allowed	Not allowed	Not allowed	Only to reconstruct the original hydro-geological conditions
B1	Not allowed	Control of the species, races, density, timing and its accessorial activities	Control of areas and timing	Control of areas and timing	Only to promote the renewal of the forests by indigenous tree species	Controlled, but only on rivers and channels	Not allowed	Not allowed, the already existing ones are planned to remove or put into the ground (electric lines)	Not allowed, the already existing ones are planned to remove	Not allowed	Only to reconstruct the original hydro-geological conditions, and also in emergency, when they can be used as storage
B2	Not allowed	Control of the species, races, density, timing and its accessorial activities	Control of areas and timing	Control of areas and timing	Only to change tree plantations into semi-natural forests	Control of timing with special respect to the water management activities of fishing, control of fish species and their amount, feeding etc.	Not allowed	Not allowed, the already existing ones are planned to remove or put into the ground (electric lines)	Not allowed, the already existing ones are planned to remove	Only on fishponds, where the quality and quantity is controlled	Control of timing according nature conservational requirements
B3	Control of timing, species, technology	Control of timing	-	-	-	-	Not allowed	Not allowed, the already existing ones are planned to remove or put into the ground (electric lines)	Not allowed, the already existing ones are planned to remove	Not allowed	Drainage is allowed in special cases, but strictly controlled
B4	-	Control of species	Without control	-	General control	-	Not allowed	Not allowed, the already existing ones are planned to remove or put into the ground (electric lines)	Not allowed, some of the existing ones are planned to remove from the area	Not allowed especially the storage of them	
C1	Not allowed	Control of species, races, density, timing and its accessorial activities	Control of areas and timing	Control of areas and timing	General control	Controlled, but only on rivers and channels	Not allowed	Allowed but strictly controlled	Allowed but strictly controlled	Not allowed	Drainage is allowed in special cases, but strictly controlled
C2	Occasional control of timing	Control of species	-	-	General control (around settlements)	-	Not allowed	Allowed but strictly controlled	Allowed but strictly controlled	Not allowed	Drainage is allowed in special cases, but strictly controlled
D	Occasional control of timing	Uncontrolled	Occasional control of timing	-	-	-	Control of species and timing	Controlled, some of the existing ones are proposed to remove or put into the ground	Allowed, but controlled	Types and amounts are controlled	Drainage, watering and local amelioration is allowed, but regional - affecting inner zones - is not desirable

Figure 1. Hortobágy National Park

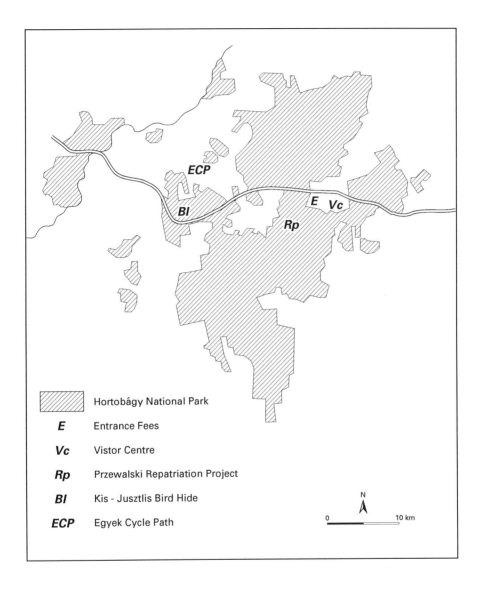

2 Introduction of an Entrance Fee System

The objective of entrance fees is to exercise full control of visitors entering the national park and at the same time to generate income for the park. The funds can be used mainly for biotope restoration, management and development of visitor facilities. In addition they make it possible to obtain more exact statistical data on the numbers of different types of visitors. Given the current status of the country, this system can be introduced only gradually. The Hortobágy National Park already provides some basic facilities such as marked routes, a visitor centre, and an information centre. In 1996 an entrance fee system was started especially with the aim of gaining experience on the responses of visitors and to get ideas for further development.

3 Visitor Centre at Hortobagy Village

Hortobágy village is the centre of Hortobágy region and the National Park. The objective is to create a new Visitor Centre in order to make the visitors more knowledgeable about the HNP values and its tourism facilities. Presently, there is an exhibition of the NP in Hortobágy but it is not developed enough to function as a real Visitor Centre. With the help of Hortobágy Nature Conservation and the Gene Preservation Public Company (HNCGPPC), which has very close connections with the HNP, feasibility studies and plans were elaborated for establishing a new Visitor Centre. At the moment we are in the fund raising stage for this project.

4 Przewalski Horse Repatriation Project

The objective of the project is to restore a few typical steppe animals already extinct in Hungary which have a constricted habitat in other parts of Europe. Thus, a park could be established which, with its unique animals, could attract a goodly number of visitors, in addition to being of outstanding importance for science and research. The national park already has designated an area of approximately 2,500-3,000 hectares of grasslands as the location of the repatriation project. The HNCGPPC started the process of acquiring the Przewalski horses. An agreement already is in place with the Association of Przewalski Breeders for four stallions, and the Public Company already owns the necessary CITES permits to import the animals. Stress is placed on starting a research program together with repatriation and eco-tourism.

5 Kis-Jusztus Bird Hide

The objective is to provide a bird watching opportunity not only for bird watchers but also for visitors with more general interests. The aim is to provide them with basic information on the characteristic bird species of the HNP. In addition the restoration activities of the park can be interpreted and explained for the visitors. With the involvement of a non-government organization (NGO), the park has already carried out wetland restoration work and the bird hide has been established. At the moment it is one of the most popular sites of HNP.

6 Western Gate -Ohat-Egyek Cycle Path

The objective is to show visitors some existing sights in the national park, which are not far from each other; thus it is possible to connect them with an approximately 30 km long cycle path, a section of which would also be part of the planned national cycle-path system. Along the route, there are various points of interest where the cyclists can stop and receive information on different subjects like old farm houses, the gene bank of the ancient pig race - mangalica - the bird rehabilitation program, the Meggyes road side inn, the boardwalk in Fekete-ret marsh and other restored wetland sites along the path. With significant financial help from the Danish Government a large scale wetland restoration program is going on in an area surrounded by the cycle path. The building of certain sections of the path and the design of interpretation exhibits along the route, is part of this project. A research and monitoring program will be started when the restoration project is finished.

7 Research Recommendations

1. Undertake economic valuation of protected areas to give area managers an effective tool, when dealing with the public and the policy makers for the acquisition of non protected lands;
2. Elaboration of simple, cost effect methodologies and systems to monitor restored sites for further planning and management.
3. Socio-economic research in protected areas in order to interpret better the benefits derived from these areas;
4. Developing more effective communication to raise public awareness, understanding and support.

Questions Related to the Rehabilitation of the Egyek-Pusztakocs Marshes

Szilvia Gori [1] and Aradi Csaba [1]

[1] Hortobágy National Park (HNP),
H-4024 Debrecen, Sumen U2, Debrecen, Hungary

Abstract. Rehabilitation of the marshes has been undertaken for conservation and ecological purposes and with the potential for ecotourism in mind. Nature-viewing opportunities and visitor facilities have been concentrated and organized to provide maximum restoration and conservation potential and provide good opportunities for tourism. Research and management recommendations include: monitoring and investigation of species tolerance to tourism; tools for concentrating the spectacle; creation of appropriate observational and other infrastructure; planning of the work schedule to minimize disturbance; and transformation of agriculture in co-operation with local people.

Keywords. National parks, Hungary, Hortobágy National Park, marshes, rehabilitation, conservation, tourism, eco-tourism, interpretation, research, planning and management, transformation of agriculture

Beyond the preservation of the complete structure and function of natural and semi-natural systems, one of the main tasks of nature conservation is to work out the management and rehabilitation of natural systems which are damaged and suffering from ecological dysfunction. Different methods of management (preservation, conservation, rehabilitation, reconstruction) have to be investigated and applied according to the characteristics and conditions of the site. Due to the natural endowment of our country and the large scale transformation of the landscape, rehabilitation and reconstruction play an important role in the practice of nature conservation.

The Egyek-Pusztakocs Marsh System once extended over approximately ten thousand hectares on the former floodplain of the River Tisza. It was inundated regularly by local waterways of the river. After water regulation works were implemented in the middle of the 19th century, the marsh area started to decrease in size and natural water ways were broken off. The dry branches received some water only from precipitation. The drainage of the marshes (1930-50s) and the cultivation of the fertile levees of the catchment area have hastened destruction. To ameliorate these effects, measures were taken to collect local run-off from precipitation, in order to reduce water loss and damages. Until recently most of these waters was accumulated in the marsh system.

Today the largest part of the remaining habitat, which is still less than half of the original area, is protected (4,073 ha). By the establishment of the Hortobágy National Park (HNP) in 1972 the entire Egyek-Pusztakocs Marsh System became a connected Nature Conservation area. Since 1993 it has been protected as part of the HNP (see Figure 1: Vegh and Gori, this volume).

The Hortobágy National Park Directorate has ranked as one of its most important tasks, the restoration work in this area, aimed at the rehabilitation and maintenance of the original water regime. The restoration works were started in 1976 with the aim of rehabilitating a permanent marsh, the Fekete-ret, designated since 1979 as a Ramsar Site. The major part of the work was implemented in 1981-82. The next step was the rehabilitation of another branch of the marsh system in 1992, this being a temporary alkaline marsh. Simultaneously a long term research program has been started to investigate the changes taking place after the restoration and in this way, to improve the implementation of similar kinds of activities. Based on the data gained during ten years of research, a feasibility study (PHARE project) was developed for the restoration of the southern wetlands and grasslands of the national park.

Today this restoration project continues with the financial help of the Danish Government. The complete water supply system will be constructed this year. Basic ecological surveys continue in the area (HNP staff and Debrecen Science University researchers: ornithological, botanical, hydrobiological survey). The effects of the rehabilitation will be followed by a monitoring program.

1 Goals of the Restoration Project

1.1 Preservation of Biodiversity

The historical transformation of the area led to the disruption of the original ecological structure. The extent of the marshes decreased, sensitive communities and species were repressed or disappeared. Edge communities (ecotones) and terrestrial connections were damaged. Management of the area is aimed at the rehabilitation of a favourable mosaic structure including edge communities and the damaged terrestrial connections of the marshes. The mosaic structure is of basic importance, especially for those species which require combined biotopes for their life. Long-term management of the site is aiming at restoration of a complete habitat system with the rehabilitation of marshes and wet meadows, alkaline grasslands, loess grasslands and forest patches. Today these marshes are functioning as a local flyway inside the Hortobagy region, connecting the current Tisza Reservoir with the fishponds and marshes of the Hortobagy and also serving as part of an international migration route.

1.2 Ecotourism Development

The restoration sites provide excellent prospects for eco-tourism. The necessary work for 'concentrating the spectacles' can be done. The required infrastructure conditions can be developed without endangering and damaging natural systems. In the project area, typical habitats and community types of the Hortobágy are concentrated in a relatively small area.

We think it is important to mention some considerations which have to be taken into account in designing paths and ecotourism development in the area. As far as possible, the interpretation of natural sites with high biodiversity has to be limited. Care has to be devoted to the psychology of interpretation of sites. More common species which occur in large numbers or spectacular species with conspicuous behaviour should be the focus rather than the rare ones. The young animals are always a big attraction. Chances for adventure or making discoveries should be given. We have planned some interesting surprises for the tourism sites and routes.

Research needed to prevent ecological damage from tourism include:
- Investigation of tolerance:
    - analysis of the reactions of the most sensitive species (distance of reaction, flight distance)
    - the structure of the community, for example for species living in colonies
    - investigation of the distribution of the values for interpretation
    - the ideal size of the visitor groups

- Tools for concentrating the spectacle:
    - creating, feeding and drinking places for the animals
    - creating advantageous mosaic structures for the species which require combined biotopes, taking advantage of the edge effect
    - resting trees, artificial nesting and reproduction places
    - plantations for interpretation (botanical garden, arboretum, wildlife park, collections with special themes)

- Creating adequate infrastructure conditions for interpretation:
    - building hidden observation points with hidden access
    - resting places for the visitors, where they can get information about the site

- good conditions for traffic (road, vehicle)

- Planning of the work schedule
    - regularity can be important for training and conditioning the species to be interpreted
    - determining the favourable periods of observation (season, part of the day)

1.3 Transformation of Agriculture on the Area

The transformation of intensive farming and its effects will take many years and can only be done in co-operation with local people. Extensive animal husbandry and traditional farming such as reed and bulrush harvesting, are compatible with nature conservation and could provide jobs for the local people. Ecotourism also can generate income for them.

2 Research Recommendations

1. Investigation of the tolerance of the area: analysis of the reactions of the most sensitive species.
2. Research in conservation biology on the effects and the application of different habitat management methods.

Bulgarian Experience in Nature Protection: Contributing to Sustainable Development

Elizaveta Matveeva[1]

1 National Nature Protection Service
   67 W. Gladstone Street, Sofia 1000, Bulgaria

Abstract. The biological diversity of Bulgaria is extremely rich and varied. Its study, as well as nature protection, have deep roots that can be traced to the late 19th century. Environmental and Biodiversity education were developed parallel with the creation of the system of protected areas in Bulgaria. The National Biological Diversity Conservation Strategy (NBDCS) and the National Action Plan for the Conservation of the Most Important Wetlands (NWCP) were developed and published in recent years, these outline the frameworks for the expansion and promotion of the system of protected areas, forming the necessary prerequisites for new educational initiatives in support of sustainable development and the conservation of valuable natural resources.

Keywords. Biological diversity, ecosystem, education, ecotourism, endemic species, nature, conservation, nature reserves, national parks, natural landmarks, protected areas, protected sites, sustainable development, wetlands

1 The Biological Diversity of Bulgaria

Bulgaria is located in Southeastern Europe, in the centre of the Balkan Peninsula, and west of the Black Sea. Being a relatively small country, with an area of 110,912 km^2, it ranks among the most biologically diverse countries in Europe, with rich wild nature. Highly varied climatic, geological, topographic and hydrological conditions allow Bulgaria to support a large composition of flora and fauna groups, including: 94 species of mammals; 383 birds; 36 reptiles; 16 amphibians; more than 27,500 insects and noninsect invertebrates ; 207 species of Black Sea and freshwater fish; about 3,750 species of vascular plants; and more than 6,700 species of nonvascular plants and fungi (Biodiversity Support Program 1994).

In terms of biodiversity, Bulgaria is the third richest area of Europe, with many rare species (including significant numbers of tertiary and glacial relict). Endemic plant species constitute about 5% of the total flora, a high proportion compared with other European countries. About 170 species and 100 subspecies of angiospermous plants are *Bulgarian endemites*. Of special significance are Bulgaria's forests, which cover over 3.2 million hectares (35% of the national total land base); 60 % of this area consists of forests of natural origin. The territory of Bulgaria is characterized by relatively small natural wetlands. At present they cover 11,000 ha, i.e. only 0.1 % of the country's territory, hence the protection of their biodiversity acquires particular significance (Ministry of Environment 1994) .

2 Biodiversity Conservation in Bulgaria. A Historical Review

Throughout the Middle Ages, the lands of the Bulgarians were known as the Sultan's wealthiest Christian province (Kanitz 1932). The territory was covered by impenetrable forests full of wildlife and game. Only two years after Bulgaria's Liberation from Ottoman domination in 1878 the first Game License Act was issued (1880) and the first Bulgarian Forest Act was published in 1883. The

Game Law of 1897 continued to provide a strong impetus to wildlife protection in the country. It placed under protection all useful animals and birds. Before the end of the last century, the Bulgarian Natural History Society, the Bulgarian Tourist Union, the Union of Foresters, and other sporting and civic organizations were established with the primary goal of advocating and disseminating the ideas and principles of nature conservation. In 1919, the issue of designating nature reserves was raised and specific legislation was advocated to save the valuable flora and fauna. During the 1920's for the first time, an appeal was issued to rally all unions and organizations whose activity has something in common with nature to protect Bulgaria's endangered and diminishing flora and fauna. The appeal for cooperation was answered by organizations such as the Bulgarian Nature Protection Society, the Society of Bulgarian Foresters, the Bulgarian Tourist Society, the Youth Tourist Union, the Botanical Society, the Geological Society, and the Speleological Society. Thus, the Bulgarian Union for Nature Protection (BUNP) was formed in 1928. The formation of the Union was in tune with the development of nature conservation in Western European countries at the time. During the 1928-1938 period, on the initiative of the Union, the establishment of the first Bulgarian nature reserves and national parks began.

The Union's major achievement in 1933-1934 was the establishment of the Vitosha National Park, the first of its kind in the Balkan Peninsula, and the designation of several nature reserves. In 1936, a Decree on the Protection of Native Nature was issued and the Rules and Regulations for the implementation of the decree were published. After realizing its primary short-term objectives, the BUNP set out new, educational and awareness goals. And this was done not so much for the organized tourists and nature lovers, but by them and through them for the future generations. The goals of the Union could be achieved only after we have achieved the necessary level of education and are fully dependent on the cultural standards and awareness level of our broad popular masses. Here we cannot refrain from noting a similarity to the goals and objective of the Rio'92 Summit and Agenda 21, formulated some 60 years later.

Despite the initial enthusiasm and success in designating protected areas, Bulgaria's nature conservation activity considerably diminished during the 1940-1960's, due to the devastating effects of World War II and the changes that followed it. Yet, between 1948-1956, the biggest reserves were registered in the country, the land area of the Vitosha National Park was extended, and the UN List included some 8 protected areas designated up to 1965. The new Law of Game Reserves and the Law of Fisheries were adopted in late 'seventies.

In 1958 a Commission for Nature Protection was established within the Bulgarian Academy of Sciences and became involved in the problems of biodiversity conservation and developing the scientific basis for the establishment of a protected area system. Several years later, the Commission strengthened and expanded its activities, becoming a National Coordination Centre for protection and restoration of nature, and a few years later came the successful outcome: the Institute of Ecology. Scientists, jointly with different institutions and institutes, worked hard to establish a foundation of Protected Areas, within a worldwide network of biosphere reserves of UNESCO. At the same time, Bulgaria maintained long-term relations with IUCN, which later became more intensive after the start of the East European Program (EEP).

In 1976 a Committee on Environmental Protection was established with the Council of Ministers. After November 1989, it was renamed as the State Committee on Environmental Protection and in the beginning of 1990 was transformed into the Ministry of Environment. The 1967 Law on Nature Protection, under which the current system of protected areas and protected plant and animal species was established and is now administered, set their status as follows:
- Nature Reserves;
- National Parks with varying characteristics: some correspond to those included in IUCN Category II, others are more similar to those of IUCN Category IV .
- Natural Landmarks (or natural sanctuaries) and Protected Sites. They correspond largely to those of IUCN Category III. More than 500 such sites have been established.
- Historic Sites (972 at present), which serve to protect lands surrounding historical and archaeological monuments.

The Bulgarian Red Data Books were published in 1984 and 1985, followed by a tremendous rise in environmental education activities and publicity. Nowadays many of the Protected Areas in Bulgaria are of International importance. Two sites (Pirin National Park and Srebarna Reserve) are recognized as World Natural Heritage Sites. As a result of the proposals made at the Council of the MAB International Committee in 1977, seventeen areas were listed as biosphere reserves, with a total area of about 60,000 ha. Four sites were designated as important wetlands areas under the Convention on Wetlands of International Importance (known as the RAMSAR Convention). In addition, 22 sites (some of which are not currently protected) have been designated by BirdLife International as Important Bird Areas in Europe.

In 1991 a new Law on Environmental Protection was adopted. Nowadays environmental issues are covered by 25 laws and 17 sets of guidelines for their application. On-going is the process of accomplishment of a package of specialized legal Acts: Protected Areas Act; Hunting Act; Fishing Act; Medicinal Plants Act; Biodiversity Conservation Act; Marine Environment Conservation Act.

Bulgaria is, or is likely to become a signatory to many international treaties and agreements that affect biodiversity conservation within the country. Among the most significant agreements are:
- The Convention of Biological Diversity;
- The Framework Convention on Climate Change;
- The Convention of Wetlands of International Importance (RAMSAR), Especially Waterfowl Habitat;
- The Convention on International Trade with the Endangered Species (CITES);
- The Convention for the Protection of the World Cultural and Natural Heritage (World Heritage Convention);
- The Man and Biosphere Program (MAB) of the United Nations Educational Scientific and Cultural Organization.

Since 1991 the Ministry of Environment has been implementing strategic planning in the fields of nature protection. USAID, the World Bank, and other international agencies undertook a series of environmental missions to Bulgaria to assist the Government in development of the Environmental Strategy Study and Action Plan, which included Bulgaria's Biological Diversity Conservation Strategy and National Wetlands Conservation Plan.

During the ensuing years a series of projects were proposed to the international donor agencies in order to lay the foundation for a comprehensive protected areas management system. The first step of this work was the development of priorities for the conservation of representative ecosystems, as well as threatened and rare species and habitats. Among the following steps was the widening of the network of protected areas and developing adequate legislative, institutional and economic mechanisms. Until 1977, the protected areas made up less than 1% of the country's territory. In the beginning of 1995, the protected areas were 4.5% of the country's territory (Figure 1).

The development of the protected areas network has been put on a biogeographic basis. Representative samples of sub biomes and ecosystems are under protection. For the purpose of their natural, long-term functioning (avoiding the effects of island biogeography), the Ministry of Environment adopted the idea of creating large national parks (three parks, namely Central Balkan, Rila and Pirin, have a total area of 220,000 ha) with a developed infrastructure of strict nature reserves (accordingly 18%, 27% and 15% of the area of these three parks).

In recent years the establishment of protected areas began after the fashion of West European nature parks where the traditional use of natural resources is combined with their protection. The establishment of the Strandja Nature Park (116,000 ha), gave a promising headstart to this category of protected areas. In terms of biodiversity, the Strandja Mountain and the Southern Black Sea coast in Bulgaria are among the pearls of Europe. They include forest communities showing many common features with the vegetation of Asia Minor and Colchis.

Figure 1. Protected Areas in Bulgaria

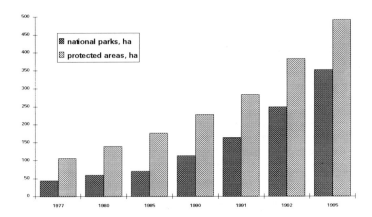

Bulgaria's biodiversity faces a broad range of anthropogenic threats. The loss and degradation of both aquatic and terrestrial habitats constitute the most significant threat to biological diversity, affecting all ecosystems from the high mountain forest and lakes to the open waters and benthic communities of the Black Sea.

As a result of human pressures, a number of Bulgarian species have in recent decades diminished to the point of extinction. These include at least 31 species of vascular plants, 7 invertebrates, 3 fishes, 2 snakes, 3 birds, 2 to 3 mammals, and 6 indigenous animal breeds.

Accelerated rates of global climate change could have far-reaching effects on Bulgaria's diversity. If global warming should result in a rise in sea level, the adverse effects along the Black Sea coast could be substantial.

The lack of knowledge and effective public policy are a less direct but no less crucial threat to biological diversity. This complex category of threats includes several general areas of special concern:
- insufficient scientific information on the status of and threats to the biological diversity;
- inadequate management and administration of protected areas;
- uncoordinated and poorly enforced conservation law and environmental regulations;
- ineffective or non-existent penalties and sanctions;
- insufficient registration and monitoring of harvested biological resources;
- lack of public understanding of biological diversity and the threats to it, and a lack of information available to the public to achieve a higher level of awareness.

While the foundation of scientific information on Bulgaria's biological diversity is one of the nation's most significant strengths, it has a number of gaps as well. In March 1994 the Ministry of Environment, which is the agency primarily responsible for developing and implementing national environmental policy, created the National Nature Protection Service (NNPS), a specialized body for the management, control and protection of biological diversity, protected natural sites and natural ecosystems.

The nature conservation policy is focused on institutional strengthening, improvement of the legislation and management of protected areas, and biodiversity. Practically, during 1993-1994 the National Action Plan for the Conservation of the most important Wetlands in Bulgaria (NWCP) was developed with the assistance of the French Government and Ramsar Convention Bureau, aiming to determine the priority areas for conservation activities and the urgent measures which have to be undertaken in short- and long-term perspective. A comprehensive National Biodiversity Conservation Strategy (NBDCS) was prepared with the assistance of USAID and the Biodiversity Support Program (WWF, the Nature Conservancy, World Resources Institute) in 1994.

The central aim of the NBDCS was to identify and coordinate a broad range of actions that will conserve biological diversity across the whole landscape, i.e. in both aquatic and terrestrial ecosystems, and on both reserved and nonreserved lands. Of primary importance was work on: analysis of the gaps in traditional knowledge; defining of conservation importance of various biological species, groups and habitats; and the identification of the threats to them.

Another basic task of NBDCS was to define areas of conservation importance. This was achieved using the method of accumulating, through the use of GIS, relevant elements of biodiversity and natural habitats, including ones that influence their vulnerability level. Identifying areas which are important for nature protection gives an opportunity for more exact estimation of the level of completion of the network of protected areas as well as for defining priorities and filling the gaps. In this context, traditional scientific and research was recognized as a solid background for development of a NBDCS. This includes a complex program for mechanisms and actions necessary for its implementation, without which biodiversity will face resistance from the enterprise sector and local government bodies.

According to the Strategy, the follow-up process should begin immediately to focus exclusively on expanding and strengthening the network of protected areas. This process should include :
- Clarification of the jurisdictional issues affecting the protected areas and full authorization of an overseeing agency to provide effective protection and management of planned areas;
- A review of the existing protected areas and identification of areas of special interest and concern (including corridors and buffer zones) that are outside the protected areas network.
- Identification and ranking of the management needs of the protected areas network, including the development of public education, information, and interpretation programs.

In addition, during recent years several pilot projects have been coordinated with international financial institutions. The objective in each of these pilot projects is similar: to provide a local model for developing a new conservation policy and financially sustainable incentive programs. Some of the projects under way are: Bulgaria-Sweden Biodiversity Conservation program; the Global Environmental Facility (GEF) Bulgaria Biodiversity Project focused on Central Balkan National Parks and Rila National Parks; the Corine Biotope project and others financed by EC-PHARE ; Sustainable tourism in Rila and Pirin mountains, with the assistance of the UK Know-How Fund; the Monaco program for preservation of wetlands of South Black Sea coast; and some others.

With the assistance of the international donor community, Bulgaria has taken several important steps towards promoting sustainable development and practical implementation of the Convention on Biological Diversity. Among the workshops held in Bulgaria is a UNEP -Ministry of Environment (MoE) Workshop on Practical Implementation of the Biodiversity Convention in Central and East European countries. To protect the unique Bulgaria nature heritage there are, several key elements that deserve immediate support both from Bulgaria and the international community. They include:
- Strengthening the scientific understanding of biological diversity in Bulgaria with focus on the gaps identified in the Strategy;
- Revision of the Bulgaria's Red Data Books and the compiling of new ones for missing taxonomic categories;
- Additional species- and community-level information;
- Encouragement of interdisciplinary research;
- Greater access to existing scientific information;
- Dissemination of existing information.

Scientific information on Bulgaria's biological diversity and its conservation is the foundation on which the National Conservation Strategy can be built. The next step in the implementation of the Strategy is to set out broad goals and to develop a specific action plan. In this action plan NNPS -

MoE will support interdisciplinary environmental education programs, including biodiversity education (BE).

One of the main efforts will be preparing an National Education and Communication Strategy, aimed at persuading decision-makers in the government to assume responsibility for biodiversity conservation. Bulgaria lacks a unified plan for biodiversity education, or even a process for designing such a plan. The legal basis for improving education on biological diversity does exist. Article 11 of the Bulgarian Environmental Protection Act of 1991 requires the authorities to publicize information on the environment through the mass media and by other means, although it does not specifically mention biological diversity. The International Convention on Biological Diversity involves an obligation on the part of states to promote and encourage understanding of biodiversity through the media and through public education. In effect, the development of this strategy has served as a first step in following these objectives.

In summarizing Bulgaria's approach to the practical implementation of the role of national parks and protected areas in sustainable development, three main goals have been recognized by the Government and the conservationalists:
- widening of the protected territory in the country;
- development of management planning and implementation processes, and demonstrating park development through site specific management programs ;
- development of financial mechanisms (i.e. nature tax and users fees) to fund a self-sustaining protected area program.

In addition to this we believe that in the long run, the conservation of Bulgaria's biological diversity will depend on general public understanding and appreciation of its value. Educational programs should not be limited to students or schools. Extension services should be organized at the national level to disseminate information to new as well as current landholders, and to communicate landholders' concerns back to the scientists and policy-makers. Educational programs are needed for the development of ecotourism as well. There is a danger that tourist agencies without proper knowledge, training and environmental education, will not adequately address key issues, including: the identification of sensitive and high pressure areas; the environmental impacts of tourism activities; and the equitable distribution of economic benefits from tourism.

The Bulgarian sense of responsibility for nature protection, the research facilities, the developed system of protected areas, the existing national strategies for biodiversity protection, and the national plan for wetlands conservation, are important prerequisites for sustainable use of nature. Increasing environmental awareness of the need to protect wildlife requires reviving of educational traditions that would guarantee the existence of unique natural wealth.

3 References

Biodiversity Support Program (1994). *Conservation of Biological Diversity in Bulgaria: The National Biological Diversity Conservation Strategy.*Washington,D.C.

Kanitz F (1932) *The Danubian Bulgaria and the Balkan.v.I,* Bulgarian Historical Library, Sofia.

Ministry of Environment (1994). *National Action Plan for the Conservation of the Most Important Wetlands in Bulgaria.* (in Bulgarian and French , 1994; in English, 1995). Ministry of Environment. Sofia.

The Emerging System of Protected Areas in Lithuania: A Review

Tadas Leoncikas[1]

[1] Department of Sociology, Central European University,
ul. Nowy Swiat, 72, PL - 00 330 Warsaw, Poland

Abstract. In post-communist years, activity in the sphere of protected areas in Lithuania has widened enormously. Nevertheless, the net is still being created. This essay describes how the work started and what features it has today. The main specific feature of Lithuania's parks is increasing emphasis on elements of ethnic culture and heritage conservation. This crystallizes into the general concept of a specially protected area/park as a representation not only of natural but of cultural values. Almost all units of the parks system work in that direction. Development of recreational services is also among the main objectives. As a specific methodological approach, functional zoning is overwhelmingly applied in protecting the parks.

Keywords. National parks and protected areas, historic parks, zoning, regional parks, ethnic culture, heritage conservation, recreation, Lithuania

1 Introduction

In Soviet times, Lithuania had just one national park in one of its regions. This was established in 1974, even though it took quite a few years before the authorities permitted it. The delay arose because of the political implications of the term, national. A number of smaller reserved units (state parks) sometimes were declared rather than really protected.

In the last decade, after the political changes, a totally new system of protected areas is emerging. In the context of a new nation-state, it has become easier to justify the need for special protection of a number of areas. Also, it is easier to create national parks.

Today, Lithuania's system of protected areas consists of National parks, Regional parks, Reservations, and Reservations (strict protection) (Table 1). These types of protected area differ according to their importance, type of protection and administration.

There are 5 national parks in Lithuania today. They were established by the decision of Parliament after the Independence Declaration in 1991. Four of them represent the main ethnographic regions of the country and are named correspondingly: Aukstaitija; Dzukija; Zemaitija; and Kursiu (Courland) peninsula. Trakai historic national park has a function of caring for the historical centre of the early Lithuanian state. This includes the castle, little town and surrounding lakes. One more historical national park is going to be established for protection of the castle complex in the capital of the country. Although the first three were established for their specific nature and landscape, now they also emphasize the general concept of combining nature and culture.

The number of regional parks is 30. 12 of them belong to Forestry system units. Others have state-budget organizations, local authorities and combined councils among their administrations. The smaller requirements and restrictions allow more grassroot initiatives in the regional parks, while national parks may be administered only by the main state institutions[1].

---

[1] i.e., national parks administration are directly affiliated to Ministeries.

Table 1. Protected Areas in Lithuania [2]

Category of the area protected	% of the protected areas
National parks	20
Regional parks	53
Reservations	24
Reservations (strict protection)	3

Nevertheless, both national and regional parks have a tendency to reflect the general concept of special parks which can be seen as having two dimensions: 1) formally, they are reflected in the laws, 2) actually, they reflect the ways in which many every-day tasks are performed. Firstly, the laws of 1991 and 1993 defined the possibility of historic parks and, among other things, stated that national parks are established in order to represent both the nature and culture of ethno-cultural regions[3].

Today, both the national parks and regional parks have become the foci of conservation and the reviving of the traditional ethnographic features and life-styles in the regions. As one of the main theoreticians of Lithuania's protected areas system P. Kavaliauskas noted, the very term 'national' [park] reflects not only recognition by the state, but also ethnic values.

Thus, activity which started mainly as the expanding of areas with special nature protection, now moves toward cultural historical heritage. Emphasis on ethnic culture elements as one of the main objects of the protected areas is a specific feature of the recently emerged system of Lithuanian parks. The concept separates them from those that concentrate on preserving wild nature.

All this leads to a special emphasis on functional zoning. The system of park planning in Lithuania is a special methodological approach. National and regional parks have been given special attention and separate financing. All park territories are analyzed and the special projects are worked out. Territory is divided into general functional zones, such as reservations, recreational zones, economic activity zones. Zones are divided into smaller elements - forest subzone, agricultural subzone, and the like. The zones determine the possible human activity in them. This is helpful in Lithuanian parks most of which are inhabited, but various problems of land reform and property restitution emerge.

It is also important to mention that the main authors and managers of the system of protected areas are geographers. In former times, people who dealt with protected areas mainly belonged to the forestry system. Today, the forestry system members take care of the areas discussed, but this is more of an executive function. The geographical scientists mainly shaped the number, types and plans/maps of the protected areas. The foregoing zoning approach is a specific outcome of the geographers work. Nevertheless, the everyday life of the parks mostly depends on their authorities, organization and finances. Also, this life depends on what image they will develop in wider society.

While most of the work on the scientific or theoretical and legal basis of the parks is done, the practical work is just getting underway. The National parks, naturally, get more attention. Not all of the rapidly proliferating regional parks have the necessary administration. They also currently tend not to differ from the ordinary reservations or simple landscape parks. Some of the regional parks find initiatives and progress easier than the national ones. Kurtuvenai regional park has been most successful in recreation because of administrative and other co-operation. There are some successful examples of new activity such as farm tourism which is likely to expand.

However, international communication and exchange of experience is essential at the level of the main coordination of parks. The biggest influence in respect to ecological activity flows from the Scandinavian countries. Scandinavians are well recognized as experts in nature use and management, and also, they have completed some concrete projects in Lithuania. Recently, the Norwegian scholars, B.P. Kaltenborn and O.I. Vistad completed an applied field survey among recreational users

---

[2] The data are taken from R.Baskyte Lietuvos regioniniai parkai: istorija, samprata, problemos in *Mokslas ir Gyvenimas*, 1996 No. 6, p.12.

[3] LithuaniaÕs regions which do have distinct landscape and ethnographic characteristics are represented by national parks; the coincidence with term Ôregional parkÕ is accidental, the latter indicates smaller scale and smaller significance of the area.

in Aukstaitija National Park. It was the first time such detailed research took place in a particular park. It also signaled the need for Lithuanian parks to develop infrastructure and services. The project was financed by the Eastern Europe Programme in Norway's Ministry of the Environment. Now, Lithuanian state parks coordinators mention the necessity for further research of this kind.

Another possible way to achieve a positive impact may be the co-operation of neighbor countries, although this has not developed to date. The Association of National Parks of the Baltic Countries has not done much yet. Recently, a proposal from Poland offered to strengthen inter-border co-operation and to join in developing the care of near-border parks. This remains a matter for the future.

Authors Note: In preparing this essay, information was used from *Girios*, as well as the articles of P.Kavaliauskas Lietuvos nacionaliniai parkai and R. Baskyte Lietuvos regioniniai parkai: istorija, samprata, problemos in *Mokslas ir Gyvenimas* 1996, No. 6-8. Interviews were completed in the Division of National Parks, Ministry of Forestry, and some park publications were consulted.

Landscape Protection in Estonia: The Case of Otepää

Monika Prede[1]

[1] Otepää Linnavalitsus
Lipuvaljak 13, EE-2513 Otepää, Estonia

Abstract. Conservation in newly independent countries presents special problems. In Estonia, the need to protect natural areas is affected by the changes introduced by a transitional economy, land reforms, and social conditions. The Landscape Protection Area in Otepää has proceeded on the assumption that the easiest way to protect nature is to prohibit all activities within the area. This approach has been unsuccessful.

Keywords. Government, economy, sustainable, development, diversity, nature, lakes, forests, agriculture, reserve, park, zone, recreation, resources, reform, community, restoration.

1 Introduction

'We have not received the Earth from our parents, we have borrowed it from our children'--the principle of the world strategy of nature conservation--should be obvious to everyone, irrespective of race, state, or religion.

Nature conservation and protection of our environment are no longer only of local importance or only the concern of a small group of specialists. Our natural environment is part of the continent's natural environment, part of everyone's common living environment. Each local land owner is responsible not just for their own homestead but must answer to all Estonians, all of Europe, and to all of the world.

In regaining Estonian independence we have stepped once again into the free world. This gives us a new opportunity while at the same time placing on us a high duty. And perhaps it is especially symbolic that NATO, which we know from peace-keeping operations, should assist us in joining the international family of nature conservation.

The following will provide a short overview of the problems facing Estonian nature conservation programs with respect to nature reserves and land ownership in Estonia's rapidly changing society (Figure 1). Otepää, in southern Estonia, is used as a case study to introduce our experiences with land owner relations, attempting to integrate the reserve with the lives of local residents at a time of transition in Estonia to a democratic market and society (Table 1 and Table 2). The References include a number of publications that are cited specifically in the following text or which can be consulted for general background and follow-up purposes.

2 Otepää Municipality and Nature Protection

2.1 History

Otepää was first mentioned in 1116 in the Novgorod Chronicle as Medvezja Golov; translated as Bear's Head. The Otepää stronghold was the centre of the ancient Estonian Ugandi county and at the crossroads of merchant routes. In 1884 the Estonian Student Society's 'blue-black-white' flag was consecrated in the Otepää church. This event made Otepää significant in Estonian culture and history.

Figure 1. Nature Protection Areas in Estonia

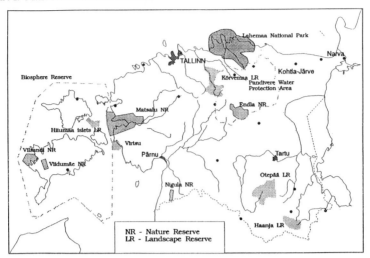

Table 1. Nature Conservation

- Traditionally we are both country people and forest people. The harsh environment has created a strong respect for nature--for preserving the living environment.
- The first traces of human settlement date back to 8,000 B.C. Fields have been cultivated here for approximately four thousand years (Ministry of Environment, 1992).
- Forests were never cleared needlessly nor wild animals needlessly killed.
- Writings from the 13th century indicate that sacred trees and groves were honoured and preserved.
- The King of Denmark prohibited the cutting of coastal woods on three islands near Tallinn as early as 1297. (Sepp, Eesti Looduskaitse, 1996)

1910: The first bird sanctuary was established on the Vaika islets.
1935: The first Conservation Act was passed.
1940: By this year there were 47 nature reserves in Estonia.
1957: The Nature Conservation Act was passed during the Soviet period.
1966: Ministry of Forest Management and Nature Protection (the present Ministry of Environment) was established.
The Nature Conservation Society was founded
1971: Lahemaa National Park was established on the north coast. to protect both nature and the traditional way of living.
This park was the first of its kind in the USSR.
1979: Endangered species are listed in the Red Data Book of Estonia (Ministry of Environment, 1992)

Table 2. Since independence, the Estonian Parliament has passed several pieces of environmental legislation:

- The Law on Environmental Fund; 1994
- The Mineral Resource Law; 1994
- The Law on Hunting Regulation; 1994
- The Water Law; 1994
- The Law on Protected Natural Objects, 1994
- The Act on the Protection of Coasts and Shores; 1995
- The Act on Sustainable Development; 1994
- The Act on Planning and Construction; 1995
- Adoption of Implementation Acts to the above mentioned laws has begun (For legislation see Ministry of Environment, 1994 and 1995).

Otepää, a typical little southeast Estonian town, was incorporated in 1936 and has since grown to about 3,000 inhabitants. It is located 227 km from Tallinn, the capital of Estonia. Otepää is Estonia's primary centre for winter sports as well as the centre for Estonian Olympic training. The town has no large industries. The majority of small industries are related to housing, provision of food, merchandising, and tourism. At the instigation of Estonia's youngest and most radical self-government, reform of the State's administrative system began, so to speak, from the bottom up. From 1988-92, Otepää played a similar role to the one Estonia had played in the former Soviet Union by initiating the dismantling of the huge centralized system and establishing autonomy. State protection reforms took place in 1992 and Otepää achieved status as an autonomous self-government.

2.2 Nature Protection

As early as 1990, Otepää's self-government declared that priority for developing the area would be based on the wise use of natural resources and geared towards an economy centred on tourism, sport, and vacations.

A Nature Protection Institution was created in 1992, associated with Otepää's municipal government and the first of its kind in Estonia. At that time, all of the decision-making power with respect to nature protection existed only at the State level--whether through the ministry, region (now county) or nature protection areas.

The first task for the Nature Protection Institution was to develop a nature protection system reform project which would address the state, county, and local self-government division of powers and the potential to work together. At the same time, a reorganization program for landscape protection was developed in response to accompanying demands for a natural park. The project's goal was the creation of an Environmental Protection Institution at the level of self-government and the integration of the Protection Area with local community life.

Today the first goal has been realized. Many direct functions are delegated to local municipalities. They are responsible for financing projects through local budgets or the Estonian Environmental Fund. They prepare applications for financing regional projects and arrange relevant tasks such as data collection, licenses on natural resources, pollution and land use permits.

2.3 Nature Protection and the Local Community

Work on the integration of the Protected Area with the community has not been so successful. To provide some historic background, during the 1970's with the continuing support of local residents, an enthusiastic forest authority together with friends of nature throughout Estonia, stood up for the establishment of landscape protection. That goal was achieved in 1979--a Landscape Protection Area was created. Until 1991 this area was formally part of the Elva Forest District. Being subject to resource uses aggravated the ability of the Protection Area to fulfill the tasks assigned to it.

During 1979-1987 staff for the Protection Area was developed including specialists and landscape protection workers. A trail centre building was erected. The task at hand was the development of a recreation based economy. Camping, boating docks, and bonfire places were part of the Protection Area. Activities geared towards a fishing economy were developed on the lakes. Fish fry were introduced. Fishing licenses were issued. A monitoring system was developed for the Pühajärv catchment area. The Otepää Landscape Protection Area Advisory Board was established, with representatives from the local self-government, the State's Environment Ministry, and the region.

By 1987-1989, the Protection Area had achieved a quantitative standard, from which a path should have developed to a qualitative standard. Unfortunately, an unrelated move towards reorganization was begun within the Ministry. A short time later, the independence process began within the State. Expectations decreased with respect to protected areas and much damage was done to work accomplished to that time.

Supervision, landscape conservation work, and development of a recreation based economy--these are the yardsticks by which the local community measures the success of the protection area. The ordinary citizens do not notice the cessation of research work or monitoring, for the meaning placed on these activities is distant for them. People want to see well kept parks, trails, information systems, and the like.

Weakening control of illegal hunting and fishing, poaching, theft from forests and defiling of the environment are obvious indications to local people that the Nature Protection Area cannot meet its responsibilities and obligations. In the eyes of local residents, the Protection Area is beginning to lose its good reputation. The same can be said for visitors--especially those who over many years spent their vacations in the same place. These people have noticed the signs of deterioration in parks, trails, and the shore line.

By 1991 the Protection Area had lost its special characteristics because of a complete collapse both for objective reasons (economic and reforms) and subjective reasons (weak administration and organization work). The situation became so critical that local residents demanded that the Protection Area be liquidated. The situation was made even more acute by state reorganization and new legislation on the basis of which the fundamental functions of protection areas were changed. This was all accompanied by a new conservative element as applied to nature protection and rigid positions with respect to returning land to private owners.

In Estonia today, protection areas do not own land (Table 3). In accordance with the new charter for protection areas, they lack the function of conservation guardianship and can only perform such duties in a minimal way. On the one side waits the general public, believing that protection areas should deal with the shaping of landscapes and guardianship; on the other side stand the land owners, new to their role and very sensitive to every kind of restriction placed on their ownership.

Because of the slowness of property reforms and the complexities involved, it is particularly difficult for property owners to understand their rights and responsibilities. Most still lack the final legal documents which make them landowners. At the same time, we must assume that lands and forests will be put to use as soon as possible because of the difficult economic situation in the area and the return of farms to private ownership.

Table 3. Protected areas in Estonia (as of 1995) (from Kulvik and Tambets, 1995)

- National Parks, 4
- Biosphere Reserve, 1
- State Nature Reserves, 5
- Hydrological Reserve, 1
- Landscape Reserves, 14
- Botanical - Zoological Reserves, 6
- Mire (bog) Reserves, 30
- Botanical Reserves, 4
- Ornithological Reserves (bird sanctuaries), 2
- Geological Reserve, 1
- Nature Park, 1.
- In total, 12% of the territory is protected by different nature protection regimes (Figure 1). A strict protection regime exists on approximately 1% of the territory of Estonia.

The Act on Protected Natural Objects determines the rights and obligations of land owners, land users and other persons in regard to protected natural objects. The Act stipulates the setting of environmental restrictions concerning property and the obligations of land owners concerning land in different protection zones. It also stipulates that economic damage or loss arising from the protection regime shall be compensated. According to the existing protection program, Statements of Protection Obligation will need to be drawn up by which every owner will be informed of the specific restrictions and obligations concerning each estate.

2.3.1 Analysis

It is against this general backdrop that Estonian nature protection areas found themselves in a deep crisis in the last few years, especially with respect to their relations with local residents and the self-government. In analyzing the opposition which has now lasted for several years, the main reasons for the conflict in Otepää are:
- different understandings of the role of the Protection Area in the life of the community
- outdated thinking by the administration of the Protected Area; rigid with respect to the community and unwilling to be flexible and in step with a changing society
- weak work with the general public
- lack of initiatives to assist in the life of the community
- weak work on the part of the Ministry in taking initiatives and at the same time in ignoring local initiatives
- isolation of the Protection Area from the local community and the activities of the municipal government

The most serious problems differ in emphasis in different protection areas. The Otepää Landscape Protection Area is strongly influenced by tourism, sport, recreation and economic demands and pressures (Table 4). Especially in such an area, it is important that a Nature Park become part of the life of the community. Maintaining a reasonable balance between pressure from a growing population and the need to preserve natural areas should be the foundation on which the area is developed. Problems must be handled not only from the perspective of nature but people's day to day economic interests must also be considered. No farmer who has just liquidated animal husbandry because it no longer was economically feasible, will begin to naturalize hay fields only because we need to conserve biological diversity. Economic interest is a primary motivational force.

Table 4. Otepää Landscape Reserve

- In 1979, the Otepää Landscape Reserve was founded, establishing a protected landscape of 232 km^2.
- In 1991, the Otepää Landscape Reserve Protection Area was separated from Elva Forestry District and become juridically independent.
- The objective of the Landscape Reserve is to preserve, protect, research and introduce for the Otepää Uplands characteristic ecosystems and traditional uses of the land that are of national importance.
- The goal is also to facilitate recreation in a natural environment, based on sustainable development.
- Research in the Reserve is linked to the preservation of rare specimens and ecosystems. (Ministry of Environment, 1994)
- All land within the Protected Area is divided into zones, as specified in the Protection Rules:
- A 'strict nature reserve zone' is an area of land or water in its natural state and free from the direct impact of human activity; an area in which the preservation of natural associations resulting only from natural processes is guaranteed.
- A 'special management zone' is a land or water area protected in order to preserve resulting or created natural and semi-natural associations.
- A 'limited management zone' is a part of a protected area used for economic purposes where restrictions, established by the authority which has responsibility for protection, must be taken into account. (Ministry of Environment, 1994)

(See Linnamagi, Erestus, 1994 and Ministry of Environment, 1994)

The Nature Park should suggest and provide alternative opportunities to local farmers in developing farm-tourism, vacation economy, and other activities. Employees of the Nature Park have a huge responsibility to persist tirelessly in explaining, advising, and assisting local residents to find new perspectives. Unfortunately, the Protection Area in Otepää lacked a good understanding of its role during the critical years. This lack of understanding was combined with a rigid belief that the easiest way to protect nature is to prohibit all kinds of activities, including tourism.

For eight years, always at the instigation of the local government, discussions were held on the subject of reorganizing the Protected Area. In all, five National Ministers have attended; all of whom acknowledged the existence of the problem and agreed with the complaints of the local government. But still the situation was never fully resolved.

The failure on the part of the Ministry of Environment to make timely staff changes had serious consequences; Estonia's largest and best known Landscape Protection Area was once again, as an experiment, made subject to the Forest District. Only two years ago there were 14 employees at the Protection Area. Now four positions are expected to remain; only two of those positions are presently filled. Because of protests by the Union of Estonian Protected Areas, the experiment has been aborted. This incident is all the more regrettable since even in 1994, the Otepää Self-Government Area Advisory Board, again at the instigation of the local government, proposed a reorganization program to the Environment Ministry.

2.4 Tailored Paths of Action For Strengthening Nature Protection Areas

Different institutions have different paths and opportunities available to them as part of a cooperative problem-solving effort. The primary question from the point of view of the development of the district is how to move forward; how to integrate the Protected Area into the life of the community.

2.4.1 Self-Government level

- start with a broad strategic planning exercise to begin the region's development directions and the role which the Protected Area (Nature Park) will have in the region
- create public opinion in support of the development of a Nature Park
- begin new projects which have nature protection as an objective, such as a parks program, and accompanying waste handling systems; seek international aid in bringing projects to life.
- work out details of an Otepää nature district project within the framework of Europe's ecological community and increase cooperative work within this district
- apply pressure to the State's institutions with the objective of creating a Nature Park which can address present day challenges

2.4.2 State level

- ban experiments which weaken protected areas and restore jurisdictional independence to the Otepää Landscape Protection Area (Nature Park)
- support local self-government initiatives which strengthen nature protection in the local area
- refrain from creating artificial official barriers; instead develop mechanisms which will permit protected areas and the small self-governments existing within their territories to address environmental problems cooperatively and possibly arrive at rational solutions

2.4.3 Protected Area level

- replenish staff through a public and open search and by taking into account the opinions of the local self government
- aside from competence, specialists should be selected on the basis of their ability to be open to new ideas, their skill at communications and their ability to take initiatives
- they should understand their mission to develop the local community and to be an initiator, leader, and coordinator in the district

3.0 Recommendations

It should be noted that, since the establishment of Tartu University in 1632, the strength of Estonian nature protection activities has always been in the area of research work. The traditionally high quality of research has been a good basis for establishing protected areas and for ongoing application in everyday work. Fifty years of occupation in Estonia by the Soviet Union have left a heavy footprint, obviously in terms of military and industry, and less visibly in terms of confusion about ownership,

economic losses, and deformed social relationships. Continuity has also been severed in the development of natural and protected areas; especially with respect to the generation of public support, relations between local communities and protected areas, and the creation of supportive economic mechanisms. Insufficient attention has been paid in Estonia to:
- analysis and improvement of the relations between protected areas, the general public and the local community
- development of optimal cooperation between self-governments and protected areas
- clarification of landowners expectations, needs, and attitudes and their collective opportunities with respect to nature protection areas
- gearing state and local economies towards welcoming opportunities within the framework of promoting conservation development
- augmenting university nature protection research programs to include self-government activities, economic problems, and communication

The aforementioned areas require more thorough research and the development of concrete mechanisms and implementation programs. Improving our ability to protect natural areas can happen only with the development of community and democracy, in the broadest sense of these terms. In large part this depends also on how quickly and painlessly Estonia is able to integrate into the European community and into free societies worldwide. Being open to new ideas, the exchange of experiences, joint work, and understanding each other, are the bases on which to change the world for the better.

4 References

Eesti NSV TEaduste Akadeemia Looduskaitse Komisjon (Estonian SSR Scientific Academic Nature Protection Commission). 1975. *Eesti Loodus-Harulduste Kaitseks.* (For the Protection of Unique Aspects of Estonian Culture). Tallinn, Estonia: Valgus (with summaries in Russian and German).

Eilart, J. 1976. Inimene, Ökosusteem ja Kultuur. (People, the Ecosystem and Culture). Tallinn, Estonia: Perioodika.

ENSV TA Kodu-uurimise Komisjon. (Estonian SSR Heritage Research Commission.) Järv, V. ed. 1966. *Otepää.* Tallinn: Estonia: Eesti Geograafia Selts. (Estonian Geographic Society).

Plancenter LTD. 1991. *Environmental Priority Action Programme for Leningrad, Leningrad Region, Karelia, and Estonia.* Helsinki, Finland: Prepared for the Ministry of the Environment of Finland. (English)

Külvik, M. and J. Tambets. eds. 1995. *Key Points of the National Biodiversity Action Plan.* Compiled for World Bank, Global Environmental Facility, and WWF Baltic Programme. Tartu-Tallinn. (unpublished paper) English.

Linnamägi, M. and T. Evestus. 1994. *Otepää Landscape Protection Reserve in Otepää Upland.* Tallinn, Estonia: As Infotrükk.

Maimets, A., H. Simm, and E. Varep. eds. 1968. *Eesti Järved.* (Estonian Lakes). Tallinn, Estonia: Eesti NSV Teaduste Akadeemia Zooloogia ja Botaanika Instituut (with summaries in English and Russian).

Margus, M. 1974. *Eesti NSV Puhkealad.* (Recreation Areas of the Estonian SSR). Tallinn, Estonia: Valgus (with summaries in Russian and German).

Ministry of Environment. 1992. *National Report of Estonia to UNCED 1992.* Tallinn, Estonia. (English).

Ministry of Environment. 1993. *Keskkond 1993.*Tallinn, Estonia.

Ministry of Environment. 1994. *Keskkond 1994.*Tallinn, Estonia.

Ministry of Environment. 1994. *Otepää Maastikukaitseala Kaitse-Eeskirjad. (Otepää's Landscape Protection Reserve Charter)*Tallinn, Estonia.

Ministry of Environment. 1994. *The Act on Protected Natural Objects.* Tallinn, Estonia. (Eng. trans.)

Ministry of Environment. 1994. *The Law on Environmental Fund.* Tallinn, Estonia. (Eng. trans.)

Ministry of Environment. 1994. *The Mineral Resource Law.* Tallinn, Estonia. (Eng. trans.)

Ministry of Environment. 1994. *The Law on Hunting Regulation.* Tallinn, Estonia. (Eng. trans.)

Ministry of Environment. 1994. *The Water Law.* Tallinn, Estonia. (Eng. trans.)

Ministry of Environment. 1995. The National Environmental Strategy for Estonia. Preliminary Draft Working Paper. *Environmental Liabilities in Estonia.*Tallinn, Estonia.
Ministry of Environment. 1995. *The Act on the Protection of Coasts and Shores.* Tallinn, Estonia. (Eng. trans.)
Ministry of Environment. 1995. *The Act on Sustainable Development.*Tallinn, Estonia. (Eng. trans.)
Ministry of Environment. 1995. *The Act on Planning and Construction.* Tallinn, Estonia. (Eng. trans.)
Punning, J-M. ed. 1994. *Ida-Virumaa Keskkonnauuringud 1980-1994.* (Eastern Virumaa Environmental Research Projects 1980-1994). Tallinn, Estonia: Stockholm Environment Institute - Tallinn (with summaries in Russian and English).
Raukas, A. ed. 1979. *Eesti NSV Saarkorguste ja Järvedenogude Kujunemine.* (The Creation of Islands and Lakes in Estonian SSR). Tallinn, Estonia: Valgus
Sepp, K. ed. 1996 Eesti Looduskaitse. (Nature Protection). Tallinn: Estonia.
Terk, E. 1993. Privatization in Estonia: Will the Economoic Approach Prevail? *The Baltic Review.* Vol 2, no. 02. March-June 1993. pp 24-25. English.
Valk, U. and J. Eilart. eds. for Eesti NSV Teaduste Akadeemia Zooloogia ja Botaanika Instituut (Academic Zoology and Botany Institute of Estonian SSR). 1974. *Eesti Metsad.* (Estonian Forests). Tallinn, Estonia:Valgus (Russian summaries).
Valk, U. and J. Eilart. eds. for Eesti NSV Teaduste Akadeemia Zooloogia ja Botaanika Instituut (Academic Zoology and Botany Institute of Estonian SSR). 1988. *Eesti Sood.* (Estonian Swamps). Tallinn, Estonia: Valgus (summaries in Russian and English).
Varep, E. and V. Maavara. 1984. *Eesti Maastikud.* (Estonian Landscapes). Tallinn, Estonia: Eesti Raamat.
Velner, H. 1993. Water Protection Progress in the Baltic Sea Area. *The Baltic Review.* Vol 2, no. 02. March-June 1993. pp 10-11. English.
Viiding, H. 1976. *Eesti NSV Maapoue Kaitsest.* (Protection Beneath the Surface of the Estonian SSR). Tallinn, Estonia: Valgus (with summaries in Russian and English).
Wilson, E. B. 1993. Tourism Action Plan: Otepää. C.E.S.O. (Canada) (English) (Prepared on behalf of the Town of Otepää) (Unpublished document).

For further reading, see also the periodical *Eesti Loodus (Estonian Nature)* English summaries. Published since 1933. Postal address: Eesti Loodus, EE2400 Tartu, pk. 110. ESTONIA.

Protected Areas and Plant Cover Diversity in the Ukrainian Carpathians: an Assessment of Representativeness

Lydia Tasenkevich[1]

[1] State Museum of Natural History, National Academy of Sciences of Ukraine
18 Teatralna St., Lviv 290008, Ukraine

Abstract. The Ukrainian Carpathians protected areas are the integral part of a Carpathian-wide protected areas network. For effective conservation, there is a need to develop a regional network of protected areas, manage them sustainably, and draw on Ukrainian scientists and park managers for research and support.

Keywords. Ukrainian Carpathians, flora, diversity, protected areas, representativeness, parks and protected areas, conservation, research

1 Introduction

The Ukrainian Carpathians are part of the Eastern Carpathians, and are divided with Poland, Slovakia and Romania. The Carpathians are a typical medium-sized mountain system with mainly dome-shaped summits, frequently united into long ranges or massives, dissected by the deep valleys of rivers. The mountain arc is about 50 km wide on its east-southern end and about 100 km on its north-western one.

2 Plant Cover Diversity

2.1 Vegetation

Five vegetation belts are distinguishable in the Ukrainian Carpathians: the oak-forests belt (100-250 m a.s.l.); the beech forests belt (250-1350 m); the spruce forests belts (700-1670 m); the subalpine belt (1300-1850 m); and alpine belt (1850-2061 m a.s.l.) (Holubets and Milkina 1988 and Malynovskyi 1968). The oak-forests belt is developed on south-western slopes only, in the warmest part of the Ukrainian Carpathians, bordering the Transcarpathian depression on the northeast part of the vast Pannonian lowland.

The largest areas in the Ukrainian Carpathians are occupied by the Carpathian beech forests. Some trees, for example, are 45 metres high and 400 years old in the Ugolka-Shyrokyi Luh, which is most likely, the largest (10,350 ha) intact part of the Central-European beech forest. It is part of the Carpathian Biosphere Reserve. The beech forests prevail on the south-western slopes, where they very often form a timberline.

On the north-eastern slopes, beginning from c. 700 m a.s.l., mixed forests occur with various proportions of fir, spruce and beech. About 1200 m a.s.l. spruce forests prevail. The highest stand occurs at the summit of Shuryn in the Chornohora Mts. at an altitude of 1,670 m.

The subalpine belt is formed by dwarf shrub associations: *Pinetum mugi, Rhodoretum myrtifolii, Juniperetum nanae, Alnetum viridi*; as well as by tall mountain grasslands with *Adenostyles alliaria (Gouan) All., Cicerbita alpina (L.) Wallr., Cirsium waldsteinii Rouy., Senecio fuchsii Gmel., Athyrium distentifolium Tausch* and high-mountain meadows with *Deschampsia cespitosa (L.) Beauv. and Calamagrostis villosa (Chaix) J.F. Gmel.* Due to burning of dwarf shrubs, and overpasturing, vast areas are occupied by *Nardus stricta* L. all over the subalpine belt.

The alpine belt is developed from 1800 m a.s.l. to the tops of the highest peaks, notably on the Chornohora Mts. There are some widely distributed associations in this belt, dominated by *Festuca airoides* Lam., *Juncus trifidus* L., *Carex sempervirens* Vill., and *Sesleria coerulans* Friv.

2.2 Statistics of Flora

The native flora of the Ukrainian Carpathians comprise over 1800 vascular plant species and subspecies. For comparison, the whole Carpathian flora consists of over 3390 taxa in an area of approximately 209,000 sq. km (Kondracki 1978).

2.3 Endemics

Among the plants composing the Ukrainian Carpathians flora, there are both common and rare species. Relic and endemic species are of special interest.

Whereas species which are relic in some areas can be very successful in other parts of their range, endemic species are restricted in their distribution to the place of their origin. Endemic species indicate the flora's originality and distinctiveness, making their conservation a special challenge.

Endemism in the Ukrainian Carpathians is still insufficiently investigated, despite numerous publications on this problem (Popov 1949; Pawlowski 1970; Chopyk 1976; Malynovskyi 1980; Stoyko & Tasenkevich 1993).

Ninety-Four species and subspecies are considered to be endemic to the flora of the Ukrainian Carpathians. They can be divided into four groups:

(1) Pan-Carpathian endemics, including 21 species
(2) West-Eastern Carpathian endemics, consisting of 12 species and subspecies
(3) East-Southern Carpathians endemics (22 taxa are known to occur in the Ukrainian Carpathians)
(4) Eastern Carpathians endemics, are the richest group of endemics in the Ukrainian Carpathians, consisting of 39 species and subspecies

The endemics are distributed irregularly in the Ukrainian Carpathians (Figure 1). In terms of endemics, the Ukrainian Carpathians territory can be divided in two parts. In the western part, the total number of endemics does not exceed 25 taxa. The eastern part has the larger number of endemics: Svydovets (40 taxa), Chornohora (51), Marmarosh (33) and Chyvchyny (56) (Figure 1).

2.4 Threatened Plant Species

The human transformation of the Ukrainian Carpathians environment has become a threat to many ecosystems and to biological diversity generally, especially in the last five decades.

Among the species composing the flora of the Ukrainian Carpathians, 168 are included in the second edition of the Red Data Book of Ukraine (Zaverukha 1992) and more than 400 species and subspecies are included in the up-dated regional Red Data List of rare and threatened plant species of the Ukrainian Carpathians.

2 Protected Areas

The protection of comparatively sizable natural areas in the Ukrainian Carpathians has been given much attention, especially among scientists, since the 1960s. Mainly due to the efforts of scientists,

Figure 1. Distribution of Endemics

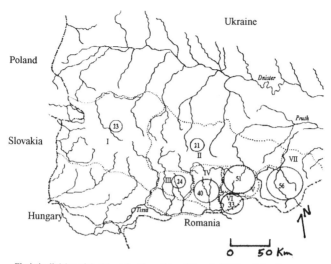

Floristic division of the Ukrainian Carpathians (Chopyk, 1976) and distribution of endemics: I — Beskydy; II — Gorgany; III — Krasna; IV — Svydovets; V — Chornohora; VI — Marmarosh; VII — Chyvchyny

more than 200 different protected areas, occupying 196,700 ha or 5.3% of the Ukrainian Carpathians' total territory have been created (Stoyko et al. 1991).

Eleven categories of protected areas are identified in Ukrainian legislation. Four of them are intended for different botanical gardens and parks and for natural areas.

With the exception of Reserved Area (small strict reserve), the categories of Strict Nature Reserve, National Park, Regional Landscape Park (protected landscape), Zakaznyk (managed resource protected area) and Natural Monument, meet the IUCN's system of Protected Areas (Guidelines...1994).

The main protected areas in the Ukrainian Carpathians are Carpathian Biosphere Reserve (32,000 ha), the Carpathian National Park (503,000 ha) and the Synevir National Park (40,400 ha). Most of the protected areas have been described in general outlines only (Stoyko 1980). Even in special monographs clear inventory data on plant species and phytosociological units are absent for the Carpathian Biosphere Reserve and the Carpathian National Park (Stoyko et al. 1982; 1993). The updating of available data and the obtaining of new data are priorities for further scientific research, if effective action plans are to be developed for protected area management. From this point of view the following principal tasks have to be solved first of all:

- Unification of the methodological base and the means of biological diversity investigation
- Development and publication of biological diversity inventories
- on the territory of the Ukrainian Carpathians
- on the territories of protected areas

Assessment of the representativeness of the Ukrainian Carpathians protected areas in terms of the biological diversity of the region is groundless without fulfilling these tasks.

3 Conclusions

The Ukrainian Carpathians protected areas are the integral part of the protected areas network of the whole Carpathian Mts. as a natural and territorial unit. The Carpathian mountain system can not be conserved effectively without the Ukrainian Carpathians link. Neglect of the Ukranian Carpathians will negatively affect natural conditions in neighbouring countries. For example, overcutting of woods in the Ukrainian Carpathians gave rise to disastrous floodings on the Romanian and Hungarian plains in the post-war times.

So, it is unfortunate that IUCN in developing its Parks for Life program, was guided by political boundaries rather than conservation considerations, in not including the Ukrainian Carpathians protected area in this plan for action. Without involving Ukraine in the Europe-wide nature conservation activities, the process of Europeanization could be very protracted. And nobody will benefit from this.

4 Acknowledgments

Support for this paper was provided by the NATO Advanced Research Workshop organizers and the Polish Academy of Sciences: Institute of Nature Conservation.

5 References

Chopyk V.I. 1976. *Altimontaine flora of the Ukrainian Carpathians.* [Vysokohirna flora Ukrainskych Karpat]. Naukova Dumka, Kyiv, 267 pp. (in Ukrainian).
Guidelines for protected areas Management Categories. 1994. IUCN, Gland. 261 pp.
Holubets M.A., Milkina L.I. 1988. Vegetation. In: *The Ukrainian Carpathians.* Nature. [Roslynnist. Ukrainski Karpaty. Pryroda.] Naukova Dumka, Kyiv, 51-63 (in Ukrainian).
Kondracki J. 1978. *Karpaty. Sydawnictwa Szkolne i Pedagogiczne.* Warszawa. 250 pp. (in Polish).
Malynovskyi K.A. 1980. *Vegetation of the altimontane part of the Ukrainian Carpathians* [Vysokohirna roslynnist' Ukrainskykh Karpat]. Naukova Dumka, Kyiv, 276 p.p. (in Ukrainian).
Pawlowski B. 1970. *Remarques sur l'endemisme dans la flora des Alpes et des Carpates.* - *Vegetatio*, 21: 181-243.
Popov M.G. 1949. *Eassay of the vegetation and flora of the Carpathians.* [Ocherk rastitel'nosti i flory Karpat]. Izdatelstvo Moscowskogo Obshchestva Ispytateley Prirody. Moskva, 302 pp. (in Russian).
Stoyko S.M. (ed.). 1980. *Nature protection in the Ukrainian Carpathians and adjacent territories* [Okhorona pryrody Ukrainskych Karpat i prylehlykh terytorij]. Naukova Dumka, Kyiv, 261 pp. (in Ukrainian).
Stoyko S.M., E. Hadach, T. Simon, S. Mikhalik. 1991. *Protected ecosystems in the Carpathians* [Zapovidni ekosystemy Karpat.] Svit, Lviv, 247 pp. (in Ukrainian).
Stoyko S.M., Milkina L.I., Tasenkevich L.O. 1993. *Nature of the Carpathian National Park.* [Pryroda Karpatskoho Natsionalnoho Parku]. Naukova Dumka, Kyiv, 212 pp. (in Ukrainian).
Stoyko S.M., Tasenkevich L. 1993. Some aspects of endemism in the Ukrainian Carpathians.- *Fragmenta Floristica et Geobotanica Suppl.* 2 (1) Krakow, 343-353
Stoyko S.M., Tasenkevich L.O., Milkina L.I. 1982. *Flora and vegetation of Carpathian Reserve.* [Flora I roslynnist Karpatskoho zapovidnyka]. Naukova Dumka, Kyiv, 218 pp. (in Ukrainian).
Zaverukha B.V. 1992. On the second edition of Red Data Book of Ukraine. [Pro druhe vydann'a Chervonoi Knyhy Ukrainy.]-*Ukr. botan. journ.*, 43, 3:82-80 (in Ukrainian).

Sultan Marshes, Turkey: A New Approach to Sustainable Wetland Management

Nilgül Karadeniz[1]

[1] Department of Landscape Architecture, Faculty of Agriculture, Ankara University, 06110, Diskapi, Ankara, Turkey

Abstract. Until recently wetlands were often regarded as wastelands. They are now generally acknowledged as being ecosystems of great value. With the recognition of wetland values, wetland protection has been emphasized by many countries. In an ecological approach to sensitive ecosystems such as wetlands, sustainability can only be achieved by rational planning decisions. In order to take rational planning decisions, the wetland resources should be evaluated according to their special ecological characteristics. In this case, Risk Analysis seemed the best solution to ensure sustainable wetland management. On the other hand the data to be used in wise management of wetlands should be realistic, reliable, accurate and should allow continuous control. The techniques which provide such data are especially Geographical Information Systems (GIS) and Remote Sensing. Sultan Marshes (called Sultansazligi in Turkish), which is one of the Ramsar Sites of Turkey, is chosen as the case area to explain how sustainable use of wetland resources can be achieved using remote sensing data, GIS techniques and the risk analysis method.

Keywords. Wetlands, marshes, protected areas, GIS, remote sensing, risk analysis, land use, assessment, Turkey.

1 The Wetland Concept

The term, wetlands, is a relatively new one to describe an environmental unit that many people knew before under different names. Wetlands occur in many different locations, many different climatic zones, and have many different soil and sediment characteristics. The term, wetland, groups together a wide range of inland, coastal and marine habitats which share common features. The Ramsar Convention defines wetland as areas of marsh, fen, peatland or water, whether natural or artificial, permanent or temporary, with water that is static or flowing, fresh, brackish or saline, including areas of marine waters, the depth of which at low tide does not exceed six meters.

In more recent times, wetlands were often regarded as wastelands. As a consequence, many natural wetlands of the world have been destroyed. Only now are they generally acknowledged as being ecosystems of great value. The recent rise in awareness of the importance of wetlands has much to do with an enhanced appreciation of their many positive ecological and environmental functions and values. Wetlands are often the only places where these functions take place. If the wetland is destroyed these functions are lost. The particular functions provided by a wetland depend on the particular ecological characteristics of the wetland. The sustainable use of wetlands implies the conservation and enhancement of these important functions, thereby safeguarding the contribution of their vulnerable elements and guaranteeing their unique qualities. Attention must be given not only to large internationally important wetlands but also to small wetlands which can also have valuable functions (Mitsch and Gosselink; 1993 and Skinner and Zalewski 1995).

Wetlands help to maintain groundwater levels and so provide important water resources for different purposes such as agriculture and industry. Wetlands play a vital role in flood control by absorbing heavy rainfall and reducing the risk of dangerous flooding downstream. By stabilizing the soil and sediment at the water's edge, they maintain shorelines of lakes, rivers and seas. By removing nutrients, such as nitrogen and phosphorus, wetlands ensure water purification and improve water quality. This helps prevent eutrophication and avoids the need to build expensive water treatment plants.

Wetlands often play a pivotal role in regional economies (Wells and Brandon 1992). And they can be seen as even more important if their functions are considered on a larger scale, ranging from sustaining the hydrological regime and chemical element cycling, to the conservation of key habitats for the survival of birds and other species.

Wetlands, which are probably known best for their waterfowl abundance, also support a large and valuable recreational hunting industry. Mitsch and Gosselink (1993) use the term industry for this function because hunters spend large sums of money in the local economy for guns, ammunition, hunting clothes, travel to hunting spots, food and lodging. Bird watching and nature tourism also are of growing economic importance in regard to wetlands.

2 The General Situation of Wetland Conservation in Turkey

Turkey is amongst the countries with the largest expanse of wetlands. Turkey is amongst the richest countries in Europe for birds, because of it's geographical position which is situated at the junction of the Asian, European and African continents and its considerable diversity of habitats, including wetlands.

Although wetlands have enormous importance, they are currently confronted with threats in Turkey. The most serious threat is the drainage of these areas. Pollution of wetlands is another source of damage. These areas are being polluted not only directly but also by the rivers that feed them. Another threat is overhunting. The collection of bird eggs, and activities such as the cutting and burning of reeds and grazing of cattle also cause extensive damage to waterfowl. In this case, Turkey merits a well planned and organized approach to environment, resource identification, evaluation, land use decisions and the creation of a balance between use and conservation.

For this purpose, the principles governing the selection and designation of protected areas were set out in the National Park Law which was enacted in 1983. This law defines the four main types of protected areas: national parks; nature parks; natural monuments; and nature reserves. These types are defined in the law as follows. A national park is a spacious land area, of national and international significance, that contains outstanding natural, historic, scientific and scenic features as well as tourism, recreation and preservation units. In Turkey, the most effective protection is being given to the national parks with a view to preserving natural and cultural wonders of the country for future generations. The creation of national parks has not only led to overall conservation of national heritage but also has helped to develop a harmonious relationship between people and their surroundings. A nature park is an expanse of land which contains a rich variety of vegetation and wildlife, that blend harmoniously in a spectacular landscape, as well as important outdoor recreation resources. Conservation and resource utilization of nature parks is practiced in accordance with the principles of the national park system.

A nature conservation area contains ecosystems that are easily endangered or have come close to extinction as well as habitats and rare specimens that have come about as acts of nature. Natural monuments are those lands or objects that have been reserved for the protection of the outstanding natural or scientific features they contain. These objects might be trees, rocks or comparable features.

3 Sultan Marshes

Sultan Marshes (called Sultansazligi in Turkish), are one of the most important and famous wetland complexes of the Central Anatolia Region (Figure 1). Sultan Marshes are situated at the junction of Develi, Yesilhisar, and Yahyali towns. These are the main settlements of the Develi closed basin which is surrounded by high mountains. The highest mountain of the Central Anatolia is Erciyes (3,917 mt) located north of Sultansazligi.

Figure 1. Sultan Marshes

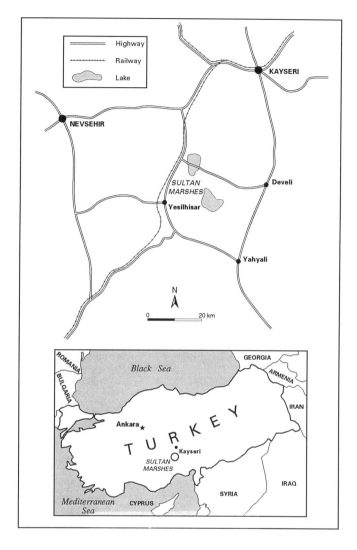

Three types of wetland occur in the area: brackish marsh; saline lakes; and freshwater marshes. All wetland types in the area are directly connected to each other. There are two main marsh complexes including the South marshes (nearly 4,000 ha) and the marshes near Kepir Marsh (nearly 2,000 ha) in the north. Originally the wetland system was fed by groundwater, water from streams such as the Yahyali, Agcasar, Develi, Yesilhisar and Dündarli: and water from springs such as the Çayirözü; Yerköy; and Soysalli. Such springs are replenished by the melting snows of the Erciyes Mountains. Yay Lake, the biggest saline lake of the complex is at the center of the Develi basin and is not more than 2 meters deep. It is very rich in microbiota and supports many water bird species especially great flamingo flocks. Its area varies greatly with seasonal changes and covers nearly 3,700 ha. In the northwest a smaller saline lake, Çöl Lake, largely dries up in summer.

Sultan Marshes, is an internationally important wetland where the variety of bird species and density of population are very high all year. Sultan Marshes satisfies the Ramsar Convention's criteria of holding more than 25,000 individuals of a species at a time for several bird species.

In spite of the fact that Sultansazligi has both national and international importance and protection statues, it is still facing the common problems of Turkish wetlands. One major threat has come from the Develi drainage project of the State Hydraulic Works (DSI). The project was planned to drain the Sultansazligi at the beginning of the 1970's. This has now been prevented by making changes in the project as a result of the efforts of the Directorate of National Parks and Wildlife and different NGO's. However, while Sultansazligi has not been drained completely, the water regime has been adversely affected because of the holding of water in dams. Another important problem is that irrigation return flow is transferred to Yay Lake. This has caused changes in the salinity of Yay lake and this in turn, will cause changes in the ecological characteristics of the lake.

As a result of conservation efforts, in 1971, Sultan Marshes was declared a Wildlife Reserve. In 1978, it was granted Biosphere Reserve status. In 1988, it was declared a Nature Conservation Area. In 1993, it was declared a Natural Site. In 1994, it was designated as one of the Ramsar Sites of Turkey.

4 Ecological Risk Analysis

In an ecological approach to sensitive ecosystems such as wetlands, sustainability can only be achieved by rational planning decisions. In order to take rational planning decisions, the data should be realistic, reliable, accurate and should allow for continuous control. The techniques which particularly provide such data are especially Geographical Information Systems (GIS) and Remote Sensing (Mulders and Epema 1992). Wetland resources should be evaluated according to their special ecological characteristics. In this case, Risk Analysis seemed the best solution to ensure sustainable wetland management (Karadeniz, 1995).

As the first step of risk analysis through GIS and remote sensing, both the natural and cultural resources of the Sultansazligi ecosystem were determined. As the second step of the process, data on the ecosystem resources were collected from conventional sources such as maps and publications. Nearly all data were loaded in the computer. And then, existing uses were evaluated on the basis of resources sensitivity and level of impacts of the uses on these resources. Finally, the results were used to evaluate for sustainable management of wetlands. Ecological Risk Analysis was made to obtain:
- Land use types and their impact on natural resources,
- Determine the assessment factors to define sensibility and density of the negative impacts,
- Ascertain sensibility to negative impacts and density of negative impacts.

Existing land use was classed as follows:

- Irrigated Agriculture,
- Dry Agriculture and Meadow,
- Dry Agriculture and Irrigated Agriculture,
- Meadow.

Ecosystem elements used for the determination of *risk analysis* were:

- Topography - Slope,
- Geology,
- Hydrogeology,
- Geomorphology,
- Soil Type,
- Land Use Capability Classes,
- Ownership.

The assessment factors used to define sensibility were determined within the framework of risk analysis (Table 1). These factors are defined as sub-units and depend on ecosystem elements such as

Table 1. Assessment factors which are used to determine sensibility and effect degrees (Karadeniz 1995).

SOURCE	ASSESSMENT FACTOR	
TOPOGRAPHY	%0-1 Slope	E1
	%2-10 Slope	E2
GEOLOGY	Alluvium: Sand, Gravel, Poor Clay Alluvial Cone(AC)	
	Alluvium: Sand, Gravel, Poor Clay Tarus(AT)	J2
	Alluvion: Sand, Gravel, Poor Clay(A)	J3
	Clay(C)	J4
	Marsh, Clay(MC)	J5
	Crystalline Limestone(CL)	J6
HYDROGEOLOGY	Flowing Artesian Aquifer(FAA)	H1
	Poor Quality Ground Water(PQG)	H2
	Salty Lake and Flood Areas(SLF)	H3
	Marsh(M)	H4
GEOMORPHOLOGY	Winter Level of Sultan Lake (1072 m)(WL)	JM1
	Flood Level of Sultan Lake (1074 m)(FL)	JM2
SOIL	Alluvial Soils(AS)	T1
	Brown Soils(BS)	T2
	Brown Soil without Lime(BSL)	T3
	Organic Soil, Salty and Muck Structure(OS)	T4
	Mixed Structure, Poor Drainage, Salty and Alcali Soils (Hydromorphic Alluvial)(HA)	T5
LANDUSE CAPABILITY CLASSES	II.Class(II.C)	A1
	III.Class(III.C)	A2
	IV.Class(IV.C)	A3
	VI.Class(VI.C)	A4
	VII.Class(VII.C)	A5
	Salty Lake and Flood Areas(SLF)	A6
	Marsh(M)	A7
OWNERSHIP	Separated Areas(SA)	M1
	Treasury Areas(TA)	M2

Brown Soils, Crystalline Limestone, and Flood Areas, Marshes or Saline Lakes. After the determination of assessment factors, a grading system was used to determine the sensibility of resources to land uses. This grading system consisted of three categories:
3 : high sensible
2 : medium sensible
1 : low sensible (Table 2).

The impacts of different land uses on the resources were graded as:
3 : high density
2 : medium density
1 : low density (Table 3).

In determining the density of the negative impacts, changes in the quality and quantity of the resources were taken into consideration.

Grading of the sensibility and impact was made by a multidisciplinary approach and an objective data determination was estimated. The analysis and determination of the data were made by workstation Arc-Info software. These analysis results were transferred to 3D views. Each GIS layer and the analysis obtained from determined factors were overlaid with the 3D views and maps. Sensibility and impact degree maps were used for risk analysis. Risks of the negative impacts on the resources are related to potential impact density and sensibility degree.

Negative impact density + sensibility of resources = RISK

Risk calculation was made by two factors. Sensibility and density were taken into consideration together and risk of negative impact was graded as follows:
3 : high risk
2 : medium risk
1 : low risk

5 Results

Although all resources of Sultan Marshes are very important, the hydrological and geomorphological elements of the ecosystem play more important roles than other elements. Saline lakes, marshes and aquifers are of high sensibility. Yay Lake, Çöl Lake, North and South Marshes are under high risk. A flowing artesian aquifer that is located in the eastern part of the area is also under high risk. Alluvial soils which contain sand, gravel and clay are under both high and medium risk. The boundary of the protected area should be expanded and so should contain all the area which is under high risk. It is necessary to conserve, and manage all the area which is under high and medium risk to ensure the sustainability of the Sultan Marshes.

6 References

Karadeniz, N., 1995. *Sultansazligi Orneginde Islak Alanlarin Cevre Acisindan Onemi Uzerinde Bir Arastirma* (Research on the Value of Wetlands in Environmental Considerations, Ph.D.,Unpublished). Doktora Tezi, Ankara Universitesi, Fen Bilimleri Enstitusu, Ankara, Turkey.

Mitsch, W. J. and Gosselink, J. G. 1993. *Wetlands*Van Nostrand Reinhold. New York, NY.

Mulders, M. A. and Epema, G. F. 1992. *Remote Sensing for Landscape Analysis*, The Netherlands: Wageningen Agricultural University, Department of Soil Science and Geology, Wageningen.

Skinner, J. and Zalewski, S., 1995. *Functions and values of Mediterranean Wetlands*. Conservation of Mediterranean Wetlands, MedWet, Tour du Valat, France.

Wells, M. and Brandon, K., 1992. *People and Parks, Linking Protected Area Management with Local Communities*. The World Bank. Washington, D.C.

Table 2. Sensibility of natural and cultural resources, according to different uses (Karadeniz 1995).

NATURAL - CULTURAL RESOURCES	USES	SENSIBILITY		
		High	Medium	Low
TOPOGRAPHY	Irrigated Agriculture	E2	E1	
	Dry Agr+Meadow			E2
	Dry Agr.+Irrigated Agr.		E2	E1
	Meadow			E2
GEOLOGY	Irrigated Agriculture	J4+J5	J3	J1
	Dry Agriculture+Meadow	J4+J5	J3	J1
	Dry Agr.+Irrigated Agr.	J4+J5	J3	J1
	Meadow	J4+J5	J3	J2
HYDROGEOLOGY	Irrigated Agriculture	H1 H2H3H4		
	Dry Agr + Meadow			H3+H4
	Dry Agr.+Irrigated Agr.			H3+H4
	Meadow			
GEOMORPHOLOGY	Irrigated Agriculture	JM1+JM2		
	Dry Agr+Meadow	JM1+JM2		
	Dry Agr.+Irrigated Agr.	JM1+JM2		
	Meadow	JM1	JM2	
SOIL	Irrigated Agriculture	T5	T4	T1
	Dry Agr+Meadow	T5	T4	T1
	Dry Agr.+Irrigated Agr.	T5	T4	T1
	Meadow			
LANDUSE CAPABILITY CLASSES	Irrigated Agriculture	A6+A7	A4	A3
	Dry Agr+Meadow	A4+A3		A2
	Dry Agr.+Irrigated Agr	A4+A3		A2
	Meadow	A4		A3+A5
OWNERSHIP	Irrigated Agriculture	M1	M2	
	Dry Agr+Meadow	M1	M2	
	Dry Agr.+Irrigated Agr.	M1	M2	
	Meadow			M1

Table 3. Effect degrees of different uses on natural and cultural resources (Karadeniz 1995).

Assesment Factor		USES			
		Irrigated Agr	Dry Agr& Meadow	Dry & Irrigated Agr	Meadow
TOPOGRAPHY	E1	3	2	2	2
	E2	2	3	3	3
GEOLOGY	J1	2	1	1	1
	J2	2	2	2	1
	J3	1	1	1	1
	J4	3	3	3	2
	J5	3	3	3	2
	J6	2	2	2	1
HYDROGEOLOGY	H1	3	1	2	1
	H2	3	1	2	1
	H3	3	2	2	2
	H4	3	2	2	2
GEOMORPHOLOGY	JM1	3	1	2	1
	JM2	3	1	2	1
SOIL	T1	1	1	1	1
	T2	1	1	1	1
	T3	1	1	1	1
	T4	3	3	3	1
	T5	3	3	3	1
LANDUSE CAPABILITY CLASSES	A1	1	1	3	2
	A2	1	1	1	2
	A3	2	2	1	1
	A4	2	2	2	1
	A5	3	3	3	3
	A6	3	3	3	3
	A7	3	3	3	3
OWNERSHIP	M1	1	1	1	1
	M2	3	3	3	1

Conservation of The Mediterranean Environment: The Vital Need to Protect and Restore the Quality of National and Cultural Resources in Libya and the Role of Parks and Protected Areas.

Amer Rghei[1] and Gordon Nelson[2]

[1] United Nations Centre for Human Settlements (Habitat)
Post Office Algeria Square, Tripoli, Libya

[2] Heritage Resources Centre, University of Waterloo
Faculty of Environmental Studies, Waterloo, Ontario N2L 3G1 Canada

1 Purposes

The aim of this paper is to highlight some challenges for the protection of the Mediterranean environment and especially for Libya. In the paper, particular stress is placed on national parks in order:
1. to describe the limited role that national parks and protected areas have played and are playing in development, settlement and land use policies and practices in Libya
2. to show the pressures or stresses that current development and land use changes are placing on important cultural and natural heritage areas.
3. to make some recommendations that would provide for more effective conservation of heritage resources through the establishment, better planning and management of parks and protected areas as a basis for sustainable development in future.

2 The Context

Libya is located on North Africa's Mediterranean coast between longitudes 9°E and 26°E and latitudes 33°N and 19°N. Libya is bordered on the East by Egypt and Sudan, on the West by Tunisia and Algeria, and on the South by Niger, Chad and Sudan. It has an area of 1,760,000 km^2; making it Africa's fourth largest state. The Libyan area has acted as a bridge between the Arab east and west and is one of the windows onto African-Mediterranean civilizations. The Mediterranean Sea and Sahara Desert are the two prominent geographical features that have affected Libya. The country includes a large part of the Sahara Desert, which extends across North Africa from the Atlantic Ocean to the Red Sea. Libya can be divided into three general climatogeographic zones as indicated by the land use and vegetation map (Figure 1):

1. The Mediterranean zone is located along the sea so that the coastal belt up to 40 km inland enjoys a mild climate. This area consists of about 45,000 square kilometers, and is the most heavily populated and most suitable for agriculture. The Mediterranean zone is backed by the mountains of Jabal Akhdar (875m) on the east side and by Tripolitania Jabal (960m) on the west side. The Jabal, an Arabic name meaning mountain, is a rocky plateau with a steep north face and a gentle south slope interrupted by some north-facing escarpments. Because of elevation, the Jabal areas receive relatively more rain and are partly cultivated.

Figure 1. Land Uses and Vegetation of Libya

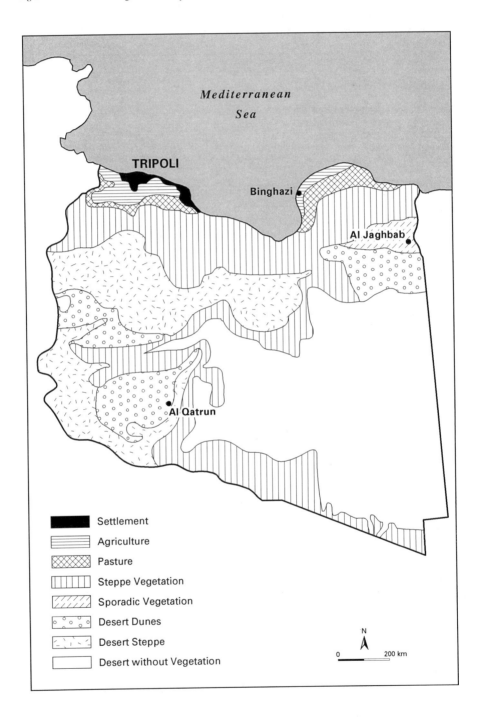

2. The inland semi-desert area of about 100,000 km^2. is chiefly steppe country and grazing land;
3. The interior desert zone contains several fertile oases. The desert and semi-desert areas are characterized by hot climatic conditions and by hospitable rugged, barren and agriculturally unproductive zones. These are still inhabited by some semi-nomads. During the last several decades the majority of the population has been concentrated along the coastal belt.

3 Brief History of Development and Land Use Change

Humans have lived in and used the Libyan landscape for thousands of years. In the process they have modified nature continuously through grazing, cultivation, settlement, the introduction of exotic plants and animals and in many other ways, as the following summary shows:

3.1 Ancient Times

About 1000 B.C., the earliest Libyans, 'The Garaments', like the modern Bedouins, had established themselves throughout North Africa and along the southern Mediterranean Sea. These people established small agricultural communities and lived in troglodytic or buried dwellings. They raised cattle.

3.2 The Phoenicians

The Phoenicians were the first to establish foreign settlements along the Libyan coast. The Phoenicians colonized Libya around the eighth century B.C. They built trading posts on the Libyan Coast as part of a chain of trading stations, including the three cities of 'Leptis Magna', 'Oea' (Tripoli), and 'Salaratha' in Tripolitania. They also cultivated the surrounding coastal areas by introducing the olive tree, thereby improving the existing agricultural system.

3.3 The Greeks

In about 600 B.C., the Greeks invaded and settled the eastern part of Libya (Cyrenaica). The Greeks established a large number of cities including Cyrene, Apollonia, Pentapolis, Ptolmais, Tauchera and Auhesperides.

3.4 The Romans

In 146 B.C., the Romans invaded the Tripolitania Region, the northwestern part of Libya along the Mediterranean Coast. The Romans with their irrigation and conservation methods brought new life to agriculture. Comfortable villas and farms spread over the countryside. They also brought also a new infusion of technical and architectural expertise to North Africa. The Roman cities resembled other provincial cities throughout the Roman Empire.

3.5 The Islamic Peoples

Their influence in Libya began about 644 A.D., soon after rulers started to establish new routes throughout the country to link all cities, villages and oasis towns across the North African Sahara. Urban and Saharan Communities were developed and adorned with gardens, market places or sugs, public baths and houses. They also built fine mosques at each centre. The Islamic faith shaped the new culture and way of life and created a special style of architecture.

Islamic settlements and towns were influenced by geographic, climatic, socio-economic and cultural factors. Settlements were also built with distinct characteristics in relation to the regional landscape they occupied. In 1551 Libya became a Turkish province. The Turks controlled the country until 1911. In that year, Italy declared war on Turkey and in 1912 Libya became an Italian colony. Fighting continued until about 1930.

3.6 Italian Colony

From 1930 to 1940 Italy launched a large scale development program and granted large tracts of land to Italian settlers. Italy wanted at that time to reclaim the history of their Roman ancestors. Italians planned to develop Libya as one part of their Empire by modernizing and developing new agricultural centres. Their development spread throughout the country, especially in the North. The Italians also established the first modern national parks and protected areas in Libya; their history is not well known and their current status uncertain. In the early years of World War II, Libya became a battleground, and in 1943 the Allied forces occupied Libya. After World War II, Libya was under the control of the United Nations and gained its independence in December 1951.

3.7 Independence Era

During the early 1950s, Libya was one of the poorest countries in the world (GNP of only US $40). By the mid 1960s, the discovery of oil had made it one of the major oil exporting countries. Since then Libya has developed rapidly from being an extremely poor to a rich oil state. In 1981 GNP per capita was US $8,450. The national economy has been greatly stimulated. Jobs became available in the major cities of the Coastal Zone.

4 Current Situation in Libya

Libya is a very large country with a comparatively small population. In 1959 the population of Libya was 1,089,000. It rose to 2,251,000 in 1973. By 1984 the population was 3,637,000. In the year 2000, Libya is expected to have 5,754,000 people.

Large scale government schemes initiated during the 1970's have channeled oil revenues into extensive construction of water canals, roads, and other infrastructure and socio-economic programs. Full attention has been given to solving the housing problem. The housing sector has been given a big push to overcome shortages of houses.

Many people migrated to the new settlements in the coastal zone from the interior steppes, deserts and oases. They were mainly rural nomads. They were attracted especially to the two large cities of Tripoli and Benghazi, where they could find better employment opportunities.

The settlement of the nomads resulted in a decline of the cultural tradition of the country. The nomads and semi-nomads (peasant and semi-peasant society) in the rural areas were a major part of the traditional Libyan economy. The Libyan economy in the recent past was based on cattle-rearing and seasonal agriculture. The nomads, who moved to highrise buildings in the urban areas of the coast, found themselves in an entirely foreign environment unable to continue their independent agricultural traditions. They adjusted to urban life by engaging in daily-wage and fixed-salaried occupations. They missed the security of the clan as they lived with thousands of other people from different areas of the country.

5 Constraints, Conflicts and the Need for National Parks and Protected Areas

The discovery of oil in Libya led to large scale development and dramatic impact on both the natural and cultural environments. Stress has been placed on scarce natural resources which are concentrated along the coast. Settlement here is very dense because the other areas are deserts or semi-deserts (Figure 2).

Figure 2. Settlements in Libya

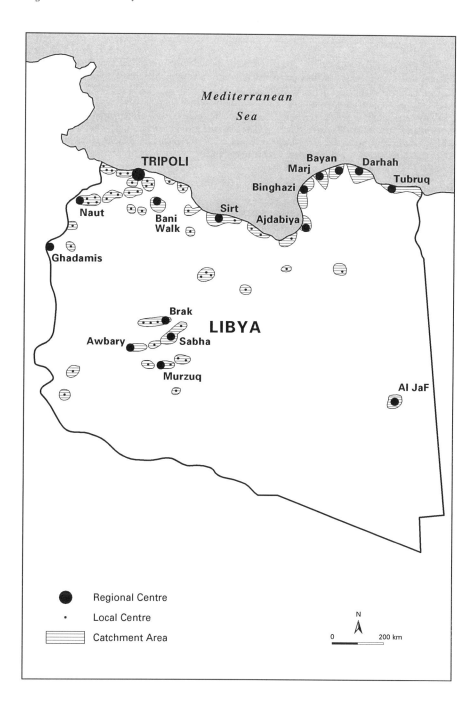

Over-population in the coastal areas has also led to a shortage of water. This problem of water is becoming a real crisis. Underground water from wells has been the main source of water supply, but it is now running down. A costly pipeline or man-made river project is nearly completed to carry the underground water from acquifers in the Sahara Desert to the densely settled regions of the Libyan Coast.

Few of Libya's natural areas and cultural resources are currently protected against uncontrolled exploitation and development. In this respect, this paper aims at strategies and research needed to establish an effective system of national parks and protected areas for the country. Such parks and protected areas are important not only in helping prevent desertification but also in maintaining water flow and other ecological processes as well as protecting ecosystems, watersheds and freshwater resources. Parks and protected areas support biodiversity and genetic materials and other bases of life. In addition, parks and protected areas have potential value for medicine, employment, recreation, tourism, education, research and other socio-economic services.

6 Summary Analysis

Environment and land use in Libya is being heavily affected by both human activities and natural changes including climatic changes and desertification. Agriculture and natural areas in the Coastal Zone are continuously converted at a high rate for urban expansion and other related uses. For example, during 1980-2000, it is estimated that about 432 km^2 of land will be converted at an annual rate of 5%.

Currently, a new phenomenon of semi-urban growth is taking place outside approved urban area boundaries. According to the studies conducted by the Urban Planning Agency, about 40% of urban land use in Tripoli and its surrounding area is semi-urban or suburban and is constructed without approval and outside urban planning boundaries. As a result, forests and natural vegetation have been damaged or destroyed. Large areas of value for grazing and agriculture are also being lost.

In this respect, conservation can be an important part of planning, especially landscape and land use planning. One of the most important tools in this regard is the use of parks and protected areas which have been neglected because of the heavy emphasis on housing, urban development, roads, water, canals and other infrastructure.

7 Recommendations

The following recommendations are made in support of sustainable development in Libya:

1. Identifying a system of natural areas or an ecological framework for the creation of a network of national parks and protected areas. Recently some proposals have been made for national parks but mainly in terms of recreational and related needs near the urbanized coastal areas. This is an unsystematic and limited approach which will not provide for the full range of park and protected area services needed for sustainable development in Libya.
2. Finding ways of linking natural and cultural heritage and conservation, for example through the use of landscape parks of the type found in Europe.
3. Studying current status and trends in wildlife, forests and other vegetation systems and identifying species, communities and areas under high stress which need protection soon.
4. Studying current status and trends in conservation and use of historic and cultural resources and areas.
5. Mapping of areas with relatively little land use or human disturbance as a basis for identifying possible sites for parks and protected areas.
6. Studying different systems of parks and protected areas in other countries in order to determine which system might be most successful in Libya.

7. Developing tourism plans that are oriented to Libyan natural and cultural heritage and which can be used to attract tourists with heritage interests in numbers that are not so great as to damage the heritage resources.
8. Undertaking a detailed assessment of the Jebel Al Akhdar or Green Mountain area adjoining the northeast coast of Libya as soon as possible. This area has many different vegetation, habitat and landscape types and has high value for biodiversity, geologic and related resources and landscapes, as well as archaeological, architectural and historic sites such as ancient Greek cities. The area has excellent potential for the establishment of parks and protected areas. These could protect the natural and cultural landscape values as well as providing excellent opportunities for education, tourism and protection of water and other resources for nearby urban and agricultural areas. The Jebel al Akhdar could be an excellent site for parks and protected areas and perhaps for a biosphere reserve. It could serve as a test and a model for other potential sites in Libya.

Many of the foregoing suggestions are in line with current thinking about sustainable development as well as the Blue Plan for the Mediterranean area.

SPEAKERS AND PARTICIPANTS

NAME	CONTACT ADDRESS
(a) Directors	
CANADA - NATO COUNTRY	
Professor Gordon Nelson	Heritage Resources Centre University of Waterloo Waterloo, Ontario N2L 3G1 CANADA Tel: +1-519-885-1211 x2072 Fax: +1-519-746-2031 Email: jgnelson@cousteau.uwaterloo.ca
POLAND - CP NATO COUNTRY	
Dr. Rafal Serafin	Polish Foundation for Nature Conservation Sebastiana 13/11 31-049 Kraków POLAND Tel: +48-12-217154 Fax: +48-12-217411 Email: kr-forum@kki.krakow.pl
(b) Key speakers - NATO Countries	
CANADA	
Mr. Harvey Locke	Canadian Parks and Wilderness Society 700 Gulf Canada Square 401-9th Avenue SW PO Box 2010 Calgary, Alberta T2P 2M2 CANADA Tel/fax: +1-403-232-0888
Mr. Paul Macnab	Terra Nova National Park Glovertown Newfoundland AOG 2LO CANADA Tel: +1-709-533-2801 Fax: +1-709-533-2706 Email: paul_mcnab@PCH.GC.CA
Ms. Anne Varangu	School of Urban and Regional Planning University of Waterloo Waterloo Ontario N2L 3G1 CANADA Email: amvarang@cousteau.uwaterloo.ca
Dr. Stephen Woodley	Natural Resources Branch Parks Canada 4th Floor, Jules Leger Building 25 Eddy Street Hull, Quebec K1A 0M5 CANADA Email: Stephen_Woodley@pch.gc.ca

FRANCE Ms. Sylvie Blangy	Consultant in Ecotourism 123 Rue de la Carriérasse 34090 Montpellier FRANCE Tel: +33-67-520994 Fax: +33-67-542567 Email: blangy@cirad.fr
ITALY Professor Almo Farina	Lunigiana Museum of Natural History Forteza della Brunzlla 540011 Aulla ITALY Tel: +39-187-400252 Fax: +39-187-420727 Email: afarina@tamnet.it
THE NETHERLANDS Professor Jan Van der Straaten	European Centre for Nature Conservation Warandelaan 2 - Y Building PO Box 1352 5004 BJ Tilburg THE NETHERLANDS Tel: + 31-13-466-3240 Fax: + 31-13-466-3250 Email: straaten@ecnc.nl www: http://www.ecnc.nl
TURKEY Professor Nilgul Karadeniz	Department of Landscape Architecture Faculty of Agriculture, Ankara University 06-110 Diskapi, Ankara TURKEY Tel/fax: +90-312-317-6467 Email: karaden@agri.ankara.edu.tr

UNITED KINGDOM Mr. David Haffey	Eastland Bank House Dipton Mill Road Hexham, Northumberland NE46 1RY UNITED KINGDOM Tel/fax: +44-143-460-7251
Ms. Jane Madgwick	Broads Authority Thomas Harvey House 18 Colegate, Norwich NR3 1BQ UNITED KINGDOM Tel: +44-1603-610734 Fax: +44-1603-765710
Professor Adrian Phillips	CNPPA Chair IUCN - The World Conservation Union UNITED KINGDOM Tel/fax: +44 1386 882094 Email: adrianp@wcmc.org. uk

UNITED STATES	
Ms. Jessica Brown	Atlantic Centre for the Environment 55 South Main Street Ipswich MA 01938 USA Tel: +1-508-356-0038 Fax: +1-508-356-7322 Email: atlantictr@igc.apc.org
Professor Gary Machlis	National Park Service Main Interior Building 1849 C Street NW (Room 3412) PO Box 37127 (470) Washington DC 20013-7127 USA Tel:+1-202-208-5391 Department of Forest Resources College of Forestry, Wildlife and Range Sciences University of Idaho Moscow, Idaho USA Tel: +1-208-885-7129 Fax: +1-208-885-6226 Email: gmachlis@uidaho.edu

c) Key speakers - Other countries	
AUSTRIA	
Mr. Philip Weller	Green Danube Programme WWF International c/o WWF Austria Ottakringer Strasse, 114-116 A-1162 Wien, Postfach 1 AUSTRIA Tel: +43-1-489-1641 Fax: +43-1-489-1641-29
LIBYA	
Dr. Amer Rghei	United Nations Centre for Human Settlements (Habitat) Post Office Algeria Square Tripoli, LIBYA Tel: +218-2132971
SLOVENIJA	
Ms. Marija Zupancic-Vicar	Commission on National Parks and Protected Areas World Conservation Union Rodine 51, 4274 Zirovnica SLOVENIJA Tel: +386-64-801035

d) Key Speakers - CP NATO Countries BULGARIA Professor Elizaveta Matveeva	National Nature Protection Service 67 W. Gladstone Street Sofia 1000 BULGARIA Tel/fax: +35-929-813384
CZECH REPUBLIC Mr. Miroslav Kundrata	Environmental Partnership for Central Europe Nadace Partnersvi Panska 9, 60200 Brno CZECH REPUBLIC Tel: +42-542-218350 Fax: +42-542-210561 Email: PSHIP@ecn.apc.org
ESTONIA Ms. Monica Prede	Otepaa Linnavalitsus Lipuvaljak 13 EE-2513 Otepaa ESTONIA Tel: +37-276-55143 Fax: +37-276-61214
HUNGARY Mr. Mihaly Vegh	Hortobagy National Park H-4024 Debrecen, Sumen U2 HUNGARY Tel: +36-2-349-922 Fax: +36-2-410-645
POLAND Mr. Andrzej Biderman	Ojców National Park Education Centre 32-047 Ojców POLAND Tel: +48-12-239063
Professor Zygmunt Denisiuk	Institute of Nature Conservation Polish Academy of Sciences Lubicz 46, 31-512 Kraków POLAND Tel/fax: +48-12-210348
Professor Anna Dyduch-Falniowska	Institute of Nature Conservation Polish Academy of Sciences Lubicz 46, 31-512 Kraków POLAND Tel/fax: +48-12-210348 Email: nodyduch@cyf.kr.edu.pl

Mr. Jerzy Sawicki	National Parks Unit Polish Ecological Club Pilsudskiego 12 31-109 Kraków POLAND Tel/fax: +48-12173578
ROMANIA Professor George Romanca	University of Bucharesti Splaiul Independenti 91-95 76201 Bucharest ROMANIA Tel/fax: +401-312-2310 Email: ROGE@BIO.BIO.UNIBUC.RO
UKRAINE Professor Stepan Stoyko	Institute of Ecology Ukrainian Academy of Sciences Chaykovski 17 290-000 Lvov UKRAINE Tel: +38-322-767623 and 726731 Fax: +380-322-728752

e) Participants - NATO Countries CANADA Mr. Tom Kovacs	Natural Resources Branch National Parks Directorate 25 Eddy Street, 4th Floor Jules Leger Building Hull, Quebec K1A 0M5 CANADA Tel: +1-819-994-2639 Fax: +1-819-997-3380 Email: tom_kovacs@pch.gc.ca
DENMARK Ms. Karin Skovhus	Department of Environmental and Land Use Management Funen County DENMARK Tel: +456-556-1000 Fax: +456-615-9160
GERMANY Ms. Krystyna Wolniakowski	German Marshall Fund Berlin GERMANY

THE NETHERLANDS Dr. Henry Baumgartl	Karkonosze Foundation PO Box 2298 NL-1620EG Hoorn THE NETHERLANDS Tel/fax: +31-229-241992 Email: H.Baumgartl@Inter.NL.net
UNITED KINGDOM Ms. Dorota Mech	Department of Agricultural Economics Wye College, University of London Wye UNITED KINGDOM

f) Participants - CP NATO Countries CZECH REPUBLIC Mr. Lada Ptacek	Adonis Advisory Centre for Sustainable Development Namesti 32, 692-01 Mikulov CZECH REPUBLIC Tel: +42-625-3239
HUNGARY Ms. Szilvia Gori	Hortobagy National Park H-4024 Debrecen, Sumen U2 HUNGARY Tel: +36-2-349-922 Fax: +36-2-410-645
LITHUANIA Mr. Tadas Leoncikas	Central European University c/o Sausio 13-osios 5-137 Vilnius 2044 LITHUANIA
POLAND Mr. Wojciech Bosak Professor Alicja Breyemeyer	Ojców National Park 32-047 Ojców POLAND Tel: +48-12-111555 Institute of Geography Polish Academy of Sciences Krakowskie Przedmieœcie 30 00-927 Warszawa POLAND Tel: +48-22-269808 Fax: +48-22-267267

Professor Tadeusz Borys	Regional Centre of Sustainable Development Nowowiejska 3 58-500 Jelenia Góra POLAND Tel/fax: +48-75-25750
Ms. Jadwiga Grabczak-Jarosz	Board of Jura Landscape Park System in Kraków Vetulaniego 1a 31-227 Kraków POLAND Tel/fax: +48-12-333088
Dr. Zbigniew Krzan	Tatra National Park Chalubińskiego 42a 34-500 Zakopane POLAND Tel: +48-165-63203 Fax: +48-165-63579 Email: dr_Krzan@tpn.zakopane.pl

Dr. Wieslaw Krzemiński	Polish Foundation for Nature Conservation Sebastiana 13/11 31-049 Kraków POLAND
Mr. Zdzislaw Krzemiński	Department of Nature Conservation Ministry of Environmental Protection, Natural Resources and Forestry Wawelska 52 Warszawa POLAND
Mr. Zbigniew Machalica	National Park Unit Polish Ecological Club Pilsudskiego 12 31-109 Kraków POLAND Tel: +48-12-173578
Mr. Rudolf Suchanek	Ojców National Park 32-047 Ojców POLAND Tel: +48-12-111555
Professor Zbigniew Witkowski	Institute of Nature Conservation Poish Academy of Sciences Lubicz 46 31-512 Kraków POLAND Tel: +48-12- 210348

SLOVAKIA	
Ms. Katarina Rajcova	Spolecnost pre trvalo udrzatelny zivot Odbocka Biele Karpaty Hviezdoslavova 1 91101 Trencin SLOVAKIA Tel: 0042 831 535763 Fax: 0042 831 37061 Email: stuztn@seps.bb.sanet.sk
Dr. Ivan Voloscuk	Sprava Narodnych Parkov Slovenskej Republiky TANAP 059-60 Tatrzanska Lomnica SLOVAKIA Tel: +969-967-3513 Fax: +969-967-124

UKRAINE	
Dr. Lidia Tasenkevych	Natural History Museum National Academy of Sciences Teatralna 18 290008 Lviv UKRAINE Tel: +38-322-728917 and 723120 Fax: +38-322-742307

NATO ASI Series G

Vol. 1: Numerical Taxonomy. Edited by J. Felsenstein. 644 pages. 1983.
(out of print)

Vol. 2: Immunotoxicology. Edited by P. W. Mullen. 161 pages. 1984.

Vol. 3: In Vitro Effects of Mineral Dusts.
Edited by E. G. Beck and J. Bignon. 548 pages. 1985.

Vol. 4: Environmental Impact Assessment, Technology Assessment, and Risk Analysis. Edited by V. T. Covello, J. L. Mumpower, P. J. M. Stallen, and V. R. R. Uppuluri. 1068 pages.1985.

Vol. 5: Genetic Differentiation and Dispersal in Plants.
Edited by P. Jacquard, G. Heim, and J. Antonovics. 452 pages. 1985.

Vol. 6: Chemistry of Multiphase Atmospheric Systems.
Edited by W. Jaeschke. 773 pages. 1986.

Vol. 7: The Role of Freshwater Outflow in Coastal Marine Ecosystems.
Edited by S. Skreslet. 453 pages. 1986.

Vol. 8: Stratospheric Ozone Reduction, Solar Ultraviolet Radiation and Plant Life.
Edited by R. C. Worrest and M. M. Caldwell. 374 pages. 1986.

Vol. 9: Strategies and Advanced Techniques for Marine Pollution Studies: Mediterranean Sea. Edited by C. S. Giam and H. J.-M. Dou. 475 pages. 1986.

Vol. 10: Urban Runoff Pollution.
Edited by H. C. Torno, J. Marsalek, and M. Desbordes. 893 pages. 1986.

Vol. 11: Pest Control: Operations and Systems Analysis in Fruit Fly Management.
Edited by M. Mangel, J. R. Carey, and R. E. Plant. 465 pages. 1986.

Vol. 12: Mediterranean Marine Avifauna: Population Studies and Conservation.
Edited by MEDMARAVIS and X. Monbailliu. 535 pages. 1986.

Vol. 13: Taxonomy of Porifera from the N. E. Atlantic and Mediterranean Sea.
Edited by J. Vacelet and N. Boury-Esnault. 332 pages. 1987.

Vol. 14: Developments in Numerical Ecology.
Edited by P. Legendre and L. Legendre. 585 pages. 1987.

Vol. 15: Plant Response to Stress. Functional Analysis in Mediterranean Ecosystems. Edited by J. D. Tenhunen, F. M. Catarino, O. L. Lange, and W. C. Oechel. 668 pages. 1987.

Vol. 16: Effects of Atmospheric Pollutants on Forests, Wetlands and Agricultural Ecosystems. Edited by T. C. Hutchinson and K. M. Meema. 652 pages. 1987.

Vol. 17: Intelligence and Evolutionary Biology.
Edited by H. J. Jerison and I. Jerison. 481 pages. 1988.

Vol. 18: Safety Assurance for Environmental Introductions of Genetically-Engineered Organisms. Edited by J. Fiksel and V.T. Covello. 282 pages. 1988.

Vol. 19: Environmental Stress in Plants. Biochemical and Physiological Mechanisms.
Edited by J. H. Cherry. 369 pages. 1989.

Vol. 20: Behavioural Mechanisms of Food Selection.
Edited by R. N. Hughes. 886 pages. 1990.

Vol. 21: Health Related Effects of Phyllosilicates.
Edited by J. Bignon. 462 pages.1990.

NATO ASI Series G

Vol. 22: Evolutionary Biogeography of the Marine Algae of the North Atlantic.
Edited by D. J. Garbary and G. R. South. 439 pages. 1990.

Vol. 23: Metal Speciation in the Environment.
Edited by J. A. C . Broekaert, Ş. Güçer, and F. Adams. 655 pages. 1990.

Vol. 24: Population Biology of Passerine Birds. An Integrated Approach.
Edited by J. Blondel, A. Gosler, J.-D. Lebreton, and R . McCleery.
513 pages. 1990.

Vol. 25: Protozoa and Their Role in Marine Processes.
Edited by P. C. Reid, C. M . Turley, and P. H. Burkill. 516 pages. 1991.

Vol. 26: Decision Support Systems.
Edited by D. P Loucks and J. R. da Costa. 592 pages. 1991.

Vol. 27: Particle Analysis in Oceanography. Edited by S. Demers. 428 pages. 1991.

Vol. 28: Seasonal Snowpacks. Processes of Compositional Change.
Edited by T. D. Davies, M . Tranter, and H . G. Jones. 484 pages. 1991.

Vol. 29: Water Resources Engineering Risk Assessment.
Edited by. J. Ganoulis. 551 pages. 1991.

Vol. 30: Nitrate Contamination. Exposure, Consequence, and Control.
Edited by I. Bogárdi and R. D. Kuzelka. 532 pages. 1991.

Vol. 31: Industrial Air Pollution. Assessment and Control.
Edited by A. Müezzinoğlu and M. L. Williams. 245 pages. 1992.

Vol. 32: Migration and Fate of Pollutants in Soils and Subsoils.
Edited by D. Petruzzelli and F. G. Helfferich. 527 pages. 1993.

Vol. 33: Bivalve Filter Feeders in Estuarine and Coastal Ecosystem Processes.
Edited by R. F. Dame. 584 pages. 1993.

Vol. 34: Non-Thermal Plasma Techniques for Pollution Control.
Edited by B. M. Penetrante and S. E. Schultheis.
Part A: Overview, Fundamentals and Supporting Technologies. 429 pages. 1993.
Part B: Electron Beam and Electrical Discharge Processing. 433 pages. 1993.

Vol. 35: Microbial Mats. Structure, Development and Environmental Significance.
Edited by L. J. Stal and P. Caumette. 481 pages. 1994.

Vol. 36: Air Pollutants and the Leaf Cuticle.
Edited by K. E. Percy, J. N. Cape, R. Jagels and C. J. Simpson. 405 pages.
1994

Vol. 37: Azospirillum VI and Related Microorganisms.
Edited by I. Fendrik, M. del Gallo, J. Vanderleyden, and M. Zamaroczy.
588 pages. 1995.

Vol. 38: Molecular Ecology of Aquatic Microbes.
Edited by I. Joint. 423 pages. 1995.

Vol. 39: Biological Fixation of Nitrogen for Ecology and Sustainable Agriculture.
Edited by A. Legocki, H. Bothe, and A. Pühler. 339 pages. 1997.

Vol. 40: National Parks and Protected Areas. Keystones to Conservation and
Sustainable Development.
Edited by J. G. Nelson and R. Serafin. 1997.